DATE DUE

JUL 2 1 2005	

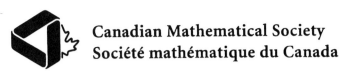

Canadian Mathematical Society
Société mathématique du Canada

Editors-in-Chief
Rédacteurs-en-chef
Jonathan M. Borwein
Peter Borwein

Springer

New York
Berlin
Heidelberg
Barcelona
Hong Kong
London
Milan
Paris
Singapore
Tokyo

CMS Books in Mathematics
Ouvrages de mathématiques de la SMC

David M. Arnold

Abelian Groups and Representations of Finite Partially Ordered Sets

 Springer

David M. Arnold
Department of Mathematics
Baylor University
Waco, TX 76798-7328
USA

Editors-in-Chief
Rédacteurs-en-chef
Jonathan M. Borwein
Peter Borwein
Centre for Experimental and Constructive Mathematics
Department of Mathematics and Statistics
Simon Fraser University
Burnaby, British Columbia V5A 1S6
Canada

Mathematics Subject Classification (2000): 16G20, 20Kxx

Library of Congress Cataloging-in-Publication Data
Arnold, David M.
  Abelian groups and representations of finite partially ordered sets / David M. Arnold.
    p. cm. — (CMS books in mathematics)
  Includes bibliographical references and indexes.
  ISBN 0-387-98982-X (hardcover : alk. paper)
  1. Abelian groups. 2. Partially ordered sets. I. Title. II. Series.
QA180 .A76 2000
512'.2—dc21   99-087081

Printed on acid-free paper.

Production managed by Michael Koy; manufacturing supervised by Jeffrey Taub.
Typeset by TechBooks, Fairfax, VA.
Printed and bound by Edwards Brothers, Inc., Ann Arbor, MI.
Printed in the United States of America.

9 8 7 6 5 4 3 2 1

ISBN 0-387-98982-X Springer-Verlag New York Berlin Heidelberg   SPIN 10751459

To my wife, Betty, for her deep affection, strong encouragement, extraordinary patience with this project, and repeated reminders of the lessons of Poseidon.

Preface

A recurring theme in a traditional introductory graduate algebra course is the existence and consequences of relationships among different algebraic structures. This is also the theme of this book, an exposition of connections between abelian groups and representations of finite partially ordered sets. Emphasis is placed throughout on classification, a description of the objects up to isomorphism, and computation of representation type, a measure of when classification is feasible.

The subject of representations of finite partially ordered sets over a field is deeply rooted in classical linear algebra. Questions such as similarity and equivalence of matrices and simultaneous similarity and equivalence of pairs of matrices can be interpreted as questions about classification of representations up to isomorphism. Fundamental theorems about representation type were originally proved in the early 1970s by matrix arguments. More recently, the subject of representations of partially ordered sets over fields has been subsumed into the more general, and more sophisticated, subject of finitely generated modules over finite-dimensional algebras.

Abelian group theory has a rich tradition of classification results. In the torsion-free case, finite direct sums of subgroups of the additive group of rational numbers were characterized by 1937. After such a promising start, by the early 1960s it became clear that torsion-free abelian groups were just too complicated to admit any reasonable classification scheme. As Irving Kaplansky wrote about torsion-free abelian groups of finite rank, "In this strange part of the subject anything that can conceivably happen actually does happen" [Kaplansky 69].

A breakthrough occurred in 1968. An equivalence between quasi-homomorphism categories of certain torsion-free abelian groups of finite rank, now called Butler groups, and representations of finite partially ordered sets over the field of rational numbers was published by M.C.R. Butler. A survey paper by the same author in 1987 contained a suggestion that the well-developed theory of representations over fields be applied to abelian group theory. Implementation of this suggestion and its natural extensions has introduced some order into the previously chaotic subject of torsion-free abelian groups of finite rank.

This book is partially motivated by the scarcity of published books and monographs on representations of partially ordered sets over fields. [Ringel 84] and [Simson 92] are refreshing exceptions but not introductory in the sense that the focus is on a more general setting. A second motivation is the absence of an expository introduction to finite rank Butler groups. Standard texts for abelian group theory were published before the development of much of the theory.

Chapter 1 is an elementary introduction to fundamental techniques for, and properties of, representations of finite partially ordered sets over a field. Basic notions of countable torsion-free abelian groups and relationships between finite rank Butler groups and representations of finite partially ordered sets over a field constitute Chapters 2 and 3, respectively.

Recently discovered relationships between Butler groups and representations of finite partially ordered sets over discrete valuation rings are included in Chapter 4. In contrast to representations over fields, very little of a theoretical nature is known about representations over discrete valuation rings. Nonetheless, representation type can be computed, primarily by the use of traditional matrix techniques.

Chapter 5 is dedicated to almost completely decomposable groups, a special case of finite rank Butler groups that is currently receiving a great deal of attention in the literature. Computations of representation type of representations of partially ordered sets over discrete valuation rings lead to explicit procedures for constructing indecomposable almost completely decomposable groups of arbitrarily large finite rank with types in a fixed finite set of types. Such constructions are quite difficult without the assistance of techniques from representation theory.

Included in Chapters 6 and 7 is a brief introductory development of Coxeter correspondences and almost split sequences within the context of representations and finite rank Butler groups, a realization theorem for endomorphism rings of finite rank Butler groups, and classification results for a special class of finite rank Butler groups called bracket groups.

Chapter 8 is devoted to applications to finite rank torsion-free modules over a discrete valuation ring and finite valuated p-groups. These applications are intriguing in that representations bridge the gap between the apparently disparate subjects of finite rank Butler groups, finite valuated groups, and p-local torsion-free abelian groups of finite rank.

No attempt has been made to be comprehensive or encyclopedic, due to the breadth of the topics discussed and the voluminous literature. Topics are selected to support the general theme of this book. A notable omission is any mention of uncountable groups, an active area of research requiring set-theoretic techniques beyond the scope of this book. Notes at the end of each chapter provide a brief guide to the published literature. In the interest of space and clarity, some long, highly technical proofs are not included, but in all cases a complete proof is referenced for the interested reader. Numerous open questions arise and some of these are mentioned at the end of the appropriate section. Explicit constructions and examples are emphasized throughout; pure existence proofs are rare.

Although the reader will witness repeated interaction between such general topics as linear algebra, rings, modules, categories, abelian groups both torsion and torsion-free, and combinatorics, the development is intended to be relatively elementary. A beginning graduate algebra text, such as [Hungerford 74], should be sufficient background. In a few instances, ancillary material is summarized and referenced.

A graduate-level course taught by the author at the University of Essen, Germany, in 1996 provided the impetus for this book. Class notes from the lectures were taken by the students. These notes provided the foundation for the first three chapters, an examination of relationships between quasi-homomorphism categories of countable Butler groups and representations of finite partially ordered sets over fields. Other graduate-level courses are possible from this book, such as "Introduction to Representations of Finite Posets," Chapters 1, 4.1–4.2, and 6, and "Introduction to Torsion-Free Abelian Groups of Countable Rank," Chapters 2.1–2.4, 3.1, 3.2, 3.4, 5.1, 5.4, and 7.

Acknowledgments

Support and facilities during the preparation of this manuscript were provided, in part, by the Ralph and Jean Storm Endowment Fund; Baylor University College of Arts and Sciences, Department of Mathematics, and Summer Sabbatical Program; German–Israeli Foundation; University of Essen, Germany; and University of Connecticut, Storrs. Special thanks to Professor Ed Oxford for his encouragement and Professors Rüdiger Göbel and Charles Vinsonhaler for their encouragement and kind hospitality.

Much of the work on the interaction of representations of partially ordered sets over discrete valuation rings and abelian groups has been done in conjunction with Professor Manfred Dugas. His insight and computational skills have been an inspiration. Motivation for an introductory book on representations of partially ordered sets was provided by the experience of former Baylor University students Rebekah Hahn, Amy Maddox, and Mary Alice Mouser, in the preparation of their master's theses. Thanks to Rebekah Hahn for permission to include some diagrams and observations from her thesis.

I am grateful to Professor E.L. Lady for the opportunity to read a preliminary manuscript on Butler modules over Dedekind domains. My appreciation to Professors S. Brenner, M.C.R. Butler, L. Fuchs, R. Göbel, H. Krause, A. Mader, K. Rangaswamy, F. Richman, C. Ringel, D. Simson, A. Skowroński, C. Vinsonhaler, and B. Zimmermann–Huisgen, among others, for illuminating conversations about abelian groups and representations of partially ordered sets.

A number of people have patiently listened to lectures about the contents of this book. Many thanks to Professors S. Files, S. Glaz, R. Göbel, J. Hurley, A. Paras, S. Pabst, J. Reid, E. Spiegel, C. Vinsonhaler, and W. Wickless for their helpful comments. My gratitude to A. Elder, G. Hennecke, A. Opendhövel, and L. Strüngmann for their assistance with the lecture notes, J. Fang and M. Kanuni for reading a preliminary version, and Roxie Ray for her skillful assistance with the preparation of this manuscript.

Contents

1
Representations of Posets over a Field

1.1 Vector Spaces with Distinguished Subspaces

Finite-dimensional vector spaces with finite sets of distinguished subspaces are illustrative examples of representations of finite posets. This provides a natural setting for equivalence and similarity of matrices, as demonstrated in the exercises.

Let k be a field and S_n denote the set $\{1, 2, \ldots, n\}$ of integers. A k-*representation* of S_n is an $(n+1)$-tuple $U = (U_0, U_1, \ldots, U_n)$ consisting of a finite-dimensional k-vector space U_0 with n distinguished subspaces $U_1, \ldots, U_n$. A *morphism* $f : U \to U'$ of representations is a k-linear transformation $f : U_0 \to U_0'$ with $f(U_i) \subseteq U_i'$ for each i.

Two representations U and U' are *isomorphic* if there is a representation morphism $f : U \to U'$ such that $f : U_0 \to U_0'$ is a vector space isomorphism. Equivalently, there is a representation morphism $g : U' \to U$ with $fg = 1_{U'}$, the identity morphism of U', and $gf = 1_U$. In this case, f is called an *isomorphism*. If f is an isomorphism from U to U', then $f : U_i \to U_i'$ is a vector space isomorphism for each i. The set of representation morphisms $U \to U$, called *endomorphisms* of U, is denoted by End U.

It follows immediately from the definition of a representation morphism that End U is a ring under the operations of addition and composition of morphisms. In particular, End U is a subring of the ring End U_0 of k-linear transformations of U_0. The field k is a subring of End U, since scalar multiplication on U_0 by an element of k is an endomorphism of U. An element f of End U is *idempotent* if $f^2 = f$. If f is idempotent, then $1 - f$ is also idempotent, since $(1 - f)^2 = 1 - 2f + f^2 = 1 - f$.

The *direct sum* of two representations $U = (U_0, U_i : 1 \leq i \leq n)$ and $U' = (U_0', U_i' : 1 \leq i \leq n)$ is the representation

$$U \oplus U' = (U_0 \oplus U_0', U_1 \oplus U_1', \ldots, U_n \oplus U_n'),$$

where $U_i \oplus U_i'$ denotes the vector space direct sum of U_i and U_i'. A representation U is *decomposable* if there is a vector space decomposition $U_0 = V_0 \oplus W_0$ with $U_i = (V_0 \cap U_i) \oplus (W_0 \cap V_i)$ for each i and both V_0 and W_0 nonzero. In this case, $V = (V_0, V \cap U_i : 1 \leq i \leq n)$ and $W = (W_0, W \cap U_i : 1 \leq i \leq n)$ are called *direct summands* of U. A representation U is *indecomposable* if $0 = (0, 0, \ldots, 0)$ and U are the only direct summands of U. Since U_0 has finite dimension, any representation U can be written as the direct sum of finitely many indecomposable representations.

Lemma 1.1.1 *A k-representation U of S_n is indecomposable if and only if 0 and 1 are the only idempotents of* End U.

PROOF. Let $U = (U_0, U_1, \ldots, U_n)$ be a representation and $f^2 = f \in$ End U. Then $U = (f(U_0), f(U_1), \ldots, f(U_n)) \oplus ((1-f)(U_0), (1-f)(U_1), \ldots, (1-f)(U_n))$ is a direct sum of representations.

To confirm this assertion, two conditions must be met. The first condition is that $U_0 = f(U_0) \oplus (1-f)(U_0)$ as vector spaces. This condition follows from the observations that $u_0 = f(u_0) + (1-f)(u_0)$ for each $u_0 \in U_0$ and $f(U_0) \cap (1-f)(U_0) = 0$, noticing that if $x = f(u_0) = (1-f)(v_0) \in f(U_0) \cap (1-f)(U_0)$, then $x = f^2(u_0) = f(x) = f(1-f)(v_0) = (f - f^2)(v_0) = 0$. The second condition is that for each i, $f(U_i) = f(U_0) \cap U_i$ and $(1-f)(U_i) = (1-f)(U_0) \cap U_i$. This condition is satisfied for the idempotent f, since $f(U_i) \subseteq f(U_0) \cap U_i$ and if $x = f(u_0) \in f(U_0) \cap U_i$, then $x = f(u_0) = f^2(u_0) = f(x) \in f(U_i)$. Since $1 - f$ is also idempotent, a similar argument shows that $(1-f)(U_i) = (1-f)(U_0) \cap U_i$, as desired.

If f is neither 0 nor 1, then U is decomposable. Consequently, if U is indecomposable, then 0 and 1 are the only idempotents of End U.

Conversely, assume that $U = V \oplus W$ is a direct sum of representations with both V_0 and W_0 nonzero. Then $U_0 = V_0 \oplus W_0$ with each $U_i = (V_0 \cap U_i) \oplus (W_0 \cap U_i)$. Let f be a vector space projection $U_0 \to V_0$ defined by $f(v_0, w_0) = v_0$. Then f is a representation morphism of U, because $f(U_i) \subseteq V_0 \cap U_i$ for each i. Furthermore, f is clearly an idempotent, and f is neither 0 nor 1, since the kernel of f is $W_0 \neq U_0$ and the image of f is $V_0 \neq U_0$. It follows that if 0 and 1 are the only idempotents of End U, then U must be indecomposable.

Representations of the form $(U_0, 0, \ldots, 0)$ with $U_0 \neq 0$ are called *trivial*. Trivial representations can be viewed as vector spaces. For instance, two trivial representations U and V are isomorphic as representations if and only if U_0 and V_0 have the same dimension as vector spaces. A trivial representation is a finite direct sum of indecomposable trivial representations, since if k^n denotes an n-dimensional vector space, then

$$(k^n, 0, \ldots, 0) = (k, 0, \ldots, 0) \oplus \cdots \oplus (k, 0, \ldots, 0)$$

is a direct sum of representations. Hence, up to isomorphism, $(k, 0, \ldots, 0)$ is the only indecomposable trivial representation.

Let $U = (U_0, U_1, \ldots, U_n)$ be a representation. Then $W_0 = U_1 + \cdots + U_n$, the subspace of U_0 generated by $U_1, \ldots, U_n$, is a vector space summand of U_0, say $U_0 = V_0 \oplus W_0$. Hence, U is the direct sum of the trivial representation $(V_0, 0, \ldots, 0)$ and the representation $(W_0, U_1, \ldots, U_n)$. In particular, if U has no trivial summands, then $U_0 = U_1 + \cdots + U_n$.

The set of isomorphism classes of indecomposable representations of S_n over a field k is denoted by $\mathrm{Ind}(n, k)$. For small n, the indecomposable representations of S_n are easily described.

Example 1.1.2 *Elements of* $\mathrm{Ind}(1, k)$ *are* $U = (k, k)$ *and* $U = (k, 0)$. *In each case,* $\mathrm{End}\, U = k$.

PROOF. As noted above, $(k, 0)$ is the only trivial indecomposable representation of S_1. If $U = (U_0, U_1)$ has no trivial summands, then $U_0 = U_1$ is a k-vector space, say of dimension m. Then U is isomorphic to $(k, k)^m$, the direct sum of m copies of the representation (k, k). Hence, U is indecomposable if and only if $m = 1$. If $U = (k, k)$ or $(k, 0)$, then $\mathrm{End}\, U = k$, since $k \subseteq \mathrm{End}\, U$ and k-endomorphisms of k are just multiplication by elements of k. But 0 and 1 are the only idempotents of the field k, whence U is indecomposable by Lemma 1.1.1.

Example 1.1.3 *Elements of* $\mathrm{Ind}(2, k)$ *are* $(k, 0, 0), (k, k, 0), (k, 0, k)$, *and* (k, k, k). *In each case, the endomorphism ring is* k.

PROOF. The representation $(k, 0, 0)$ is the only trivial indecomposable representation of S_2. Assume that U is a nontrivial indecomposable representation of S_2. Then $U = (U_0, U_1, U_2)$ with $U_0 = U_1 + U_2$. If $U_1 \cap U_2 = 0$, then $U_0 = U_1 \oplus U_2$ and $U = (U_1, U_1, 0) \oplus (U_2, 0, U_2)$. Since U is indecomposable, U is isomorphic to either $(k, k, 0)$ or $(k, 0, k)$.

Next, assume that $U_1 \cap U_2 \neq 0$ and write $U_0 = (U_1 \cap U_2) \oplus V$ for some subspace V of U_0. Since $U_1 \cap U_2 \subseteq U_i$, $U_i = (U_1 \cap U_2) \oplus (U_i \cap V)$ for $i = 1, 2$. Consequently,

$$U = (U_0, U_1, U_2) = (U_1 \cap U_2, U_1 \cap U_2, U_1 \cap U_2) \oplus (V, U_1 \cap V, U_2 \cap V).$$

As U is indecomposable and $U_1 \cap U_2 \neq 0$, it follows that $V = 0$ and U is isomorphic to (k, k, k). The argument that each of the U's listed has endomorphism ring k, hence is indecomposable, is the same as that for Example 1.1.2.

The first three indecomposable representations of S_2 listed in Example 1.1.3 have at least one subspace equal to 0. Deleting a zero subspace gives an indecomposable representation of S_1. Thus, (k, k, k) is the only indecomposable representation of S_2 that does not arise from a representation of S_1. This is an illustration of the next proposition.

Proposition 1.1.4 *Given positive integers $m \leq n$, there is a correspondence F from k-representations of S_m to k-representations of S_n such that if U and U' are k-representations of S_m, then* End $U =$ End $F(U)$ *and U is isomorphic to U' if and only if $F(U)$ is isomorphic to $F(U')$. Moreover, $F :$ Ind$(m, k) \rightarrow$ Ind(n, k) is a one-to-one function.*

PROOF. The correspondence F is defined by

$$U = (U_0, U_1, \ldots, U_m) \rightarrow F(U) = (U_0, U_1, \ldots, U_m, 0, \ldots, 0).$$

It is easy to confirm that End $U =$ End$F(U)$ and if U and U' are k-representations of S_m, then U is isomorphic to U' if and only if $F(U)$ is isomorphic to $F(U')$. As an application of Lemma 1.1.1, U is indecomposable if and only if $F(U)$ is indecomposable. Thus, $F :$ Ind$(m, k) \rightarrow$ Ind(n, k) is a well-defined one-to-one function.

Elements of Ind$(3, k)$ can also be listed, but the proof of completeness of the list is slightly more complicated than for the case $n \leq 2$. Two arguments are given. The first argument, Example 1.1.5, is technical but does illustrate the solution of a matrix problem, while the second argument, a reduction to Ind$(2, k)$, is Corollary 1.2.4.

Given a finite-dimensional k-vector space U, let $(1+1)U = \{(x, x) : x \in U\}$, the image of the diagonal embedding of U in $U \oplus U$. Write $U \oplus 0 = \{(x, 0) : x \in U\}$ and $0 \oplus U = \{(0, x) : x \in U\}$ for the coordinate spaces of $U \oplus U$. Then $(1+1)U$, $U \oplus 0$, and $0 \oplus U$ are subspaces of $U \oplus U$ and $(U \oplus U, U \oplus 0, 0 \oplus U, (1+1)U)$ is a k-representation of S_3.

Example 1.1.5 *Elements of* Ind$(3, k)$ *are*

$$(k, 0, 0, 0), \quad (k, k, 0, 0), \quad (k, 0, k, 0), \quad (k, 0, 0, k), \quad (k, k, k, 0),$$
$$(k, 0, k, k), \quad (k, k, 0, k), \quad (k, k, k, k), \quad (k \oplus k, k \oplus 0, 0 \oplus k, (1+1)k).$$

In each case, the endomorphism ring is isomorphic to k.

PROOF. Each of the representations listed has endomorphism ring k, hence is indecomposable by Lemma 1.1.1. In particular, for each of the first eight representations, $U_0 = k$, so that End $U = k$. For the last case, let $U = (k \oplus k, k \oplus 0, 0 \oplus k, (1+1)k)$ and $f \in$ End U. Since f preserves the two coordinate spaces, $f = (f_1, f_2)$ with each f_i a k-endomorphism of k. But f also preserves the diagonal embedding, whence $f_1 = f_2 \in k$. It follows that $k = \text{End}\, U$, recalling that k is contained in End U via scalar multiplication.

The remainder of the proof consists in confirming that the list is complete. In view of Proposition 1.1.4, $\text{Ind}(2, k)$ can be embedded in $\text{Ind}(3, k)$ for each two-element subset S_2 of S_3. This, together with Example 1.1.3, accounts for the first 7 representations in the list, a complete list of those indecomposables $U = (U_0, U_1, U_2, U_3)$ with some $U_i = 0$.

Suppose $U = (U_0, U_1, U_2, U_3)$ is an indecomposable representation of S_3 with each $U_i \neq 0$. Then $U_0 = U_1 + U_2 + U_3$, since U is nontrivial and indecomposable. First, assume that $W = U_1 \cap U_2 \cap U_3 \neq 0$. Write $U_0 = W \oplus V$ for some V so that $U_i = W \oplus (V \cap U_i)$ for each i, since $W \subseteq U_i$. Consequently,

$$U = (W, W, W, W) \oplus (V, V \cap U_1, V \cap U_2, V \cap U_3)$$

is a direct sum of representations. Since U is indecomposable with $W \neq 0$, it follows that $V = 0$ and U is isomorphic to (k, k, k, k).

Next, assume that $U_1 \cap U_2 \cap U_3 = 0$. Then $U_i \cap U_j = 0$ for each $1 \leq i \neq j \leq 3$. To see this, suppose, for example, that $U_1 \cap U_2 \neq 0$. Then $U_0 = (U_1 \cap U_2) \oplus V$ for some $V \supseteq U_3$, using the assumption that $U_1 \cap U_2 \cap U_3 = 0$. Hence, $U_i = (U_1 \cap U_2) \oplus (U_i \cap V)$ for $i = 1, 2$, and so

$$U = (U_0, U_1, U_2, U_3) = (U_1 \cap U_2, U_1 \cap U_2, U_1 \cap U_2, 0) \oplus (V, U_1 \cap V, U_2 \cap V, U_3).$$

But U is indecomposable with $U_1 \cap U_2 \neq 0$. Thus, $0 = V \supseteq U_3$, a contradiction to the assumption that $U_3 \neq 0$. The arguments for the remaining cases are similar.

At this stage, the assumptions are that $U = (U_0, U_1, U_2, U_3)$ is indecomposable, $U_0 = U_1 + U_2 + U_3$ with each $U_i \neq 0$, and $U_i \cap U_j = 0$ for $1 \leq i \neq j \leq 3$. Write $U_3 = (U_3 \cap (U_1 \oplus U_2)) \oplus V$ for some vector space V and observe that $U_0 = U_1 + U_2 + U_3 = (U_1 \oplus U_2) \oplus V$. Then

$$U = (U_0, U_1, U_2, U_3) = (U_1 \oplus U_2, U_1, U_2, U_3 \cap (U_1 \oplus U_2)) \oplus (V, 0, 0, V).$$

Since U is indecomposable with each $U_i \neq 0$, it follows that $V = 0$ and $U_1 \oplus U_2 = U_0 \supseteq U_3$.

The only case remaining is the case that $U = (U_0, U_1, U_2, U_3)$ is indecomposable, $U_0 = U_1 \oplus U_2$, each $U_i \neq 0$, and $U_i \cap U_3 = 0$ for $i = 1, 2$. This case can be interpreted as a matrix problem.

Let $B_1 = \{x_1, \ldots, x_r\}$ be a k-basis for U_1, $B_2 = \{y_1, \ldots, y_s\}$ a k-basis for U_2, and $B_3 = \{z_1, \ldots, z_t\}$ a k-basis for U_3, a subspace of $U_0 = U_1 \oplus U_2$. For each i,

$$z_i = \Sigma a_{ij} x_j + \Sigma b_{ij} y_j \quad \text{for some} \quad a_{ij}, b_{ij} \in k.$$

In particular, U_3 is the row space of a $t \times (r + s)$ k-matrix

$$M = (A \mid B),$$

where $A = (a_{ij})$ is a $t \times r$ k-matrix and $B = (b_{ij})$ is a $t \times s$ k-matrix. Columns of A are labeled by B_1, columns of B are labeled by B_2, and rows of M are labeled by B_3. The matrices A and B are referred to as *block matrices*.

Obviously, M depends on a choice of bases for the U_i's, but U does not. The following invertible matrix operations on M do not change U:

(a) Elementary column operations within A (a basis change for U_1),
(b) Elementary column operations within B (a basis change for U_2),
(c) Elementary row operations on M (a basis change for U_3).

Write $M \approx N$ if the matrix N can be obtained from M by a sequence of operations (a), (b), (c). Using (a) and (c), the block matrix A can be reduced to echelon form

$$\begin{pmatrix} I & 0 \\ 0 & 0 \end{pmatrix},$$

with I denoting an identity matrix. This echelon matrix must actually be of the form $(I \ 0)$ because the block matrix A cannot have a row of zeros. Otherwise, there is some row in M having nonzero entries only in the columns indexed by a basis of U_2. This row would then denote a basis element of U_3 that is also an element of $U_2 \cap U_3 = 0$, a contradiction.

In fact, A also cannot have a column of zeros. To see this, suppose that the jth column of A is all zeros. Then $U_1 = kx_j \oplus V_1$ for some V_1 with $U_3 \subseteq V_1 \oplus U_2$, and so

$$U = (kx_j, kx_j, 0, 0) \oplus (V_1 \oplus U_2, V_1, U_2, U_3).$$

Since U is indecomposable and $kx_j \neq 0$, U is isomorphic to $(k, k, 0, 0)$, a contradiction to the assumption that each $U_i \neq 0$.

Consequently,

$$M \approx (I \mid B)$$

for some B with I a $t \times t$ identity matrix, and so $t = r$. An elementary row operation E on $(I \mid B)$ yields

$$M \approx (E \mid EB).$$

However, using (a),

$$(E \mid EB) \approx (I = EE^{-1} \mid EB).$$

Hence,

$$M \approx (I \mid EB)$$

for any elementary row operation E on M.

Now use (b) and (c) to reduce B to echelon form I, noting that B cannot have a row or column of zeros for the same reasons that A cannot. As a consequence of the preceding remarks, it follows that

$$M \approx (I \mid I).$$

In particular, $t = r = s$ and U_3 is isomorphic to $\oplus\{k(x_j + y_j): 1 \leq j \leq t\}$, corresponding to the row space of $(I \mid I)$. Thus, U is isomorphic to $\oplus\{(kx_j \oplus ky_j, kx_j, ky_j, k(x_j + y_j)): 1 \leq j \leq t\}$, a direct sum of representations. Since U is indecomposable, U must be isomorphic to $(k \oplus k, k \oplus 0, 0 \oplus k, (1+1)k)$. This completes the proof that an indecomposable representation U of S_3 is isomorphic to one of the nine representations listed.

The preceding examples confirm that $\text{Ind}(n, k)$ is finite and the endomorphism ring of each indecomposable k-representation of S_3 is isomorphic to k for $n \leq 3$. The following two examples demonstrate that neither of these properties holds for $n \geq 4$. Let $k[x]$ denote the polynomial ring with coefficients in k.

Example 1.1.6
(a) *Ind(n, k) is infinite for $n \geq 4$.*
(b) *For each irreducible polynomial $g(x)$ in $k[x]$ and each integer $e \geq 1$, there is an indecomposable k-representation U of S_4 with $\text{End}\, U = k[x]/\langle g(x)^e \rangle$.*

PROOF. Let A be an $m \times m$ k-matrix and define

$$U_A = (k^m \oplus k^m, k^m \oplus 0, 0 \oplus k^m, (1+1)k^m, (1+A)k^m),$$

where $(1 + A)k^m = \{(x, Ax): x \in k^m\}$. Then U_A is a k-representation of S_4.

Given another $m \times m$ k-matrix B, U_A is isomorphic to U_B if and only if A and B are *similar*, i.e., there is an invertible k-matrix M with $MAM^{-1} = B$. To see this, let $f : U_A \to U_B$ be an isomorphism. Since f preserves the first three subspaces, it follows that $f = (h, h)$ with h a k-automorphism of k^m. Moreover,

$$f((1 + A)k^m) = \{(h(x), h(Ax)): x \in k^m\} = \{(y, By): y \in k^m\} = (1 + B)k^m,$$

whence $f(Ax) = h(Ax) = Bh(x) = Bf(x)$ for each x in k^m. Letting M be the matrix of h relative to the standard basis for k^m gives an invertible matrix M with $MA = BM$. Conversely, the existence of such an M gives rise to a representation isomorphism $f = (h, h)$ from U_A to U_B.

Let $\text{Mat}_m(k)$ denote the ring of all $m \times m$ k-matrices. An argument like that of the preceding paragraph shows that $\text{End}\, U_A$ is isomorphic to the ring $C(A) = \{M : M \in \text{Mat}_m(k), MA = AM\}$, called the *centralizer* of the matrix A.

The remainder of the argument is an exercise in linear algebra [Hungerford, 74]. Given an $m \times m$ k-matrix A, define a $k[x]$-module structure on $V_A = k^m$ by $xy = Ay$ for each $y \in V_A$. Since $k[x]$ is a principal ideal domain, V_A is an indecomposable $k[x]$-module if and only if V_A is isomorphic to $k[x]/\langle g(x)^e \rangle$ for some irreducible polynomial $g(x) \in k[x]$ and integer $e \geq 1$. In this case, m is the degree of $g(x)^e$, and $g(x)^e$ is the minimal polynomial of A.

Now assume that V_A is an indecomposable $k[x]$-module. As x acts on V_A by A, $C(A)$ is isomorphic to $\text{End}_{k[x]}(k[x]/\langle g(x)^e \rangle)$. But $k[x]/\langle g(x)^e \rangle$ is isomorphic to $\text{End}_{k[x]}(k[x]/\langle g(x)^e \rangle)$ via $b \mapsto$ multiplication by b. It follows that End U_A is isomorphic to $k[x]/\langle g(x)^e \rangle$. Hence, U_A is an indecomposable representation by Lemma 1.1.1, since 0 and 1 are the only idempotents of $k[x]/\langle g(x)^e \rangle$.

(a) In view of Proposition 1.1.4, it is sufficient to assume that $n = 4$. Since $m \geq 1$ is arbitrary, there are infinitely many indecomposable k-representations U_A of S_4 with A an $m \times m$ k-matrix. In particular, for each $\lambda \in k$ and $m \geq 1$, there is an $m \times m$ Jordan block matrix $A(m, \lambda)$ with minimal polynomial $(x - \lambda)^m$, where $A(m, \lambda)$ is the matrix with λ's on the diagonal, 1's on the superdiagonal, and 0's elsewhere. Notice that $V_{A(m,\lambda)}$ is an indecomposable $k[x]$-module and $A(m, \lambda)$ and $A(m', \lambda')$ are similar if and only if $m = m'$ and $\lambda = \lambda'$. This shows that

$$\{U_A : A = A(m, \lambda), m \geq 1, \lambda \in k\}$$

is an infinite subset of $\text{Ind}(4, k)$.

(b) Suppose $g(x)$ is an irreducible polynomial and $V = k[x]/\langle g(x)^e \rangle$ has dimension m as a k-vector space. Pick a basis for V and let A be the $m \times m$ k-matrix determined by multiplication by x relative to this basis. As above, End U_A is isomorphic to $k[x]/\langle g(x)^e \rangle$, and so U_A is an indecomposable representation.

Even though there are infinitely many elements in $\text{Ind}(4, k)$, they have all been classified; see Example 6.2.7. In fact, each U in $\text{Ind}(4, k)$ has endomorphism ring isomorphic to either k or $k[x]/\langle g(x)^e \rangle$ for some irreducible $g(x)$ in $k[x]$.

The situation is quite different for $n \geq 5$. Given a field k, a ring R is a *k-algebra* if $k = k1_R$ is a subring of $CR = \{x \in R : xy = yx \text{ for all } y \in R\}$, a subring of R called the *center* of R. A k-algebra R is a k-vector space and is said to be a *finite-dimensional k-algebra* if R has finite k-dimension.

Example 1.1.7 *If $n \geq 5$ and R is a finite-dimensional k-algebra, then there is a k-representation U of S_n with End U isomorphic to R.*

PROOF. Via Proposition 1.1.4, it is sufficent to assume that $n = 5$. Let A and B be two $m \times m$ k-matrices. Define the centralizer of A and B, $C(A, B)$, to be $\{M : M \in \text{Mat}_m(k), MA = AM, \text{ and } MB = BM\}$, a subring of $\text{Mat}_m(k)$. Define

$$U_{A,B} = (k^m \oplus k^m, k^m \oplus 0, 0 \oplus k^m, (1 + 1)k^m, (1 + A)k^m, (1 + B)k^m),$$

a k-representation of S_5.

Then End $U_{A,B}$ is isomorphic to $C(A, B)$. To see this, first observe that if $f \in$ End $U_{A,B}$, then $f = (g, g)$ for some endomorphism g of k^m, since f preserves the first three subspaces of $k^m \oplus k^m$. Preservation of the last two subspaces yields $g(Ax) = Ag(x)$ and $g(Bx) = Bg(x)$ for each $x \in k^m$. Choose M to be the matrix of g relative to the standard basis for k^m so that $MA = AM$ and $MB = BM$.

It now follows that the correspondence $f \mapsto M$ is an isomorphism from
End $U_{A,B}$ to $C(A, B)$, noticing that if $M \in C(A, B)$, then M induces a endomorphism of $U_{A,B}$.

Let R be a finite-dimensional k-algebra and $\{x_1, \ldots, x_t\}$ a k-basis for R with
$x_1 = 1$. Define $(t + 2) \times (t + 2)$ R-matrices

$$A = \begin{pmatrix} 0 & 1 & 0 & \cdots & 0 & 0 \\ 0 & 0 & 1 & \cdots & 0 & 0 \\ & & & \cdots & & \\ 0 & 0 & 0 & \cdots & 0 & 1 \\ 0 & 0 & 0 & \cdots & 0 & 0 \end{pmatrix} ; \quad B = \begin{pmatrix} 0 & 0 & 0 & \cdots & 0 & 0 & 0 \\ 1 & 0 & 0 & \cdots & 0 & 0 & 0 \\ 1 & 1 & 0 & \cdots & 0 & 0 & 0 \\ 0 & x_2 & 1 & \cdots & 0 & 0 & 0 \\ & & & \cdots & & & \\ 0 & 0 & 0 & \cdots & 1 & 0 & 0 \\ 0 & 0 & 0 & \cdots & x_t & 1 & 0 \end{pmatrix}.$$

It suffices to prove that R is isomorphic to $C(A, B)$. In this case, $U_{A,B}$ is a k-
representation of S_5 with End $U_{A,B}$ isomorphic to R, as desired.

Suppose $M = (a_{ij}) \in C(A, B)$ is a $(t + 2) \times (t + 2)$ matrix with each a_{ij} a $t \times t$
k-matrix representing a k-endomorphism of the t-dimensional vector space R.
Since $MA = AM$, it follows that M is of the form

$$\begin{pmatrix} a_1 & a_2 & a_3 & \cdots & a_{t+2} \\ 0 & a_1 & a_2 & \cdots & a_{t+1} \\ & & & \cdots & \\ 0 & 0 & 0 & \cdots & a_2 \\ 0 & 0 & 0 & \cdots & a_1 \end{pmatrix}.$$

Since $MB = BM$, it follows that $M = aI$ for some $a \in \mathrm{End}_k R$ with $ax_i = x_i a$
for each i. Because the x_i's are a k-basis for R, $ar = ra$ for each $r \in R$. In
particular, a is an element of $\mathrm{End}_R(R)$, the R-endomorphism ring of R regarded
as a left R-module. This shows that $C(A, B)$ is contained in $\mathrm{End}_R(R)$. Conversely,
let $f \in \mathrm{End}_R(R)$ and let a denote the $t \times t$ k-matrix of f relative to the k-basis
$\{x_1, \ldots, x_t\}$ of R. Then $M = aI \in C(A, B)$, and so $C(A, B) = \mathrm{End}_R(R)$. Finally,
R is isomorphic to $\mathrm{End}_R(R)$, the isomorphism given by $r \mapsto \mu_r$, where μ_r is right
multiplication by r. The proof is now complete.

EXERCISES

1. Two $m \times m$ k-matrices A and B are *equivalent* if there are invertible $m \times m$ k-matrices
   M and N with $MAN = B$. Define a k-representation W_A of S_3 by

   $$W_A = (k^m \oplus k^m, k^m \oplus 0, 0 \oplus k^m, (1 + A)k^m).$$

   (a) Prove that A and B are equivalent if and only if the representations W_A and W_B are
       isomorphic.
   (b) Use Example 1.1.5 to prove that a square k-matrix is equivalent to a diagonal matrix
       with entries either 0 or 1.

2. Two pairs of $m \times m$ k-matrices (A, B) and (A', B') are *simultaneously equivalent* if there are invertible k-matrices P and Q with $PA = A'Q$ and $PB = B'Q$. Define a k-representation $W_{A,B}$ of S_4 by

$$W_{A,B} = (k^m \oplus k^m, k^m \oplus 0, 0 \oplus k^m, (1+A)k^m, (1+B)k^m).$$

   (a) Prove that (A, B) and (A', B') are simultaneously equivalent if and only if $W_{A,B}$ and $W_{A',B'}$ are isomorphic as representations.
   (b) All elements of $\text{Ind}(S_4, k)$ are listed in Example 6.2.8. What are the consequences of this classification for pairs of square k-matrices up to simultaneous equivalence?

3. Two pairs (A, B) and (A', B') of $m \times m$ k-matrices are *simultaneously similar* if there is an invertible $m \times m$ k-matrix P with $PAP^{-1} = A'$ and $PBP^{-1} = B'$. Let $U_{A,B}$ be the k-representation of S_5 defined in the proof of Example 1.1.7.
   (a) Prove that $U_{A,B}$ and $U_{A',B'}$ are isomorphic as representations if and only if (A, B) and (A', B') are simultaneously similar.
   (b) What are the consequences of Example 1.1.7, Example 1.4.1, and Proposition 1.4.2 for the problem of classifying pairs of matrices up to simultaneous similarity?

1.2 Representations of Posets and Matrix Problems

A representation of a finite partially ordered set (poset) over a field can be written uniquely up to isomorphism and order as a finite direct sum of indecomposable representations (Theorem 1.2.2). In particular, classification of representations up to isomorphism reduces to classification of indecomposable representations. With each poset, there is an associated matrix problem as illustrated in Example 1.1.5 for S_3. Interpretation of representations as matrices gives a procedure for computing endomorphism rings and constructing indecomposable representations (Theorem 1.2.6 and Example 1.2.7).

A *partially ordered set* is a set S together with a binary relation $\leq$ on S that is *reflexive*, $s \leq s$ for each $s \in S$; *antisymmetric*, if $s \leq t$ and $t \leq s$, then $s = t$; and *transitive*, if $s \leq t$ and $t \leq u$, then $s \leq u$. An *antichain* is a partially ordered set such that no two elements are related. For example, S_n is an antichain.

Fundamental properties of representations of posets are conveniently expressed in a categorical setting. Following is a brief summary of standard categorical terminology that will be used henceforth. A *category* is a class $\mathbf{X}$ of objects together with a set of morphisms $\text{Hom}(A, B)$ for each pair A, B of objects in $\mathbf{X}$ and a composition $\text{Hom}(B, C) \times \text{Hom}(A, B) \mapsto \text{Hom}(A, C)$, written $(g, f) \mapsto gf$, for each triple A, B, C of objects of $\mathbf{X}$ such that

   (i) composition is associative and
   (ii) for each object A there is 1_A in $\text{Hom}(A, A)$ with $f1_A = f$ and $1_A g = g$ for $f \in \text{Hom}(A, B)$, $g \in \text{Hom}(C, A)$, and objects B and C of $\mathbf{X}$.

The notation $A \in \mathbf{X}$ stands for "A is an object of $\mathbf{X}$," a mild abuse of notation, since $\mathbf{X}$ need not be a set. A category $\mathbf{X}$ is *additive* if, in addition,

(iii) for each A, $B \in \mathbf{X}$, there is a binary operation $+$ such that $(\mathrm{Hom}(A, B), +)$ is an abelian group with $g(f_1 + f_2) = gf_1 + gf_2$ and $(f_1 + f_2)h = f_1h + f_2h$ for $f_i \in \mathrm{Hom}(A, B)$, $g \in \mathrm{Hom}(B, C)$, $h \in \mathrm{Hom}(D, A)$, and C, $D \in \mathbf{X}$ and
(iv) given $A_1, \ldots, A_n \in \mathbf{X}$ there is an object A and morphisms $i_j \in \mathrm{Hom}(A_j, A)$ such that if $f_j \in \mathrm{Hom}(A_j, B)$, then there is a unique $f \in \mathrm{Hom}(A, B)$ with $fi_j = f_j$ for each $1 \le j \le n$.

The object A of (iv) is called a *direct sum* of $A_1, \ldots, A_n$, denoted up to isomorphism by $A_1 \oplus \cdots \oplus A_n$. The i_j's are called *injections*. If $\mathbf{X}$ is an additive category and A is an object of $\mathbf{X}$, then it follows from (iii) that $\mathrm{Hom}(A, A)$ is a ring with identity 1_A, denoted by End A.

As an example, the class of finite-dimensional vector spaces over a field k is an additive category; morphisms are k-linear transformations. In this case, $A = A_1 \oplus \cdots \oplus A_n$ is the usual vector space direct sum with injections $i_j : A_j \to A$ given by $i_j(a_j) = (0, \ldots, 0, a_j, 0, \ldots, 0)$.

Let k be a field and S a finite poset. Define rep(S, k) to be the category with objects $U = (U_0, U_i : i \in S)$, where U_0 is a finite-dimensional k-vector space, each U_i is a subspace of U_0, and if $i \le j$ in S, then U_i is contained in U_j. Morphisms in this category, called *representation morphisms*, are k-linear transformations $f : U_0 \to U_0'$ with $f(U_i)$ a subset of U_i' for each i in S. If U is an object of rep(S, k), then End U is a k-algebra, since endomorphisms of U commute with multiplication by elements of k. Objects of rep(S_n, k) are precisely the k-representations of S_n, as defined in Section 1.1.

Let R be a ring with identity 1_R. An element r of R is a *unit* of R if there is an $s \in R$ such that $rs = 1_R = sr$. The ring R is *local* if $r + s$ is a nonunit for each pair r and s of nonunits of R.

Lemma 1.2.1 *Suppose k is a field and S is a finite poset.*

(a) *The category* rep(S, k) *is an additive category. Furthermore, if $U \in$ rep(S, k) and e is an idempotent in* End U, *then $U = e(U) \oplus (1 - e)(U)$ in* rep(S, k).
(b) *A representation U in* rep(S, k) *is indecomposable if and only if* End U *is a local ring.*

PROOF. (a) A routine verification shows that rep(S, k) is an additive category with a direct sum of $U = (U_0, U_i : i \in S)$ and $V = (V_0, V_i : i \in S)$ given by

$$U \oplus V = (U_0 \oplus V_0, U_i \oplus V_i : i \in S).$$

This follows directly from the fact that the category of finite-dimensional vector spaces is an additive category. Injections for this direct sum are representation morphisms $i_U : U \to U \oplus V$ defined by $i_U(u_0) = u_0 \oplus 0$ for each $u_0 \in U_0$ and $i_V(v_0) = 0 \oplus v_0$ for each $v_0 \in V_0$. Moreover, if U is in rep(S, k) and $e^2 = e$ in End(U), then

$$U = e(U) \oplus (1 - e)(U) = (e(U_0), e(U_i) : i \in S) \oplus ((1 - e)(U_0), (1 - e)(U_i) : i \in S)$$

is a direct sum decomposition of U in rep(S, k). The proof is just as in Lemma 1.1.1 for the case that $S = S_n$.

(b) A representation U is indecomposable if and only if 0 and 1 are the only idempotents of End U, a consequence of (a). The ring End U is a finite-dimensional k-algebra, since U_0 is a finite-dimensional k-space and End U is contained in $\mathrm{End}_k U_0$. Hence, End U is a local ring if and only if 0 and 1 are the only idempotents of End U (Exercise 1.2.3). The proof of (b) is now complete.

Suppose $A_1, \ldots, A_n \in \mathrm{rep}(k, S)$ with each $A_j = (A_{j0}, A_{js} : s \in S)$. Then $U = (U_0, U_s : s \in S) = (\oplus_j A_{j0}, \oplus_j A_{js} : s \in S) = A_1 \oplus \cdots \oplus A_n$ is a direct sum in rep(k, S) with injections $i_j : A_j \to U$ defined by $i_j(a_{j0}) = (0, \ldots, 0, a_{j0}, 0, \ldots, 0)$ for each $1 \le j \le n$. Define $p_j : U_0 \to A_{j0}$ by $p_j(a_{10}, \ldots, a_{n0}) = a_{j0}$ for each j. Each p_j is a representation morphism such that $p_j i_t = 0$ if $j \ne t$, $p_j i_j$ the identity on A_j, $e_j = i_j p_j$ is an idempotent of End U, and $1_U = e_1 + \cdots + e_n$. The p_j's are called *projections* for the direct sum $U = A_1 \oplus \cdots \oplus A_n$.

Conversely, if $U, A_1, \ldots, A_n \in \mathrm{rep}(k, S)$ with morphisms $i_j : A_j \to U$ and $p_j : U \to A_j$ for each $1 \le j \le n$ such that $p_j i_t = 0$ if $j \ne t$, $p_j i_j$ is the identity of A_j, $e_j = i_j p_j$ is an idempotent of End U, and $1_U = e_1 + \cdots + e_n$, then U is isomorphic to $A_1 \oplus \cdots \oplus A_n$ (Lemma 1.2.1(a) and Exercise 1.2.6(b)).

Theorem 1.2.2 *Let S be a finite poset and k a field,*

(a) *Each $U \in \mathrm{rep}\,(S, k)$ is a finite direct sum of indecomposable representations.*

(b) *A decomposition of U into indecomposable representations is unique up to isomorphism and order of summands.*

PROOF. (a) The proof is by induction on dim U, the k-dimension of U_0. If dim $U = 1$, then End $U = k$, and so U is indecomposable. Now assume that dim $U > 1$ and that U is not indecomposable. Write $U = V \oplus W$ as a direct sum of nonzero representations. Then dim $V < $ dim U and dim $W < $ dim U. By induction, both V and W are finite direct sums of indecomposable representations. Since dim U is finite, the proof is complete.

(b) Suppose $U = A_1 \oplus \cdots \oplus A_m = B_1 \oplus \cdots \oplus B_n$, a direct sum of representations with each $A_j = (A_{j0}, A_{js} : s \in S)$ and $B_i = (B_{i0}, B_{is} : s \in S)$ indecomposable representations. Since $U = A_1 \oplus \cdots \oplus A_m$, there are injections $i_j : A_j \to U$ and projections $p_j : U \to A_j$ as defined above with $p_j i_t = 0$ if $j \ne t$, $p_j i_j$ the identity of A_j, $e_j = i_j p_j$ an idempotent of End U, and $1_U = e_1 + \cdots + e_m$. Similarly, there are injections $i'_j : B_j \to U$ and projections $p'_j : U \to B_j$ for each $1 \le j \le n$ with $p'_j i'_t = 0$ if $j \ne t$, $p'_j i'_j$ the identity on B_j, $e'_j = i'_j p'_j$ an idempotent of End U, and $1_U = e'_1 + \cdots + e'_n$.

Notice that the identity of A_1 is equal to $p_1(1_U)i_1 = p_1(e'_1 + \cdots + e'_n)i_1 = \Sigma_t p_1(i'_t p'_t)i_1 = \Sigma_t(p_1 i'_t)(p'_t i_1)$. Since End A_1 is a local ring, $(p_1 i'_t)(p'_t i_1) = u$ is a unit of End A_t for some t. Assume, for notational convenience, that $t = 1$. This involves a permutation of the subscripts of the B_j's. Then $f = p_1 i'_1 : B_1 \to A_1$

and $g = p'_1 i_1 u^{-1} : A_1 \to B_1$ with $fg = uu^{-1} = 1$. Now, $e = gf \in \text{End } B_1$ with $e^2 = fgfg = fg = e$. By Lemma 1.2.1(a), $B_1 = e(A_1) \oplus (1 - e)(A_1)$. As B_1 is indecomposable, $B_1 = e(A_1)$ and $eg : A_1 \to B_1$ is an isomorphism.

It follows that $U = A_1 \oplus B_2 \oplus \cdots \oplus B_n$, where the injection for A_1 is $i_1 u^{-1}$, the projection for A_1 is $fp'_1 = (p_1 i'_1) p'_1 = p_1 (i'_1 p'_1) = p_1 e'_1$, the injection for each B_j is i'_j, and the projection for each B_j is p'_j. Since $U = A_1 \oplus A_2 \oplus \cdots \oplus A_m$, $A_2 \oplus \cdots \oplus A_m$ is isomorphic to $B_2 \oplus \cdots \oplus B_n$. The proof is completed by an induction on m.

The point of Theorem 1.2.2 is that all representations in $\text{rep}(S, k)$ can be determined uniquely up to isomorphism from the indecomposable ones. Write $\text{Ind}(S, k)$ for the set of isomorphism classes of indecomposables in $\text{rep}(S, k)$. If $S = S_n$, then $\text{Ind}(S, k) = \text{Ind}(n, k)$, as defined in Section 1.1.

The next theorem shows that classification of nontrivial elements of $\text{Ind}(S_n, k)$ can be interpreted as a matrix problem of the kind discussed in Section 1.1. Define $\Delta(S_n, k)$ to be the full subcategory of $\text{rep}(S_{n+1}, k)$ with objects of the form

$$U = (U_0 = \oplus_i U_i, U_i, U_* : i \in S_n)$$

such that $U_i \cap U_* = 0$ for each $i \in S_n$. The term *full subcategory* means that objects of $\Delta(S_n, k)$ are objects of $\text{rep}(S_{n+1}, k)$ and morphism sets of objects in $\Delta(S_n, k)$ are exactly as in $\text{rep}(S_{n+1}, k)$.

Theorem 1.2.3 *Let k be a field.*

(a) *There is a correspondence $H : \Delta(S_n, k) \to \text{rep}(S_n, k)$ with $\text{Hom}(U, U')$ isomorphic to $\text{Hom}(H(U), H(U'))$ for each U, U' in $\Delta(S_n, k)$. The image of H consists of all representations of S_n with no trivial summands.*

(b) *The correspondence H induces an injection $\text{Ind}(\Delta(S_n, k)) \to \text{Ind}(S_n, k)$ with nontrivial indecomposable representations as the image.*

PROOF. (a) Define $H : \Delta(S_n, k) \to \text{rep}(S_n, k)$ by

$$H(U) = (V_0, V_i : i \in S) \text{ with } V_0 = U_0 / U_* \text{ and } V_i = (U_i + U_*) / U_*,$$

where $U = (U_0 = \oplus_i U_i, U_i, U_* : i \in S_n) \in \Delta(S_n, k)$. If $f : U \to U'$ is a morphism in $\Delta(S_n, k)$, then f induces a representation morphism $H(f) : H(U) \to H(U')$, since $f(U_*) \subseteq U'_*$ and $f(U_i) \subseteq U'_i$ for each i in S_n. The correspondence H induces a group homomorphism $H : \text{Hom}(U, U') \to \text{Hom}(H(U), H(U'))$. This homomorphism is one-to-one, since if $H(f) = 0$, then $f(U_0) \subseteq U'_*$ and $f(U_i) \subseteq U'_i \cap U'_* = 0$ for each i.

To confirm that $H : \text{Hom}(U, U') \to \text{Hom}(H(U), H(U'))$ is onto, let $g : H(U) \to H(U')$. Then $g(V_i) \subseteq V'_i$ with $V_i = (U_i + U_*) / U_*$ isomorphic to U_i, since $U_i \cap U_* = 0$. In particular, g induces a k-linear transformation $f : \oplus_i U_i \to \oplus_i U'_i$ with $f(U_i) \subseteq U'_i$ and $\pi' f = g\pi$, where $\pi : \oplus U_i \to V_0$ is the k-linear transformation with $\ker \pi = \{x \in \oplus U_i \mid \pi(x) = 0\} = U_*$. Therefore, $f : U_* = \ker \pi \to$

$U'_* = \ker \pi'$, from which it follows that $f : U \to U'$ is a representation morphism with $H(f) = g$.

To find the image of H, suppose $V = (V_0, V_i : i \in S) \in \text{rep}(S_n, k)$ with no trivial summands. Then $V_0 = \Sigma_i V_i$, and so inclusion of each V_i in V_0 induces an epimorphism $\pi : \oplus_i V_i \to V_0$ with $\ker \pi = V_*$ and $V_i \cap V_* = 0$ for each i. Consequently, $(\oplus_i V_i, V_i, V_*) \in \Delta(S_n, k)$ with $H(\oplus_i V_i, V_i, V_*) = V$, as desired.

(b) By (a), $H : \text{Hom}(U, V) \to \text{Hom}(H(U), H(V))$ is a group isomorphism for each U and V in $\Delta(S_n, k)$. In particular, U is isomorphic to V if and only if $H(U)$ is isomorphic to $H(V)$. Notice that $H(fg) = H(f)H(g)$ for representation morphisms f and g. Thus, e is an idempotent of End U if and only if $H(e)$ is an idempotent of End $H(V)$, and if f is an idempotent of End $H(V)$, then there is an indempotent e in End U with $H(e) = f$. In view of Lemma 1, U is indecomposable if and only if 0 and 1 are the only idempotents of End U. Since $H(0) = 0$ and $H(1_U) = 1_{H(U)}$, the correspondence sending the isomorphism class of U to the isomorphism class of $H(U)$ induces an injection $\text{Ind}(\Delta(S_n, k)) \to \text{Ind}(S_n, k)$. Finally, apply (a) to see that the image consists of the isomorphism classes of nontrivial indecomposables in $\text{rep}(S_n, k)$.

In Example 1.1.5, classification of indecomposable representations in $\text{rep}(S_3, k)$ is reduced to classification of indecomposable representations in $\Delta(S_2, k)$. These indecomposables were then described by solving a matrix problem. As an alternative proof, the correspondence H of Theorem 1.2.3 can be used to translate indecomposables in $\Delta(S_2, k)$ into indecomposables in $\text{rep}(S_2, k)$, which are then easily classified as in Example 1.1.3.

Corollary 1.2.4 *If* $U = (U_1 \oplus U_2, U_1, U_2, U_3) \in \text{rep}(S_3, k)$ *is indecomposable with each* $U_i \neq 0$ *and* $U_1 \cap U_3 = U_2 \cap U_3 = 0$, *then* U *is isomorphic to* $(k \oplus k, k \oplus 0, 0 \oplus k, (1 + 1)k)$.

PROOF. Notice that $U \in \Delta(S_2, k)$. By Theorem 1.2.3, $H(U) = (V_0, V_1, V_2) \in \text{rep}(S_2, k)$ is indecomposable with V_i isomorphic to $U_i \neq 0$ for $i = 1, 2$ and $V_0 = (U_1 \oplus U_2)/U_3$. As a consequence of Example 1.1.3, $H(U)$ is isomorphic to (k, k, k). However, $H(k \oplus k, k \oplus 0, 0 \oplus k, (1 + 1)k) =$

$$((k \oplus k)/(1 + 1)k, (k \oplus 0)/(1 + 1)k, (0 \oplus k)/(1 + 1)k)$$

is isomorphic to (k, k, k). Apply Theorem 1.2.3 to conclude that U must be isomorphic to $(k \oplus k, k \oplus 0, 0 \oplus k, (1 + 1)k)$.

In view of Theorem 1.2.3, indecomposable representations in $\text{rep}(S_n, k)$ are determined by indecomposable representations in $\Delta(S_n, k)$. Each representation U in $\Delta(S_n, k)$ can be interpreted as a partitioned matrix M_U. Let $U = (U_0, U_i, U_* : i \in S_n) \in \Delta(S_n, k)$, recalling that $U_0 = \oplus_i U_i$ and $U_i \cap U_* = 0$ for each i. Pick a basis B_i for each U_i and a basis B of U_*. Write each element of B as a k-linear combination of elements of the B_i's, and use the resulting coefficients to represent

U_* as the row space of a k-matrix

$$M_U = (\cdots |A_i| \cdots)$$

with one block matrix A_i for each $i \in S_n$. The columns of each A_i are labeled by B_i, and the rows of M_U are labeled by B, with the convention that an A_i-block is empty if $U_i = 0$.

The condition that $U_i \cap U_* = 0$ for each i means that each row of M_U must have nonzero entries in at least two of the blocks. Moreover, if U has no 1-dimensional summands, then each column of M_U must be nonzero. A detailed example of the construction and properties of M_U is given in the proof of Example 1.1.5, wherein $n = 3$.

The representation U remains unchanged by the following invertible matrix operations on M_U:

(a) Elementary column operations within each block A_i for each $i \in S_n$ (a basis change of U_i),
(b) Elementary row operations on M_U (a basis change of U_*).

Representations U and V in $\Delta(S_n, k)$ are isomorphic if M_U can be reduced to M_V by a series of matrix operations (a) and (b). This is because U and V are isomorphic if and only if there is an isomorphism $f : U_0 \to V_0$ with $f(U_i) = V_i$ for each i. Hence, classification of indecomposables in $\Delta(S_n, k)$ up to isomorphism amounts to the solution of a *matrix problem*, i.e., find canonical forms for the matrices M_U subject to the matrix operations (a) and (b).

The following example demonstrates how to interpret the representation U_A of S_4 given in Example 1.1.6 as a matrix.

A square matrix A is *indecomposable* if the vector space V_A, as defined in Example 1.1.6, is an indecomposable $k[x]$-module.

Example 1.2.5 *Let A be an indecomposable $n \times n$ k-matrix. If $U = (U_0 = k^n \oplus k^n \oplus k^n \oplus k^n, U_1, U_2, U_3, U_4, U_*) \in \Delta(S_4, k)$, where U_i is the ith coordinate space of U_0 for $1 \le i \le 4$ and U_* is the row space of the matrix*

$$M_U = \left(\begin{array}{c|c|c|c} I & I & I & 0 \\ I & A & 0 & I \end{array} \right),$$

then $H(U)$ is isomorphic to $(k^n \oplus k^n, k^n \oplus 0, 0 \oplus k^n, (1+1)k^n, (1+A)k^n) = U_A$.

PROOF. Notice that $U_0 = U_1 \oplus \cdots \oplus U_4$ and that $U_i \cap U_* = 0$ for each i, since no linear combination of the rows of M_U will result in a vector with a nonzero entry in exactly one block. The dimension of U_0 is $4n$, the dimension of each U_i is n, and the dimension of U_* is $2n$. With H as defined in Theorem 1.2.3, $H(U) = (V_0, V_1, V_2, V_3, V_4)$, where $V_0 = V_1 \oplus V_2 = ((U_1 + U_*)/U_*) \oplus ((U_2 + U_*)/U_*) = k^n \oplus k^n$. The first row of M_U gives $V_3 = (U_3 + U_*)/U_* = (1+1)k^n$, and the second row yields $V_4 = (U_4 + U_*)/U_* = (1+A)k^n$.

There is a generalization of the notion of a matrix problem for $\mathrm{rep}(S_n, k)$ to an arbitrary finite poset S, but the construction is more complicated than it is for antichains.

Let $U = (U_0, U_i : i \in S) \in \mathrm{rep}(S, k)$ and assume that U has no trivial summands. For each i in S, $U_i = U_i^* \oplus \Sigma\{U_j : j < i\}$ for some subspace U_i^* of U_i with U_i^* isomorphic to $U_i / \Sigma\{U_j : j < i\}$. In general, the subspace U_i^* is unique only up to isomorphism. Since U has no trivial summands, $U_0 = \Sigma\{U_i : i \in S\}$, whence $U_0 = \Sigma\{U_i^* : i \in S\}$ and $U_i = \Sigma\{U_i^* : j \leq i\}$ for each i in S. Let U_* denote the kernel of the k-linear transformation $\oplus\{U_i^* : i \in S\} \to U_0$ induced by inclusion of each U_i^* in U_0. Then $U_i^* \cap U_* = 0$ for each i.

To find a matrix associated with U, pick a basis B_i for each U_i^* and a basis B of U_*. Write each element of B as a k-linear combination of elements of the B_i's and use the resulting coefficients to represent U_* as the row space of a matrix

$$M_U = (\cdots | A_i | \cdots)$$

with one block matrix A_i for each $i \in S$. The columns of A_i are labeled by B_i, the rows of M_U are labeled by B, and the block A_i is empty if $U_i^* = 0$. If $S = S_n$, then $U_i^* = U_i$ for each i so that this construction agrees with the previous construction for antichains.

Theorem 1.2.3 can be generalized to arbitrary finite posets (Exercise 1.2.7). As part of this generalization, the next theorem shows how the endomorphism ring of a representation U can be computed from the ring of endomorphisms R_U of the partitioned matrix M_U. This, together with Lemma 1.2.1, gives a computational procedure to determine whether or not U is an indecomposable representation. Let $\mathrm{rs}(M_U)$ denote the row space of the matrix M_U.

Theorem 1.2.6 *Let S be a finite poset, k a field, and $U = (U_0, U_i : i \in S) \in \mathrm{rep}(S, k)$ with no trivial summands. For each $i \in S$, choose a summand U_i^* of U_i with U_i^* isomorphic to $U_i / \{\Sigma U_j : j < i\}$. Then $\mathrm{End}\, U$ is isomorphic to the ring R_U consisting of all $f = \oplus\{f_i : i \in S\} \in \Pi\{\mathrm{End}\, U_i^* : i \in S\}$ with $f(\mathrm{rs}(M_U)) \subseteq \mathrm{rs}(M_U)$.*

PROOF. Recall that U_* is the kernel of the onto linear transformation $\oplus\{U_i^* : i \in S\} \to U_0$ with $U_i = \Sigma\{U_j^* : j \leq i \text{ in } S\}$ for each i in S and $U_* = \mathrm{rs}(M_U)$. Each g in $\mathrm{End}\, U$ induces $\phi(g) = \oplus\{g_i : i \in S\}$ with $g_i : U_i / (\Sigma\{U_j : j < i\}) \to U_i / (\Sigma\{U_j : j < i\})$ defined by $g_i(u_i + \Sigma_{j<i} U_j) = g(u_i) + \Sigma_{j<i} U_j$. This is well-defined, since $g(U_i) \subseteq U_i$ for each i.

Identify U_i^* with $U_i / (\Sigma\{U_j : j < i\})$ for each $i \in S$. Then $\phi(g)(\mathrm{rs}(M_U)) = \phi(g)(U_*) \subseteq U_* = \mathrm{rs}(M_U)$. Hence, $\phi : \mathrm{End}\, U \to R_U$ is a 1-to-1 ring homomorphism. To show that ϕ is onto, let $f = \oplus\{f_i : i \in S\} \in R_U$. Then $f = \oplus_i f_i$ is an endomorphism of $\oplus\{U_i^* : i \in S\}$ with $f(U_*) \subseteq U_*$. Since $U_0 = (\oplus_i U_i^*)/U_*$, f induces an endomorphism g of U_0 with $g(U_i) \subseteq U_i$ for each i, recalling that $U_i = \Sigma\{U_j^* : j \leq i \text{ in } S\}$. Hence, $g \in \mathrm{End}\, U$ with $\phi(g) = f$, as desired.

Indecomposable matrices can be used to construct representations U that can be shown to be indecomposable by computing their endomorphism rings via R_U.

Example 1.2.7 *Let S be the poset $(1 < 2, 3 < 4, 5 < 6)$. Given an $n \times n$ k-matrix A, define $(U_0, U_1, U_2, U_3, U_4, U_5, U_6) \in \mathrm{rep}(S, k)$ by*

$$U_0 = k^n \oplus k^n \oplus k^n,$$

$$U_1 = 0 \oplus 0 \oplus k^n \subseteq U_2 = (1 + A)k^n \oplus k^n,$$

$$U_3 = (1 + 1)k^n \oplus 0 \subseteq U_4 = k^n \oplus k^n \oplus 0, \text{ and}$$

$$U_5 = 0 \oplus (1 + 1)k^n \subseteq U_6 = k^n \oplus (1 + 1)k^n.$$

Then

$$
M_U = \left(
\begin{array}{c|c|c|c|c|c}
0 & I & 0 & A & 0 & I \\
0 & 0 & I & I & 0 & I \\
I & 0 & 0 & I & I & 0
\end{array}
\right),
$$

End U is isomorphic to $C(A)$, and U is indecomposable if A is an indecomposable matrix.

PROOF. A direct calculation shows that candidates for the (U_i^*)'s with $U_i = U_i^* \oplus \Sigma\{U_j : j < i\}$ are $U_1^* = 0 \oplus 0 \oplus k^n$, $U_2^* = (1 + A)k^n \oplus 0$, $U_3^* = (1 + 1)k^n \oplus 0$, $U_4^* = 0 \oplus k^n \oplus 0$, $U_5^* = 0 \oplus (1 + 1)k^n$, and $U_6^* = k^n \oplus 0 \oplus 0$. Recall that the row space of M_U is the kernel U_* of $\oplus\{U_i^* : i \in S\} \to U_0$. Notice that U_6^*, U_4^*, and U_1^* are the respective coordinate spaces of $k^n \oplus k^n \oplus k^n$. The first row of M_U arises from $U_2^* = (1 + A)k^n \oplus 0$, the second row from $U_3^* = (1 + 1)k^n \oplus 0$, and the third row from $U_5^* = 0 \oplus (1 + 1)k^n$.

By Theorem 1.2.6, End U is isomorphic to R_U. To compute R_U, let $f = f_1 \oplus \cdots \oplus f_6$ with $f(\mathrm{rs}(M_U))$ contained in $\mathrm{rs}(M_U)$. View each f_i as an $n \times n$ k-matrix. Then $f_3 = f_4 = f_6$ (from the second row of M_U), $f_1 = f_4 = f_5$ (from the third row of M_U), and $f_2 = f_6$ with $f_4 A = A f_4$ (from the first row of M_U). Thus, $f = (h, \ldots, h)$ with $h \in C(A)$, i.e., $hA = Ah$. Conversely, each h in $C(A)$ determines an element $f = (h, \ldots, h) \in R_U$. This shows that R_U may be identified with $C(A)$, whence End U is isomorphic to $C(A)$ by Theorem 1.2.6.

Finally, if A is an indecomposable matrix, then $C(A)$ has only 0 and 1 as idempotents, recalling from Section 1.1 that $C(A)$ is isomorphic to $k[x]/\langle g(x)^e \rangle$ with $g(x)$ an irreducible polynomial and $g(x)^e$ the minimal polynomial of A. This shows that U is indecomposable.

The matrix M_U is defined in terms of a choice of the U_i^*'s. The following invertible matrix operations on $M_U = (\cdots |A_i| \cdots)$ preserve U, recalling that the columns of A_i are labeled by a basis of U_i^* and the rows of M_U by a basis of U_*.

(a) Elementary column operations within each block A_i (a basis change of the subspace U_i^*),
(b) Elementary row operations on M_U (a basis change of the subspace U_*),
(c) Elementary column operations from block A_i to block A_j if $i < j$ in S (a different choice of an embedding of the subspace U_i^* into U_j).

Two representations U and V are isomorphic if M_U can be reduced to M_V by a series of matrix operations (a), (b), and (c). Classification of indecomposables in $\text{rep}(S, k)$ amounts to finding a canonical form for matrices M_U subject to the matrix operations (a), (b), and (c). This procedure is practical only for posets that are relatively uncomplicated.

Example 1.2.8 *Suppose* $S = \{1 < 2 < \cdots < n\}$ *is a poset. The elements of* $\text{Ind}(S, k)$ *are*

$$(k, 0, \ldots, 0), \quad (k, 0, \ldots, 0, k), \quad (k, 0, \ldots, 0, k, k), \quad \ldots,$$

$$(k, 0, k, \ldots, k), \quad (k, k, \ldots, k, k).$$

In each case, the endomorphism ring is k.

PROOF. Let $U = (U_0, U_1 \subseteq \cdots \subseteq U_n) \in \text{rep}(S, k)$ be an indecomposable representation and write $M_U = (A_1| \cdots |A_n)$. First assume that U_1 is nonzero. Use elementary row and column operations (a) and (b) to reduce A_1 to a matrix of the form $\begin{pmatrix} I & 0 \\ 0 & 0 \end{pmatrix}$. Next use (c) to see that

$$M_U \approx \left(\begin{array}{cc|c|c|c} I & 0 & 0 & \cdots & 0 \\ 0 & 0 & B_2 & \cdots & B_n \end{array} \right)$$

for some k-matrices B_i. But $(I \quad 0 \mid 0 \mid \cdots \mid 0)$ determines a representation summand $V = (V_0, V_i)$ of U with $0 \neq V_1^*$, and $V_i^* = 0$ for each $i \geq 2$. Since S is linearly ordered with 1 as the least element, $V_1^* = V_1 = V_2 = \cdots = V_n$. But U is indecomposable and $V_1 \neq 0$, so that U must be isomorphic to $(k, k, \ldots, k)$.

Now assume that $U_1 = 0$. Then A_1 is empty, and the proof is completed by an induction on n.

EXERCISES

1. Solve a matrix problem to find all elements of $\text{Ind}(S, k)$, where S is the poset $\{1, 2 < 3\}$.

2. Let A and B be $m \times m$ k-matrices.
    (a) Find a representation W isomorphic to the representation given in Example 1.2.7 with
    $$M_W = \left(\begin{array}{c|c|c|c|c|c} I & 0 & 0 & 0 & I & A \\ 0 & I & 0 & 0 & I & I \\ 0 & 0 & I & I & 0 & I \end{array} \right),$$

    (b) Define $U = (k^m \oplus k^m, k^m \oplus 0, 0 \oplus k^m, (1+1)k^m, (1+A)k^m, (1+B)k^m) \in \text{rep}(S_5, k)$. Show that $M_U = \left(\begin{array}{c|c|c|c|c} I & I & I & 0 & 0 \\ I & A & 0 & I & 0 \\ I & B & 0 & 0 & I \end{array} \right)$ and $\text{End } U = C(A, B)$.

(c) Define a matrix $M = \left(\begin{array}{c|c|c|c|c} I & I & I & I & 0 \\ I & A & B & 0 & I \end{array} \right)$.

   Find a $V \in \text{rep}(S_5, k)$ with $M_V = M$ and show that $\text{End } V = C(A, B)$.

(d) Explain why the representations U and V are not isomorphic.

3. Let R be a finite-dimensional k-algebra, k a field. Prove that R is a local ring if and only if 0 and 1 are the only idempotents of R.

4. Let k be a field of characteristic 0 and define $U = (U_0, U_1, U_2, U_3) \in \text{rep}(S_3, k)$, where $U_0 = k^4$, $U_1 = k \oplus k \oplus 0 \oplus 0$, $U_2 = 0 \oplus 0 \oplus k \oplus k$, and $U_3 = (1, 1, 1, 1)k \oplus (2, 3, 3, 3)k$. Find M_U and solve a matrix problem to write M_U as a direct sum of indecomposable representations. Describe these indecomposable representations. What changes if U_3 is replaced by $(1, 1, 1, 1)k \oplus (2, 3, 3, 3)k \oplus (1, 0, 1, 0)k$?

5. Let S be a finite poset. Define a poset $T = S \cup \{*\}$ with $s < *$ for each s in S. Prove that there is a one-to-one correspondence between nontrivial indecomposables in $\text{Ind}(T, k)$ and nontrivial indecomposables in $\text{Ind}(S, k)$.

6. Let $\mathbf{X}$ be an additive category. *Idempotents split* in $\mathbf{X}$ if whenever $e \in \text{End } A$ is an idempotent for some object A of $\mathbf{X}$, then there is an object B in $\mathbf{X}$, $q \in \text{Hom}(B, A)$, and $p \in \text{Hom}(A, B)$ with $qp = e$ and $pq = 1_B$.

(a) Prove that idempotents split in $\text{rep}(S, k)$.

(b) Assume that idempotents split in $\mathbf{X}$ and that $U, A_1, \ldots, A_n$ are objects of $\mathbf{X}$. Prove that $U = A_1 \oplus \cdots \oplus A_n$ if and only if there are morphisms $i_j : A_j \to U$ and $p_j : U \to A_j$ for each $1 \leq j \leq n$ such that $p_j i_t = 0$ if $j \neq t$, $p_j i_j$ is the identity of A_j, $e_j = i_j p_j$ is an idempotent element of $\text{End } U$, and $1_U = e_1 + \cdots + e_n$.

(c) ([Bass 68] or see [Arnold 82]) Suppose $\mathbf{X}$ is an additive category such that idempotents split and $A = A_1 \oplus \cdots \oplus A_n$ is a direct sum of objects A_i of $\mathbf{X}$ with each $\text{End } A_i$ a local ring. Generalize the proof of Theorem 1.2.2 to show that if $A = B_1 \oplus \cdots \oplus B_t$ with each B_i an indecomposable object of $\mathbf{X}$, then $t = n$ and there is a permutation σ of $\{1, 2, \ldots, n\}$ such that A_i is isomorphic to $B_{\sigma(i)}$ for each i.

7. Given a finite poset S, define $\Delta(S, k)$ and a correspondence $H : \Delta(S, k) \to \text{rep}(S, k)$ generalizing Theorem 1.2.3 such that $R_{H(U)} = \text{End } U$ for each $U \in \Delta(S, k)$.

1.3   Finite Representation Type

Finite posets having only finitely many isomorphism classes of indecomposable k-representations can be described (Theorem 1.3.6). This description depends only on the poset and is independent of the field k. In this case there is a systematic procedure for finding all indecomposable k-representations up to isomorphism (Theorem 1.3.7 and references).

A finite poset S has *finite representation type* over a field k if $\text{Ind}(S, k)$ is finite and has *infinite representation type* if $\text{Ind}(S, k)$ is infinite. For example, S_n has finite representation type for $n \leq 3$ (Example 1.1.5) and infinite representation type for $n \geq 4$ (Example 1.1.6(a)).

A *subposet* of a poset S is a subset T of S equipped with the ordering on S; specifically, if $i, j \in T$, then $i \leq j$ in S if and only if $i \leq j$ in T. Part (a) of the following proposition extends the correspondence F of Proposition 1.1.4 to arbitrary finite posets S; part (b) is an alternative correspondence.

Proposition 1.3.1 *Suppose T is a subposet of a finite poset S.*

(a) *There is a correspondence $F^+ : \text{rep}(T, k) \to \text{rep}(S, k)$ with $\text{Hom}(U, V)$ isomorphic to $\text{Hom}(F^+U, F^+V)$ for each U, V in $\text{rep}(T, k)$.*
(b) *There is a correspondence $F^- : \text{rep}(T, k) \to \text{rep}(S, k)$ with $\text{Hom}(U, V)$ isomorphic to $\text{Hom}(F^-U, F^-V)$ for each U, V in $\text{rep}(T, k)$.*
(c) *If S has finite representation type, then so does T.*
(d) *If T has infinite representation type, then so does S.*

PROOF. (a) The correspondence is given by $F^+(U) = W = (W_0, W_i : i \in S) \in \text{rep}(S, k)$ for $U = (U_0, U_i : i \in T)$, with $W_0 = U_0$, $W_i = U_i$ if $i \in T$, $W_i = \Sigma\{U_j : j \in T, j < i\}$ if $i \in S \backslash T$ and there is $j \in T$ with $j < i$, and $W_i = 0$ otherwise. If $f : U \to V$ is a morphism in $\text{rep}(T, k)$, define $F^+(f) = f$, a morphism in $\text{rep}(S, k)$. Then $F^+ : \text{Hom}(U, V) \to \text{Hom}(F^+(U), F^+(V))$ is an isomorphism for each pair U, V of objects of $\text{rep}(T, k)$.

(b) The correspondence is given by $F^-(U) = W = (W_0, W_i : i \in S) \in \text{rep}(S, k)$ for $U = (U_0, U_i : i \in T)$ with $W_0 = U_0$, $W_i = U_i$ if $i \in T$, $W_i = \cap\{U_j : j \in T, j > i\}$ if $i \in S \backslash T$ and there is $j \in T$ with $j > i$, and $V_i = U_0$ otherwise. If $f : U \to V$ is a morphism in $\text{rep}(T, k)$, define $F^-(f) = f$, a morphism in $\text{rep}(S, k)$. Then $F^- : \text{Hom}(U, V) \to \text{Hom}(F^-(U), F^-(V))$ is an isomorphism for each pair U, V of objects of $\text{rep}(T, k)$.

(c), (d) Just as in Proposition 1.1.4, both F^+ and F^- induce an injection $\text{Ind}(T, k) \mapsto \text{Ind}(S, k)$.

The *width* of a poset S, denoted by $w(S)$, is the largest number of pairwise incomparable elements of S.

Corollary 1.3.2 *If S is a finite poset with $w(S) \geq 4$, then $\text{rep}(S, k)$ has infinite representation type.*

PROOF. Apply Proposition 1.3.1 and the fact that $\text{rep}(S_4, k)$ has infinite representation type (Example 1.1.6(a)).

A finite poset S is a *chain* if S is linearly ordered.

Lemma 1.3.3 *Suppose S is a finite poset with $w(S) = 1$.*

(a) *Then S is a chain and $\text{rep}(S, k)$ has finite representation type. Elements of $\text{Ind}(S, k)$ are*

$$(k, 0, \ldots, 0), \quad (k, 0, \ldots, 0, k), \quad (k, 0, \ldots, 0, k, k), \quad \ldots,$$
$$(k, 0, k, \ldots, k), \quad (k, k, \ldots, k, k).$$

(b) *If $U = (U_0, U_i : i \in S)$ and $V = (V_0, V_i : i \in S) \in \mathrm{rep}(S, k)$ with $U_0 \subseteq V_0$ and $U_i = U_0 \cap V_i$ for each $i \in S$, then $V = W \oplus U$ for some $W \in \mathrm{rep}(S, k)$.*

PROOF. Since $w(S) = 1$, any two elements of S are comparable. Thus, S is linearly ordered, say $S = \{1 < 2 < \cdots < m\}$.

(a) This is proved in Example 1.2.8 using a matrix argument. Following is an alternative proof. Let $U = (U_0, U_1 \subseteq \cdots \subseteq U_m)$ be an indecomposable representation of S and write $U_0 = U_m \oplus V_0$ for some V_0. There is a representation direct sum

$$U = (V_0, 0, \ldots, 0) \oplus (U_m, U_1, \ldots, U_m).$$

Since U is indecomposable, either $U_m = 0$ and U is isomorphic to $(k, 0, \ldots, 0)$, or $V_0 = 0$ and $U = (U_m, U_1, \ldots, U_{m-1}, U_m)$. Next write $U_m = U_{m-1} \oplus V_m$ so that

$$U = (V_m, 0, \ldots, 0, V_m) \oplus (U_{m-1}, U_1, \ldots, U_{m-1}, U_{m-1}).$$

Since U is indecomposable, either $U_{m-1} = 0$ and U is isomorphic to $(k, 0, \ldots, 0, k)$, or else $V_m = 0$ and $U = (U_{m-1}, U_1, \ldots, U_{m-2}, U_{m-1}, U_{m-1})$. Repeating this process completes the list of indecomposables as given.

(b) Let $X = (V_0/U_0, (V_1 + U_0)/U_0, \ldots, (V_m + U_0)/U_0) \in \mathrm{rep}(S, k)$. First assume that X is indecomposable. Then X is isomorphic to one of the representations listed in (a). There is some t and $x \in V_t \backslash U_0$ such that $V_0/U_0 = k(x + U_0)$, $(V_i + U_0)/U_0 = 0$ for $i < t$, and $(V_i + U_0)/U_0 = k(x + U_0)$ for $i \geq t$. Hence, $V_0 = kx \oplus U_0$ with $kx \subseteq V_i$ for $i \geq t$. Then $V_i = V_i \cap V_0 = kx \oplus (U_0 \cap V_i) = kx \oplus U_i$ for each $i \geq t$ and $V_i = U_0 \cap V_i = U_i$ for each $i < t$. This shows that

$$V = (V_0, V_1, \ldots, V_m) = (kx, 0, \ldots, 0, kx, \ldots, kx) \oplus (U_0, U_1, \ldots, U_m)$$

$$= W \oplus U$$

is a direct sum of representations, as desired.

For the general case, write X as a direct sum of n indecomposable representations in $\mathrm{rep}(S, k)$. Use the above argument on an indecomposable summand of X and induction on n to see that U is a summand of V.

Given $U = (U_0, U_i : i \in S) \in \mathrm{rep}(S, k)$, define the *dimension of U* to be the k-dimension of U_0.

Lemma 1.3.4 *If S is a finite poset with $w(S) = 2$, then $\mathrm{rep}(S, k)$ has finite representation type. Moreover, each indecomposable $U \in \mathrm{rep}(S, k)$ has dimension 1.*

PROOF. Let a be a minimal element of S and partition S into three sets $A = \{a\}$, $B = \{i \in S : i > a\}$, and $C = \{i \in S : i \not\geq a\}$. Observe that C is a chain, since $w(S) = 2$ and the minimality of a in S implies that any two elements of C are comparable.

The proof is by induction on the cardinality $|S|$ of S. If $|S| = 2$, then $S = S_2$. By Example 1.1.3, S_2 has finite representation type and each indecomposable representation has dimension 1. Now assume that $|S| > 2$, rep(T, k) has finite representation type, and each indecomposable $U \in$ rep(T, k) has dimension 1 for each proper subposet T of S. Let $U = (U_0, U_i : i \in S)$ be an indecomposable representation of S.

If $U_a = 0$, then $T = B \cup C$ is a proper subposet of S, and U is in the image of $F^+ :$ rep$(T, k) \rightarrow$ rep(S, k) given in Proposition 1.3.1(b). By the induction hypothesis, it follows that there are, up to isomorphism, only finitely many $U \in$ Ind(S, k) with $U_a = 0$.

Next assume that $U = (U_0, U_i : i \in S)$ is indecomposable with $U_a \neq 0$. Then $W = (U_a, U_a \cap U_j : j \in C)$ and $U_C = (U_0, U_j : j \in C)$ are in rep(C, k). Since C is a chain and $U_a \leq U_0$, it follows from Lemma 1.3.3(b) that $U_C = W \oplus V$ for some $V = (V_0, V_j : j \in C)$. Specifically, write $C = \{1 < 2 < \cdots < m\}$ so that

$$U_0 = U_a \oplus V_0,$$
$$U_j = (U_a \cap U_j) \oplus V_j \text{ for } 1 \leq j \leq m,$$
$$V_1 \subseteq \cdots \subseteq V_m \subseteq V_0.$$

The next step is to verify that $U = X \oplus Y$, where $X = (X_0, X_i : i \in S)$ is defined by

$$X_0 = U_a, X_i = U_a \text{ for } i \geq a, \text{ and } X_i = U_a \cap U_i \text{ for } i \in C$$

and $Y = (Y_0, Y_i : i \in S)$ is defined by

$$Y_0 = V_0, Y_a = 0, Y_i = V_0 \cap U_i \text{ for } i > a, \text{ and } Y_i = V_i \text{ for } i \in C.$$

Notice that $X, Y \in$ rep(S, k). From the above decomposition of U_C,

$$U_0 = U_a \oplus V_0 = X_0 \oplus Y_0,$$
$$U_i = (U_a \cap U_i) \oplus V_i = X_i \oplus Y_i \text{ for } i \in C,$$
$$U_a = U_a \oplus 0 = X_a \oplus Y_a.$$

Finally, assume that $i \geq a$. Then

$$U_i = U_a \oplus V_0 \cap U_i = X_i \oplus Y_i,$$

since $U_a \subseteq U_i \subseteq U_0$ and $U_0 = U_a \oplus V_0$.

Since U is indecomposable and $U_a \neq 0$, it must be the case that $U = X$. Therefore, U is in the image of $F^- :$ rep$(C, k) \rightarrow$ rep(S, k) given in Proposition 1.3.1(b). As a consequence of Lemma 1.3.3(a), there are, up to isomorphism, only finitely many indecomposable U's with $U_a \neq 0$, and each such U has dimension 1. This completes the proof.

There is a more general version of Lemma 1.3.4. A *splitting decomposition* of a finite poset S is a partition of S into three subsets A, B, C such that C is either a chain or empty and $a < b$ for each $a \in A, b \in B$. For example, if A and B are finite posets and S is the disjoint union of A and B subject to the additional relations $a < b$ for each a in A and b in B, then the partition $A \cup B$ is a splitting decomposition of S with C empty.

Proposition 1.3.5 *If a finite poset S has a splitting decomposition A, B, C, then each U in $\mathrm{Ind}(S, k)$ is in the image of either $\mathrm{Ind}(A \cup C, k)$ or $\mathrm{Ind}(B \cup C, k)$.*

PROOF. The proof is essentially the same as the proof of Lemma 1.3.4 and is left as an exercise (Exercise 1.3.2).

The only remaining case for determination of finite representation type is for posets of width 3. This case is significantly more difficult.

There is a standard notation for posets. A chain with n elements is denoted by (n). If $T_1, \ldots, T_m$ are posets, then the disjoint union of the T_i's is written as $(T_1, \ldots, T_m)$. As an example, $(1, 1, 1, 1, 1)$ is an alternative notation for the antichain S_5, and $(2, 2, 2)$ denotes the poset $\{1 < 2, 3 < 4, 5 < 6\}$. Define N to be the poset $\{1 < 2 > 3 < 4\}$.

The following theorem shows that finite representation type for $\mathrm{rep}(S, k)$ does not depend on the field k.

Theorem 1.3.6 [Kleiner 75A] *Let S be a finite poset. Then S does not contain $S_4, (2, 2, 2), (1, 3, 3), (1, 2, 5)$, or $(N, 4)$ as a subposet if and only if $\mathrm{rep}(S, k)$ has finite representation type.*

PROOF. ($\Leftarrow$) In view of Proposition 1.3.1, it suffices to show that if S is one of $S_4, (2, 2, 2), (1, 3, 3), (1, 2, 5)$, or $(N, 4)$, then S has infinite representation type. These five posets are called *critical posets*.

Following is a list of representations for each critical poset S, demonstrating that $\mathrm{rep}(S, k)$ has infinite representation type. The idea is to show that if A is an indecomposable $n \times n$ k-matrix, then there is a representation $U \in \mathrm{rep}(S, k)$ with $\dim U \geq n$ and $\mathrm{End}\, U = C(A)$, the centralizer of A. Since Jordan block matrices provide examples of $n \times n$ indecomposable matrices, ranks of indecomposable representations in $\mathrm{rep}(S, k)$ are unbounded. This shows that $\mathrm{rep}(S, k)$ has infinite representation type.

The only difficulty is, given the list, to show that each of these representations has endomorphism ring $C(A)$. The case that $S = S_4$ is proved in Example 1.1.6 by computing $\mathrm{End}\, U$ directly. The case that $S = (2, 2, 2)$ is Example 1.2.7, confirmed by computing R_U from the matrix M_U. The remainder of the computations are given as exercises.

(i) $S = S_4$

$U = (U_0, U_1, U_2, U_3, U_4)$

$U_0 = k^n \oplus k^n,$

$U_1 = k^n \oplus 0, \quad U_2 = 0 \oplus k^n, \quad U_3 = (1+1)k^n, \quad U_4 = (1+A)k^n.$

(ii) $S = (2, 2, 2) = $
$$
\begin{array}{ccc}
2 & 4 & 6 \\
| & | & | \\
1 & 3 & 5
\end{array}
$$

$U = (U_0, U_1 \subseteq U_2, U_3 \subseteq U_4, U_5 \subseteq U_6),$

$U_0 = k^n \oplus k^n \oplus k^n,$

$U_1 = 0 \oplus 0 \oplus k^n, \qquad U_2 = (1+A)k^n \oplus k^n,$

$U_3 = (1+1)k^n \oplus 0, \quad U_4 = k^n \oplus k^n \oplus 0,$

$U_5 = 0 \oplus (1+1)k^n, \quad U_6 = k^n \oplus (1+1)k^n.$

(iii) $S = (1, 3, 3) = $
$$
\begin{array}{ccc}
 & 4 & 7 \\
 & | & | \\
1 & 3 & 6 \\
 & | & | \\
 & 2 & 5
\end{array}
$$

$U = (U_0, U_1, U_2 \subseteq U_3 \subseteq U_4, U_5 \subseteq U_6 \subseteq U_7),$

$U_0 = k^n \oplus k^n \oplus k^n \oplus k^n,$

$U_1 = (1+1)k^n \oplus (1+1)k^n,$

$U_2 = 0 \oplus (1+1)k^n \oplus 0, \quad U_3 = 0 \oplus k^n \oplus k^n \oplus 0, \quad U_4 = 0 \oplus k^n \oplus k^n \oplus k^n,$

$U_5 = k^n \oplus 0 \oplus 0 \oplus 0, \qquad U_6 = k^n \oplus 0 \oplus 0 \oplus k^n, \quad U_7 = k^n \oplus (1+A)k^n \oplus k^n.$

(iv) $S = (N, 4) = $
$$
\begin{array}{ccc}
 & & 8 \\
 & & | \\
 & & 7 \\
 & & | \\
1 & 3 & 6 \\
\diagdown & | & | \\
2 & 4 & 5
\end{array}
$$

$U = (U_0, U_2 \subseteq U_1 \supseteq U_4 \subseteq U_3, U_5 \subseteq U_6 \subseteq U_7 \subseteq U_8),$

$U_0 = k^n \oplus k^n \oplus k^n \oplus k^n \oplus k^n,$

$U_2 = (1+1)k^n \oplus (1+1)k^n \oplus 0, \quad U_1 = k^n \oplus k^n \oplus k^n \oplus k^n \oplus 0,$

$U_4 = 0 \oplus (1+1)k^n \oplus 0 \oplus 0, \qquad U_3 = 0 \oplus k^n \oplus k^n \oplus (1+1)k^n,$

$U_5 = 0 \oplus 0 \oplus 0 \oplus 0 \oplus k^n, \qquad\qquad U_6 = k^n \oplus 0 \oplus 0 \oplus 0 \oplus k^n,$

$U_7 = k^n \oplus 0 \oplus 0 \oplus k^n \oplus k^n, \qquad U_8 = k^n \oplus (1+A)k^n \oplus k^n \oplus k^n.$

$$
\begin{array}{c}
8 \\
| \\
7 \\
\end{array}
$$

$$
\begin{array}{ccc}
1 & 3 & 6 \\
| & | & | \\
2 & 5 \\
\end{array}
$$

(v) $S = (1, 2, 5) = \qquad 4$

$U = (U_0, U_1, U_2 \subseteq U_3, U_4 \subseteq U_5 \subseteq U_6 \subseteq U_7 \subseteq U_8)$,

$U_0 = k^n \oplus k^n \oplus k^n \oplus k^n \oplus k^n \oplus k^n$,

$U_1 = (0 + 1 + 1 + 0 + 0 + 0)k^n + (0 + 0 + 1 + 0 + 0 + 1)k^n$

$\qquad + (0 + 0 + 0 + 1 + 1 + 0)k^n$,

$U_2 = (1 + 1)k^n \oplus (1 + 1)k^n \oplus 0 \oplus 0, \qquad U_3 = k^n \oplus k^n \oplus k^n \oplus k^n \oplus 0 \oplus 0,$

$U_4 = 0 \oplus 0 \oplus 0 \oplus 0 \oplus 0 \oplus k^n, \qquad U_5 = 0 \oplus 0 \oplus 0 \oplus 0 \oplus k^n \oplus k^n,$

$U_6 = k^n \oplus 0 \oplus 0 \oplus 0 \oplus k^n \oplus k^n, \qquad U_7 = k^n \oplus 0 \oplus 0 \oplus k^n \oplus k^n \oplus k^n,$

$U_8 = k^n \oplus (1 + A)k^n \oplus k^n \oplus k^n \oplus k^n$.

($\Rightarrow$) Suppose S has infinite representation type. Then $w(S) \geq 3$ by Lemmas 1.3.3 and 1.3.4. Moreover, any poset S containing S_4 as a subposet has infinite representation type by Corollary 1.3.2.

It now suffices to assume that $w(S) = 3$ and prove that S must contain (2,2,2), (1,3,3), (1,2,5), or $(N,4)$ as a subposet. There are several proofs of this result. All are lengthy and complicated and as such are beyond the scope of this book. A proof in [Ringel 84, Section 2.6] uses the theory of integral quadratic forms. A more traditional proof is given, for example, in [Simson 92, Theorem 10.1]. This proof uses the notion of a Zavadiskij derivative of a poset, as defined below. An elementary outline of this proof is given in [Hahn 97].

In view of Proposition 1.3.1, elements of $\mathrm{Ind}(S, k)$ include those elements of $\mathrm{Ind}(T, k)$ for each proper subposet T of S. Proposition 1.3.5 illustrates that for certain finite posets S, all elements of $\mathrm{Ind}(S, k)$ are of this form. These observations motivate the notion of sincere representations and posets.

Given $U = (U_0, U_i : i \in S) \in \mathrm{rep}(S, k)$, define $\mathrm{cdn}\, U = (u_0, u_i : i \in S)$, where u_0 is the k-dimension of U_0, u_i is the k-dimension of $U_i/U_i^{\#}$, and $U_i^{\#}$ is the subspace of U_i generated by $\{U_j : j < i \text{ in } S\}$. Call U a *sincere* representation if each $u_i \neq 0$. If U is not sincere, say $u_i = 0$ for some minimal i, then $T = S\backslash\{i\}$ is a proper subposet of S with U in the image of $F^+ : \mathrm{rep}(T, k) \rightarrow \mathrm{rep}(S, k)$, as given in Proposition 1.3.1. Moreover, if $\mathrm{rep}(S, k)$ has finite representation type, then so does $\mathrm{rep}(T, k)$.

A finite poset S is *sincere* if there is a sincere indecomposable representation U in $\mathrm{rep}(S, k)$.

Theorem 1.3.7 [Kleiner 75B] *A finite poset S with finite representation type is sincere if and only if S is one of the posets* (1), (1, 1), (1, 1, 1), (1, 1, 2), (1, 2, 2), (1, 2, 3), (1, 2, 4), $(N, 2)$, $(N, 3)$,

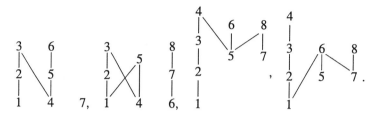

PROOF. [Simson 92, Theorem 10.2] or [Ringel 84, Section 2.6].

A complete list of all of the indecomposable sincere representations for each sincere poset, a total of 41 representations, is given in [Kleiner 75B]. There are several minor typographical errors in this list, repeated in [Arnold 89] for Butler groups. A corrected list can be found in [Ringel 84, Section 2.6], [Simson 92, Section 10.7], or [Arnold, Richman 92]. In the latter paper, an algorithm was implemented confirming that each of these representations is in fact indecomposable.

Here is an example of the representations in this list:

Example 1.3.8

(a) *If* $S = (1, 1, 2)$, *then the only sincere element of* Ind(S, k) *is*

$$(k \oplus k, k \oplus 0, (1 + 1)k, 0 \oplus k, k \oplus k).$$

(b) *If* $S = (1, 2, 2)$, *then the sincere elements of* Ind(S, k) *are*

$$(k \oplus k \oplus k, (1 + 1 + 1)k, k \oplus 0 \oplus 0, k \oplus k \oplus 0, 0 \oplus 0 \oplus k, 0 \oplus k \oplus k),$$

$$(k \oplus k \oplus k, (1 + 0 + 1)k + (1 + 1 + 0)k, k \oplus 0 \oplus 0, k \oplus k \oplus 0, 0 \oplus 0 \oplus k, 0 \oplus k \oplus k),$$

$$(k \oplus k, (1 + 1)k, k \oplus 0, k \oplus k, 0 \oplus k, k \oplus k).$$

The following corollary is striking. It can be proved by examining the list of 41 sincere indecomposable representations, together with the observation that if U is indecomposable in rep(S, k), then U may be regarded as a representation of T for some sincere subposet T of S. There is a more theoretical proof in [Ringel 84] using integral quadratic forms.

Corollary 1.3.9 *If S is a finite poset, k is a field, and* rep(S, k) *has finite representation type, then each indecomposable U in* rep(S, k) *has rank less than or equal to 6 and* End $U = k$.

All the representations given for critical posets in the proof of Theorem 1.3.6($\Leftarrow$) can be constructed from the fundamental representation U_A of S_4 defined in Example 1.1.6. These constructions employ the Zavadskij derivative on a finite poset relative to a suitable pair of elements of S. This derivative is also used in the proof of Theorems 1.3.6 and 1.4.4 given in [Simson 92]. Although the proofs are not included, definitions and some examples are given in the remainder of this section.

An ordered pair of elements (a, b) in S is called *suitable* if a is not less than or equal to b, and the subposet, $S_a^b = S\backslash(a^\triangleleft \cup b^\triangleright)$ of S is either empty or a chain $C = \{c_1 < c_2 < \cdots < c_m\}$, where $a^\triangleleft$ is the set of all elements in S greater than or equal to a, and $b^\triangleright$ is the set of all elements in S less than or equal to b. Define two copies of C,

$$C^- = \{c_1^- < c_2^- < \cdots < c_m^-\} \text{ and } C^+ = \{c_1^+ < c_2^+ < \cdots < c_m^+\}$$

if S_a^b is not empty. Otherwise, let C^- and C^+ be empty. The *derivative of S with respect to the suitable pair* (a, b), $S' = \partial_{(a,b)}S$, is the disjoint union of posets $a^\triangleleft \cup b^\triangleright \cup C^- \cup C^+$ with additional order relations given below.

1. If $x, y \in a^\triangleleft \cup b^\triangleright$, then $x \leq y$ in S' if and only if $x \leq y$ in S.
2. $c_i^- < c_i^+$ for all i.
3. If $x \in b^\triangleright\backslash\{b\}$, then $x < c_i^-$ in S' if and only if $x < c_i$ in S.
4. If $x \in a^\triangleleft\backslash\{a\}$, then $x > c_i^+$ in S' if and only if $x > c_i$ in S.
5. $a < b$, $a < c_1^+$, and $c_m^- < b$.
6. If the above relations result in $x < y$ and $y < x$ for some $x, y \in S'$, then set $x = y$.

The following example demonstrates, in terms of derivatives, an interrelationship among the five critical posets $(1, 1, 1, 1)$, $(2, 2, 2)$, $(1, 3, 3)$, $(N, 4)$, and $(1, 2, 5)$ listed in Theorem 1.3.6.

Example 1.3.10

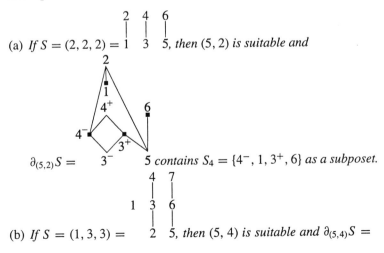

(a) *If $S = (2, 2, 2) =$ ... , then $(5, 2)$ is suitable and $\partial_{(5,2)}S = $ ... 5 contains $S_4 = \{4^-, 1, 3^+, 6\}$ as a subposet.*

(b) *If $S = (1, 3, 3) = $ ... , then $(5, 4)$ is suitable and $\partial_{(5,4)}S = $*

contains $(2, 2, 2) = \{2 < 3, 1^- < 1^+, 6 < 7\}$ *as a subposet.*

(c) *If* $S = (N, 4) =$ 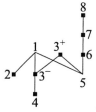 *, then* $(5, 1)$ *is suitable and* $\partial_{(5,1)} S =$

contains $(1, 3, 3) = \{1, 4 < 3' < 3^+, 6 < 7 < 8\}$ *as a subposet.*

(d) *If* $S = (1, 2, 5) =$ 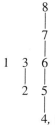 *, then* $(4, 3)$ *is suitable and* $\partial_{(4,3)} S =$

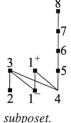 *contains* $(N, 4) = \{2 < 3 > 1^- < 1^+, 5 < 6 < 7 < 8\}$ *as a subposet.*

PROOF. Only (a) is confirmed; the other three cases are left as exercises. Recall that $C = S_5^2 = \{3 < 4\}$, so that $C^- = \{3^- < 4^-\}$ and $C^+ = \{3^+ < 4^+\}$. Then $\partial_{(5,2)}(2, 2, 2) = \{1, 2, 3^- < 4^-, 3^+ < 4^+, 5, 6\}$ with relations

1. $1 < 2, 5 < 6$;

2. $3^- < 3^+, 4^- < 4^+$;
5. $5 < 2, 5 < 3^+, 4^- < 2$.

The numbering corresponds to that found in the definition of $\delta_{(a,b)}S$, and 3, 4, and 6 of the definition add no new relations. Thus, $\delta_{(5,2)}(2,2,2)$ is as pictured in (a).

A derivative $\delta_{(a,b)}$ on a finite poset S with suitable pair (a,b) induces a correspondence $\delta_{(a,b)} : \mathrm{rep}(S,k) \to \mathrm{rep}(S'k)$. Let S be a finite poset with suitable pair (a,b) and $S_a^b = \{c_1 < \cdots < c_n\}$. Given $U = (U_0, U_i : i \in S) \in \mathrm{rep}(S,k)$, define $\delta_{(a,b)}U = W \in \mathrm{rep}(S',k)$ as follows. Choose a subspace V of $U_a = U_b$ with

$$U_a + U_b = V \oplus U_b$$

and let $W = (W_0, W_i : i \in S') \in \mathrm{rep}(S',k)$, where

$$W_0 = U_0/V, \quad W_{c(j)-} = \left((U_b \cap U_{c(j)}) + V\right)/V \text{ for each } j \leq n,$$
$$W_{c(j)+} = \left(U_a + U_{c(j)}\right)/V \text{ for each } j \leq n,$$
$$W_t = (U_t + V)/V \text{ for each } t \in a^\triangleleft \cup b^\triangleright.$$

Theorem 1.3.11 *Suppose S is a finite poset with suitable pair (a,b), $S_a^b = \{c_1 < \cdots < c_n\}$, and $S' = \delta_{(a,b)}S$.*

(a) *The correspondence $\delta_{(a,b)} : \mathrm{rep}(S,k) \to \mathrm{rep}(S',k)$ yields an onto correspondence $\delta_{(a,b)} : \mathrm{Ind}(S,k) \to \mathrm{Ind}(S',k)$.*

(b) *$|\mathrm{Ind}(S,k)| = |\mathrm{Ind}(S',k)| + 1 + n$. In particular, $\mathrm{rep}(S,k)$ has finite representation type if and only if $\mathrm{rep}(S',k)$ has finite representation type.*

PROOF. [Simson 92, Theorem 9.10].

It appears from the definition that the derivative of a poset is more complicated than the poset. However, as a consequence of Theorem 1.3.11(b), if S is a finite poset, then S has finite representation type if and only if a repeated application of the derivative eventually results in a 1-element poset. There is another "derivative" of a finite poset with this same property that can be used to classify those finite posets S such that $\mathrm{rep}(S,k)$ has finite representation type [Gabriel 73A].

Corollary 1.3.12 *There are embeddings $\mathrm{Ind}(S_4,k) \hookrightarrow \mathrm{Ind}((2,2,2),k) \hookrightarrow \mathrm{Ind}((1,3,3),k) \hookrightarrow \mathrm{Ind}((N,4),k) \hookrightarrow \mathrm{Ind}((1,2,5),k)$.*

PROOF. Theorem 1.3.11(a), Example 1.3.10, and Proposition 1.3.1.

As an illustration, the last example of this section shows how the representation for $S = (2,2,2)$ in Theorem 1.3.6(ii) can be constructed from the one for $(1,3,3)$

in Theorem 1.3.6(iii) and vice versa. The inverse of the correspondence $\delta_{(a,b)}$ of Theorem 1.3.11 is spelled out in [Simson 92, Section 9.3].

Example 1.3.13 *Let* $S = (1, 3, 3) = $ *with suitable pair* $(5, 4)$ *and* $U \in \text{rep}(S, k)$ *as given in Theorem 1.3.6(iii). Then* $\delta_{(5,4)}S \supseteq (2, 2, 2)$ *and* $\delta_{(5,4)}U$ *contains the representation for* $(2, 2, 2)$ *listed in Theorem 1.3.6(ii).*

PROOF. The representation $U \in \text{rep}((1, 3, 3), k)$ in Theorem 1.3.6(iii) is

$$U_0 = k^n \oplus k^n \oplus k^n \oplus k^n, \qquad U_1 = (1+1)k^n \oplus (1+1)k^n,$$

$$U_2 = 0 \oplus (1+1)k^n \oplus 0, \qquad U_3 = 0 \oplus k^n \oplus k^n \oplus 0,$$

$$U_4 = 0 \oplus k^n \oplus k^n \oplus k^n, \qquad U_5 = k^n \oplus 0 \oplus 0 \oplus 0,$$

$$U_6 = k^n \oplus 0 \oplus 0 \oplus k^n, \qquad U_7 = k^n \oplus (1+A)k^n \oplus k^n.$$

Notice that $U_5 + U_4 = U_5 \oplus U_4$, so choose $V = U_5$ in the definition of $\delta_{(5,4)}U$. Then $\delta_{(5,4)}U$ is given by

$$W_0 = U_0/V = k^n \oplus k^n \oplus k^n, \qquad W_{1-} = ((U_4 \cap U_1) + V)/V = 0 \oplus (1+1)k^n,$$

$$W_{1+} = (U_5 + U_1)/V = k^n \oplus (1+1)k^n, \qquad W_2 = (U_2 + V)/V = (1+1)k^n \oplus 0,$$

$$W_3 = (U_3 + V)/V = k^n \oplus k^n \oplus 0, \qquad W_4 = (U_4 + V)/V = k^n \oplus k^n \oplus k^n,$$

$$W_5 = (U_5 + V)/V = 0 \oplus 0 \oplus 0, \qquad W_6 = (U_6 + V)/V = 0 \oplus 0 \oplus k^n,$$

$$W_7 = (U_7 + V)/V = (1+A)k^n \oplus k^n.$$

By Example 1.3.10(b), $(2, 2, 2) = \{2 < 3, 1^- < 1^+, 6 < 7\}$ is a subposet of $\delta_{(5,4)}S$. Then $(W_0, W_6 \subseteq W_7, W_2 \subseteq W_3, W_{1-} \subseteq W_{1+})$ is the representation of $(2, 2, 2)$ given in Theorem 1.3.6(ii). $\qquad \blacksquare$

EXERCISES

1. Let $S = (1, 2, 2)$.
   (a) Find all subposets of S.
   (b) List all elements of $\text{Ind}(S, k)$.

2. Prove Proposition 1.3.5.

3. Show that End $U = C(A)$ for each of the representations U in Theorem 1.3.6, (iii), (iv), and (v).

4. Confirm the assertions of Example 1.3.10.

5. Prove that if $U = (U_0, U_i : i \in S) \in \text{rep}(S, k)$ is indecomposable and $i \in S$ with $0 \neq U_i \neq U_0$, then there is a 3-element antichain T that is a subposet of S with $i \in T$.

6. Prove, without reference to Theorem 1.3.7, that if S is a sincere poset with $w(S) = 3$, then S has 3 maximal elements.

1.4 Tame and Wild Representation Type

Infinite representation type is typically divided into two kinds. Heuristically, if $\text{rep}(S, k)$ has tame representation type, then there is a potential classification scheme for indecomposables in $\text{rep}(S, k)$. On the other hand, if $\text{rep}(S, k)$ has wild representation type, then it is unrealistic to attempt a complete classification of indecomposable representations.

Representations of finite posets over a field can be generalized to an arbitrary ring R. It is assumed herein that a ring has an identity and that R-modules are left unitary R-modules unless otherwise specified.

Given a finite poset S, define $\text{rep}(S, R)$ to be the category with objects $U = (U_0, U_i : i \in S)$, where U_0 is a finitely generated free R-module, each U_i is a finitely generated free R-summand of U_0, and if $i \leq j$ in S, then $U_i \subseteq U_j$. Morphisms in this category, called *representation morphisms*, are R-module homomorphisms $f : U_0 \to U_0'$ with $f(U_i) \subseteq U_i'$ for each i in S. Then $\text{rep}(S, R)$ is an additive category just as in Lemma 1.2.1, since R-modules with R-homomorphisms form an additive category.

Let $\mathbf{X}$ and $\mathbf{X}'$ be additive categories. An (additive) *functor* $F : \mathbf{X} \to \mathbf{X}'$ is a correspondence F from the objects of $\mathbf{X}$ to the objects of $\mathbf{X}'$ and a group homomorphism $F : \text{Hom}(U, V) \to \text{Hom}(F(U), F(V))$ for each pair of objects U, V of $\mathbf{X}$ such that $F(fg) = F(f)F(g)$ for each $g \in \text{Hom}(U, V)$, $f \in \text{Hom}(V, W)$, and $F(1_U) = 1_{F(U)}$. For example, the correspondences H of Theorem 1.2.3 and F^+, F^- of Proposition 1.3.1 are functors. If F is a functor, then $F(U \oplus V)$ is isomorphic to $F(U) \oplus F(V)$ as a consequence of the definition of a direct sum in an additive category.

A functor F *preserves indecomposables* if $F(U)$ is indecomposable whenever U is indecomposable and *reflects isomorphisms* if U is isomorphic to V whenever $F(U)$ is isomorphic to $F(V)$. Moreover, F is a *faithful* functor if $F : \text{Hom}(U, V) \to \text{Hom}(F(U), F(V))$ is one-to-one for each pair U, V of objects of $\mathbf{X}$ and a *full* functor if $F : \text{Hom}(U, V) \to \text{Hom}(F(U), F(V))$ is onto for each pair U, V of objects of $\mathbf{X}$. Notice that a fully faithful functor preserves indecomposables, since $F(e)$ is an idempotent if and only if e is an idempotent, and reflects isomorphisms, since $F(fg) = F(f)F(g) = 1$ if and only if $fg = 1$.

Let $k\langle x, y \rangle$ denote the k-algebra of polynomials in noncommuting variables x and y with coefficients in a field k, and let $\text{mod}\, k\langle x, y \rangle$ denote the category of $k\langle x, y \rangle$-modules that have finite k-dimension. The endomorphism ring of $N \in \text{rep}(S, k \langle x, y \rangle)$ is denoted by $\text{End}_{k\langle x, y \rangle} N$. Write $\text{End}_k N$ for the endomorphism ring of N regarded as a k-representation, albeit one with possibly infinite dimension.

The category $\text{rep}(S, k)$ has *wild representation type* if there is an $N \in \text{rep}(S, k\langle x, y\rangle)$ such that the functor $F_N : \text{mod } k\langle x, y\rangle \rightarrow \text{rep}(S, k)$ given by $F_N(M) = W \otimes_{k\langle x,y\rangle} M$ preserves indecomposables and reflects isomorphisms and $\text{End}_{k\langle x,y\rangle} N = \text{End}_k N$. The category $\text{rep}(S, k)$ has *fully wild representation type* if F_N is a fully faithful functor for some N. If $\text{rep}(S, k)$ has fully wild representation type, then $\text{rep}(S, k)$ has wild representation type, since a fully faithful functor preserves indecomposables and reflects isomorphisms.

Example 1.4.1 *The category* $\text{rep}(S_5, k)$ *has fully wild representation type.*

PROOF. Define $N \in \text{rep}(S_5, R)$ for $R = k\langle x, y\rangle$ by

$$N = (R \oplus R, R \oplus 0, 0 \oplus R, (1 + 1)R, (1 + x)R, (1 + y)R).$$

As in Example 1.1.7, $\text{End}_k N = \{r \in \text{End}_k R : xr = rx, yr = ry\}$. Then $\text{End}_k N$ is isomorphic to $R = \text{End}_R N$, the endomorphism ring of R as an R-module. If $M \in \text{mod } R$, then

$$N \otimes_R M = ((R \oplus R) \otimes_R M, (R \oplus 0) \otimes_R M, (0 \oplus R) \oplus_R M,$$

$$(1 + 1)R \otimes_R M, (1 + x)R \otimes_R M, (1 + y)R \otimes_R M)$$

$$= (M \oplus M, M \oplus 0, 0 \oplus M, (1 + 1)M, (1 + x)M, (1 + y)M).$$

Hence, $N \otimes_R M \in \text{rep}(S, k)$, since M has finite k-dimension.

For $M, M' \in \text{mod } k\langle x, y\rangle$, define $\phi : \text{Hom}(M, M') \rightarrow \text{Hom}(N \otimes_R M, N \otimes_R M')$ by $\phi(f) = 1 \otimes f$. Then ϕ is a one-to-one group homomorphism. Furthermore, ϕ is onto because a representation morphism $g : N \otimes_R M \rightarrow N \otimes_R M'$ is of the form $g = (f, f)$, f a k-homomorphism from $M \rightarrow M'$ with $fx = xf$ and $fy = yf$. It follows that f is an R-homomorphism from M to M' with $1 \otimes f = g$.

A salient property of $\text{rep}(S_5, R)$ confirmed in Example 1.1.7 is that each finite-dimensional k-algebra can be realized as the endomorphism ring of some $U \in \text{rep}(S_5, k)$. As an immediate consequence of the following proposition, if $\text{rep}(S, k)$ has wild representation type, then $\text{rep}(S, k)$ contains a copy of each finite-dimensional indecomposable R-module for each finite-dimensional k-algebra R. This is why it is unreasonable to expect an explicit classification of all indecomposable representations in a category with wild representation type. Let $\text{Ind}(\text{mod } R)$ denote the set of isomorphism classes of indecomposable modules in $\text{mod } R$.

Proposition 1.4.2 [Brenner 74A] *Let R be a finite-dimensional k-algebra.*

(a) *There is a fully faithful functor $F : \text{mod } R \rightarrow \text{mod } k\langle x, y\rangle$.*
(b) *If $\text{rep}(S, k)$ has wild representation type, then there is an embedding of $\text{Ind}(\text{mod } R)$ into $\text{Ind}(S, k)$.*

PROOF. Let $\{x_1, \ldots, x_n\}$ be a k-basis for R with $x_1 = 1$ and let A and B be $(n+2) \times (n+2)$ R-matrices as defined in Example 1.1.7. Specifically,

$$A = \begin{pmatrix} 0 & 1 & 0 & \cdots & 0 & 0 \\ 0 & 0 & 1 & \cdots & 0 & 0 \\ & & & \cdots & & \\ 0 & 0 & 0 & \cdots & 0 & 1 \\ 0 & 0 & 0 & \cdots & 0 & 0 \end{pmatrix}, \quad B = \begin{pmatrix} 0 & 0 & 0 & \cdots & 0 & 0 & 0 \\ 1 & 0 & 0 & \cdots & 0 & 0 & 0 \\ 1 & 1 & 0 & \cdots & 0 & 0 & 0 \\ 0 & x_2 & 1 & \cdots & 0 & 0 & 0 \\ & & & \cdots & & & \\ 0 & 0 & 0 & \cdots & 1 & 0 & 0 \\ 0 & 0 & 0 & \cdots & x_n & 1 & 0 \end{pmatrix}.$$

Suppose M is an R-module with finite k-dimension. Define $F(M) = M^{n+2}$, a $k\langle x, y\rangle$-module in mod $k\langle x, y\rangle$ via the multiplication $xm = Am$, and $ym = Bm$ for each m in M^{n+2}. If $f : M \to M'$ is an R-homomorphism, define $F(f) = fI, I$ an $(n+2) \times (n+2)$ identity matrix. It is routine to verify that $F : \mathrm{Hom}_R(M, M') \to \mathrm{Hom}_{k\langle x,y\rangle}(F(M), F(M'))$ is a one-to-one group homomorphism. To see that F is onto, let $g : F(M) \to F(M')$ be a $k\langle x, y\rangle$-homomorphism. A matrix computation, just as in Example 1.1.7, shows that $g = (f, \ldots, f)$ with $f \in \mathrm{Hom}_k(M, M')$ and $fx_i = x_i f$ for each $1 \le i \le n$. Thus, $g = fI$ with $f \in \mathrm{Hom}_R(M, M')$ as desired.

Call rep(S, k) *endowild* if for each finite-dimensional k-algebra R, there is $U \in$ rep(S, k) with End U isomorphic to R.

Corollary 1.4.3 *Let S be a finite poset and k a field. If* rep(S, k) *has fully wild representation type, then* rep(S, k) *is endowild.*

PROOF. Let R be a finite-dimensional k-algebra. Then $R \in \mathrm{mod}\, R$ with $\mathrm{End}_R(R) = R$. By Proposition 1.4.2, $F(R) = M \in \mathrm{mod}\, k\langle x, y\rangle$ with $\mathrm{End}_{k\langle x,y\rangle}(M) = R$, because F is fully faithful. Since rep(S, k) has fully wild representation type, $U = N \otimes_{k\langle x,y\rangle} M \in \mathrm{rep}(S, k)$ with End U isomorphic to R.

Theorem 1.4.4 [Nazarova 75] *Let S be a finite poset and k a field. Then* rep(S, k) *has wild representation type if and only if S contains S_5, $(1, 1, 1, 2)$, $(2, 2, 3)$, $(1, 3, 4)$, $(N, 5)$, or $(1, 2, 6)$ as a subposet.*

PROOF. ($\Leftarrow$) By Proposition 1.3.1, if T is a subposet of S, then there is a fully faithful functor $F^+ : \mathrm{rep}(T, k) \to \mathrm{rep}(S, k)$. Since the composition of fully faithful functors is fully faithful, it is sufficent to show that if T is one of S_5, $(1, 1, 1, 2)$, $(2, 2, 3)$, $(1, 3, 4)$, $(N, 5)$, or $(1, 2, 6)$, then rep(T, k) has fully wild representation type. In particular, rep(S, k) would then have wild representation type.

The case that $T = S_5$ is presented in Example 1.4.1. Following is a list of representations $N_T = (U_0, U_i : i \in T)$ with $F_{N_T} : \mathrm{mod}\, k\langle x, y\rangle \to \mathrm{rep}(S, k)$ a fully faithful functor. The proof is a computation like that of Example 1.4.1 to show that $\mathrm{End}_k N_T = R = \mathrm{End}_R(R)$ and $\mathrm{Hom}_{k\langle x,y\rangle}(M, M')$ is isomorphic to

$\mathrm{Hom}(N_T \otimes_{k\langle x,y\rangle} M, N_T \otimes_{k\langle x,y\rangle} M')$ for each M, M' in mod $k\langle x, y\rangle$. Write $R = k\langle x, y\rangle$.

(i) $T = (S_5)$ or $T = (1, 1, 1, 1, 2) = 1 \quad 2 \quad 3 \quad \overset{\displaystyle 5}{\underset{\displaystyle 4}{|}}$

$$U_0 = R \oplus R \oplus R \oplus R, \qquad U_1 = R \oplus R \oplus 0 \oplus 0,$$

$$U_2 = 0 \oplus 0 \oplus R \oplus R, \qquad U_3 = (1 + 0 + 1 + 0)R \oplus (0 + 1 + 0 + 1)R,$$

$$U_4 = (1 + 0 + x + 1)R, \qquad U_5 = (1 + 0 + x + 1)R \oplus (0 + 1 + y + 0)R.$$

(ii) $T = (2, 2, 3) = \begin{matrix} & & & 5 \\ & & & | \\ & 2 & 4 & 6 \\ & | & | & | \\ 1 & 3 & 7 \end{matrix}$

$$U_0 = R \oplus R \oplus R \oplus R \oplus R \oplus R,$$

$$U_1 = (y + x + 0 + 1 + 0 + 1)R \oplus (0 + 1 + 1 + 0 + 1 + 0)R,$$

$$U_2 = (1 + 0 + 1 + 0 + 0 + 0)R \oplus (0 + 1 + 0 + 1 + 0 + 0)R$$

$$\oplus (y + x + 0 + 1 + 0 + 1)R \oplus (0 + 1 + 1 + 0 + 1 + 0)R,$$

$$U_3 = 0 \oplus 0 \oplus 0 \oplus 0 \oplus R \oplus R, \qquad U_4 = 0 \oplus 0 \oplus R \oplus R \oplus R \oplus R,$$

$$U_7 = R \oplus 0 \oplus 0 \oplus 0 \oplus 0 \oplus 0, \qquad U_6 = R \oplus R \oplus 0 \oplus 0 \oplus 0 \oplus 0,$$

$$U_5 = R \oplus R \oplus R \oplus R \oplus 0 \oplus 0.$$

(iii) $T = (1, 3, 4) = \begin{matrix} & & & 8 \\ & & & | \\ & 4 & 7 \\ & | & | \\ & 3 & 6 \\ & | & | \\ 1 & 2 & 5 \end{matrix}$

$$U_0 = R \oplus R \oplus R \oplus R \oplus R \oplus R \oplus R \oplus R,$$

$$U_1 = (0 + 1 + 1 + 0 + 1 + 0 + 0 + 0)R$$

$$\oplus (y + x + 0 + 1 + 0 + 1 + 0 + 0)R$$

$$\oplus (1 + 0 + 1 + 0 + 0 + 0 + 1 + 0)R$$

$$\oplus (0 + 1 + 0 + 1 + 0 + 0 + 0 + 1)R,$$

$$U_2 = 0 \oplus 0 \oplus 0 \oplus 0 \oplus 0 \oplus 0 \oplus R \oplus R,$$

$$U_3 = 0 \oplus 0 \oplus 0 \oplus 0 \oplus R \oplus R \oplus R \oplus R,$$

$$U_4 = 0 \oplus 0 \oplus R \oplus R \oplus R \oplus R \oplus R \oplus R,$$

$$U_5 = R \oplus 0 \oplus 0 \oplus 0 \oplus 0 \oplus 0 \oplus 0 \oplus 0,$$

$$U_6 = R \oplus R \oplus R \oplus 0 \oplus 0 \oplus 0 \oplus 0 \oplus 0,$$

$$U_7 = R \oplus R \oplus R \oplus R \oplus R \oplus 0 \oplus 0 \oplus 0,$$

$$U_8 = R \oplus R \oplus R \oplus R \oplus R \oplus R \oplus R \oplus 0.$$

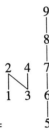

(iv) $T = (N, 5) =$

$$U_0 = R \oplus R \oplus R \oplus R \oplus R \oplus R \oplus R \oplus R \oplus R \oplus R,$$

$$U_1 = 0 \oplus 0 \oplus 0 \oplus 0 \oplus 0 \oplus 0 \oplus R \oplus R \oplus R \oplus R,$$

$$U_2 = 0 \oplus 0 \oplus R \oplus R \oplus 0 \oplus 0 \oplus R \oplus R \oplus R \oplus R$$

$$\oplus (0+0+0+0+1+0+0+0+1+0)R$$

$$\oplus (0+0+0+0+0+1+0+0+0+1)R,$$

$$U_3 = (0+0+0+0+1+0+0+0+1+0)R$$

$$\oplus (0+0+0+0+0+1+0+0+0+1)R,$$

$$U_4 = (0+1+1+0+1+0+0+0+0+0)R$$

$$\oplus (y+x+0+1+0+1+0+0+0+0)R$$

$$\oplus (1+0+1+0+0+0+1+0+0+0)R$$

$$\oplus (0+1+0+1+0+0+0+1+0+0)R$$

$$\oplus (0+0+0+0+1+0+0+0+1+0)R$$

$$\oplus (0+0+0+0+0+1+0+0+0+1)R,$$

$$U_5 = R \oplus 0 \oplus 0 \oplus 0 \oplus 0 \oplus 0 \oplus 0 \oplus 0 \oplus 0 \oplus 0,$$

$$U_6 = R \oplus R \oplus R \oplus 0 \oplus 0 \oplus 0 \oplus 0 \oplus 0 \oplus 0 \oplus 0,$$

$$U_7 = R \oplus R \oplus R \oplus R \oplus R \oplus 0 \oplus 0 \oplus 0 \oplus 0 \oplus 0,$$

$$U_8 = R \oplus R \oplus R \oplus R \oplus R \oplus R \oplus R \oplus 0 \oplus 0 \oplus 0,$$

$$U_9 = R \oplus R \oplus R \oplus R \oplus R \oplus R \oplus R \oplus R \oplus 0 \oplus 0.$$

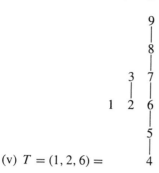

(v) $T = (1, 2, 6) =$

$$U_0 = R \oplus R \oplus R \oplus R \oplus R \oplus R \oplus R \oplus R \oplus R \oplus R \oplus R \oplus R,$$

$$U_1 = (0 + 0 + 0 + 1 + 0 + 0 + 0 + 0 + 0 + 1 + 0 + y)R$$

$$\oplus (0 + 0 + 1 + 0 + 0 + 0 + 0 + 0 + 1 + 0 + 1 + x)R$$

$$\oplus (0 + 1 + 0 + 0 + 0 + 1 + 0 + 0 + 0 + 1 + 1 + 0)R$$

$$\oplus (1 + 0 + 0 + 0 + 1 + 0 + 0 + 0 + 1 + 0 + 0 + 1)R$$

$$\oplus (0 + 0 + 0 + 0 + 0 + 0 + 0 + 1 + 0 + 0 + 1 + 0)R$$

$$\oplus (0 + 0 + 0 + 0 + 0 + 0 + 1 + 0 + 0 + 0 + 0 + 1)R,$$

$$U_2 = 0 \oplus 0 \oplus 0 \oplus 0 \oplus 0 \oplus 0 \oplus 0 \oplus 0 \oplus R \oplus R \oplus R \oplus R,$$

$$U_3 = 0 \oplus 0 \oplus 0 \oplus 0 \oplus R \oplus R \oplus R \oplus R \oplus R \oplus R \oplus R \oplus R,$$

$$U_4 = R \oplus R \oplus 0 \oplus 0 \oplus 0 \oplus 0 \oplus 0 \oplus 0 \oplus 0 \oplus 0 \oplus 0 \oplus 0,$$

$$U_5 = R \oplus R \oplus R \oplus 0 \oplus 0 \oplus 0 \oplus 0 \oplus 0 \oplus 0 \oplus 0 \oplus 0 \oplus 0,$$

$$U_6 = R \oplus R \oplus R \oplus R \oplus 0 \oplus 0 \oplus 0 \oplus 0 \oplus 0 \oplus 0 \oplus 0 \oplus 0,$$

$$U_7 = R \oplus R \oplus R \oplus R \oplus R \oplus R \oplus 0 \oplus 0 \oplus 0 \oplus 0 \oplus 0 \oplus 0,$$

$$U_8 = R \oplus R \oplus R \oplus R \oplus R \oplus R \oplus R \oplus R \oplus 0 \oplus 0 \oplus 0 \oplus 0,$$

$$U_9 = R \oplus R \oplus R \oplus R \oplus R \oplus R \oplus R \oplus R \oplus R \oplus R \oplus 0 \oplus 0.$$

($\Rightarrow$) This argument is long and technical and as such is omitted. A complete proof can be found in [Simson 92, Theorem 15.3].

Define $\text{Rep}(S, k)$ to be the category with objects $U = (U_0, U_i : i \in S)$, where U_0 is a k-vector space with countable dimension, each U_i is a subspace of U_0, and if

$i \leq j$ in S, then $U_i \subseteq U_j$. Morphisms in this category are k-linear transformations $f : U_0 \to U'_0$ with $f(U_i) \subseteq U'_i$ for each i in S. Then $\mathrm{rep}(S, k)$ is the full subcategory of $\mathrm{Rep}(S, k)$ with objects those $U \in \mathrm{Rep}(S, k)$ such that U_0 has finite k-dimension. For a ring R, $\mathrm{Mod}\, R$ denotes the category of all countably generated R-modules.

A category $\mathrm{rep}(S, k)$ has *strongly wild representation type* if there is an N in $\mathrm{rep}(S, k\langle x, y \rangle)$ such that $\mathrm{End}_{k\langle x,y \rangle} N = \mathrm{End}_k N$ and the functor $F_N : \mathrm{Mod}\, k\langle x, y \rangle \to \mathrm{Rep}(S, k)$ preserves indecomposables and reflects isomorphisms. Clearly, if $\mathrm{rep}(S, k)$ has strongly wild representation type, then it has wild representation type, since $\mathrm{mod}\, k\langle x, y \rangle$ is a full subcategory of $\mathrm{Mod}\, k\langle x, y \rangle$.

All of the variations on the definitions of wild representation type are equivalent for $\mathrm{rep}(S, k)$.

Proposition 1.4.5 *Let S be a finite poset and k a field. The following statements are equivalent:*

(a) $\mathrm{rep}(S, k)$ *has fully wild representation type;*
(b) $\mathrm{rep}(S, k)$ *has wild representation type;*
(c) $\mathrm{rep}(S, k)$ *has strongly wild representation type.*

In this case, $\mathrm{rep}(S, k)$ *is endowild.*

PROOF. (a) $\Leftrightarrow$ (b) is a consequence of the definition of fully wild representation type and the proof of Theorem 1.4.4.

(c) $\Rightarrow$ (b) follows from the definition of strongly wild representation type.

(b) $\Rightarrow$ (c) Assume that $\mathrm{rep}(S, k)$ has wild representation type, T is one of the posets S_5, $(1, 1, 1, 2)$, $(2, 2, 3)$, $(1, 3, 4)$, $(N, 5)$, or $(1, 2, 6)$, and let N_T be the associated representation given in (i)–(v) of Theorem 1.4.4. The same proof as that of Theorem 1.4.4 shows that $\mathrm{Hom}_{k\langle x,y \rangle}(M, M')$ is isomorphic to $\mathrm{Hom}(N_T \otimes_{k\langle x,y \rangle} M, N_T \otimes_{k\langle x,y \rangle} M')$ for each M, M' in $\mathrm{Mod}\, k\langle x, y \rangle$. This shows that $\mathrm{rep}(T, k)$ has strongly wild representation type for each such T. Finally, if T is a subposet of S, then there is a fully faithful functor $F^+ : \mathrm{Rep}(T, k) \to \mathrm{Rep}(S, k)$ defined just as in Proposition 1.3.1 for $\mathrm{rep}(S, k)$. The conclusion is that $\mathrm{rep}(S, k)$ has strongly wild representation type. For the last statement, apply Corollary 1.4.3.

As a consequence of Theorems 1.4.4 and 1.3.6, $\mathrm{rep}(S, k)$ has exactly one type from among finite representation type, wild representation type, and infinite but not wild representation type. The category $\mathrm{rep}(S, k)$ has *tame representation type*, provided that it has infinite representation type and for each $w = (w_0, w_i : i \in S)$ there are finitely many $N_1, \ldots, N_m \in \mathrm{rep}(S, k[x])$ with $\mathrm{End}_k N_i = \mathrm{End}_{k[x]} N_i$ for each i such that if $U \in \mathrm{rep}(S, k)$ is indecomposable with cdn $U = w$, then U is isomorphic to $N_i \otimes_{k[x]} B$ for some i and indecomposable cyclic $k[x]$-module B. In other words, if $\mathrm{rep}(S, k)$ has tame representation type, then the elements of $\mathrm{Ind}(S, k)$ are characterized up to isomorphism by cdn U, finitely many $k[x]$-representations of S, and indecomposable $k[x]$-modules of the form $k[x]/\langle g(x)^e \rangle$ for some irreducible polynomial $g(x)$.

Theorem 1.4.6 *Let S be a finite poset and k a field. The following statements are equivalent:*

(a) *The category* rep(S, k) *has tame representation type;*
(b) *The category* rep(S, k) *has infinite representation type but does not have wild representation type;*
(c) *The poset S contains S_4, $(2, 2, 2)$, $(1, 3, 3)$, $(1, 2, 5)$, or $(N, 4)$ as a subposet but does not contain S_5, $(1, 1, 1, 2)$, $(2, 2, 3)$, $(1, 3, 4)$, $(N, 5)$, or $(1, 2, 6)$ as a subposet.*

PROOF. (b) $\Leftrightarrow$ (c) Theorems 1.4.4 and 1.3.6.
(a) $\Leftrightarrow$ (b) is a complicated argument, proved in [Simson 92, Theorems 15.54 and 15.57] by induction and the Zavadskij derivative.

The next lemma relates representations of a finite poset $S = (S, \leq)$ and the opposite poset $S^{\mathrm{op}} = (S, \geq)$. A *contravariant functor* $F : X \to X'$ between two additive categories X and X' is a correspondence that satisfies all of the properties of a functor except that $F(fg) = F(g)F(f)$ for each f and g, i.e., the order of composition is interchanged by F.

For example, if $\mathbf{X}$ is the category of finite-dimensional vector spaces, then $F : \mathbf{X} \to \mathbf{X}$, defined by $F(V) = \mathrm{Hom}(V, k)$, is a contravariant functor. In particular, if $f : V \to W$ is a k-linear transformation, then $F(f) : \mathrm{Hom}(W, k) \to \mathrm{Hom}(V, k)$ is defined by $F(f)(\alpha) = \alpha f$ for each $\alpha \in \mathrm{Hom}(W, k)$. Moreover, for each finite-dimensional vector space V, there is an isomorphism $\phi_V : V \to F^2(V) = \mathrm{Hom}(\mathrm{Hom}(V, k), k)$ given by $\phi_V(v)(\alpha) = \alpha(v)$ for each $v \in V$ and $\alpha \in \mathrm{Hom}(V, k)$. This isomorphism is a *natural* isomorphism from the identity functor to F^2 in the sense that if $f : V \to V'$ is a k-linear transformation, then $F^2(f)\phi_V = \phi_{V'} f$. A contravariant functor F with a natural isomorphism from the identity functor to F^2 is called a *contravariant duality*.

Proposition 1.4.7 *Let S be a finite poset and k a field. There is a contravariant duality $\sigma : \mathrm{rep}(S, k) \to \mathrm{rep}(S^{\mathrm{op}}, k)$. Moreover, $\sigma : \mathrm{Ind}(S, k) \to \mathrm{Ind}(S^{\mathrm{op}}, k)$ is one-to-one and onto, and $\mathrm{rep}(S, k)$ has finite representation type if and only if $\mathrm{rep}(S^{\mathrm{op}}, k)$ has finite representation type.*

PROOF. Given $U = (U_0, U_i) \in \mathrm{rep}(S, k)$, define $\sigma(U) = V$, where $V_0 = \mathrm{Hom}_k(U_0, k)$ and $V_i = \{f : U_0 \to k : f(U_i) = 0\}$. The proof is an exercise in vector space duality (Exercise 1.4.4).

The converse to the last conclusion of Proposition 1.4.5 remains unresolved.

Open Question: *If S is a finite poset, k a field, and $\mathrm{rep}(S, k)$ is endowild, does $\mathrm{rep}(S, k)$ have wild representation type?*

EXERCISES

1. Let $(S(n), m) = \{a_{0m}, a_{1m}, \ldots, a_{nm}\}$ be a copy of the poset $\{a_0, a_1, \ldots, a_n\}$ consisting of an antichain $\{a_1, \ldots, a_n\}$ with $a_0 > a_i$ for each $1 \leq i \leq n$. Given positive integers n and j, define $S(n, j)$ to be the poset consisting of a disjoint union of the posets $(S(n), 0), \ldots, (S(n), j-1)$ with the additional relation $a_{i,m} \leq a_{i,m+1}$ for each $0 \leq i \leq n, 0 \leq m \leq j-2$. Determine, in terms of n and j, exactly when $\text{rep}(S(n, j), k), k$ a field, has finite, tame, or wild representation type.

2. Given positive integers n and j, define $P(n, j)$ to be the poset $\{a_0 > b_1 > \cdots > b_{j-1}, a_1, \ldots, a_n\}$ with $\{a_1, \ldots, a_n\}$ an antichain and $a_0 > a_i$ for each $1 \leq i \leq n$. Determine, in terms of n and j, exactly when $\text{rep}(S(n, j), k), k$ a field, has finite, tame, or wild representation type.

3. Provide the missing computations for the proof of Theorem 1.4.4($\Leftarrow$).

4. Complete the proof of Proposition 1.4.7.

5. Let $S = (1, 2, 2)$.
   (a) Describe S^{op}.
   (b) Find a sincere indecomposable in $\text{rep}(S, k)$ that is decomposable when viewed as an element of $\text{rep}(S^{\text{op}}, k)$.
   (c) Use the duality σ to compare the indecomposable representations of $\text{rep}(S, k)$ and $\text{rep}(S^{\text{op}}, k)$ (see Exercise 1.3.1).

6. Let T be a subposet of a finite poset S and let F^+ and F^- be the functors defined in Section 1.3. Find a functor $G : \text{rep}(S, k) \rightarrow \text{rep}(T, k)$ such that both $GF^+(U)$ and $GF^-(U)$ are isomorphic to U for each $U \in \text{rep}(T, k)$.

1.5   Generic Representations

There are infinite-dimensional k-representations of a finite poset S, called generic representations, that determine the representation type of $\text{rep}(S, k)$ (Theorems 1.5.3 and 1.5.4). Generic representations give rise to, and arise from, finite-dimensional representations.

For a ring R and a finite poset S, define a category of R-representations $\text{Rep}(S, R)$ with objects $U = (U_0, U_i : i \in S)$, where U_0 is a countably generated free R-module, each U_i is a countably generated free R-summand of U_0, and $U_i \subseteq U_j$ if $i \leq j$ in S. A morphism $f : U \rightarrow V$ is an R-homomorphism $f : U_0 \rightarrow V_0$ with $f(U_i) \subseteq V_i$ for each i in S.

An R-module M is *simple* if 0 and M are the only submodules of M. The R-module M has a *composition series* if there is a finite ascending chain $M_0 = 0 \subset M_1 \subset \cdots \subset M_n = M$ of R-submodules of M such that each M_i/M_{i-1} is simple. In this case, M is said to have R-*length* n. The R-length of M is independent of the composition series for M as a consequence of the classical Jordan–Hölder theorem [Anderson, Fuller 74].

If $U = (U_0, U_i : i \in S) \in \text{Rep}(S, R)$, then U_0 is a left module over $\text{End}\, U$, the endomorphism ring of U, defined by $fu = f(u)$ for $u \in U_0$ and $f \in \text{End}\, U$. This operation is well-defined, since End U is a subring of the R-endomorphism ring of U_0. The length of U_0 as an End U-module is called the *endolength* of U.

Given a field k, a representation $U = (U_0, U_i : i \in S) \in \text{Rep}(S, k)$ is a *generic representation* if U is indecomposable, U_0 has countably infinite k-dimension, and U_0 has finite length as an End U-module. If U is indecomposable and U_0 has finite k-dimension, then U satisfies all of the conditions for a generic representation except for having infinite k-dimension. A number of examples of generic representations are given below.

Generic representations have special endomorphism rings. Following is a brief review of some ring-theoretic notions and properties that are used; see, for example, [Anderson, Fuller 74] or [Hungerford 74]. The *Jacobson radical JR* of a ring R is the intersection of the maximal right ideals of R. Then JR is a two-sided ideal of R with $J(R/JR) = 0$ and if $x \in JR$ and $r \in R$, then $1 - rx$ is a unit of R. Moreover, R is a local ring if and only if R/JR is a division ring. A ring R is *semisimple* if $JR = 0$ and *left Artinian* if it has the descending chain condition on left ideals. An ideal I is *nilpotent* if I^n, the ideal of R generated by $\{x_1 x_2 \cdots x_n : x_i \in I\}$, is equal to zero for some positive integer n. If R is left Artinian, then JR is nilpotent. A semisimple left Artinian ring R is a finite product of full matrix rings over division rings. In this case, R is also right Artinian. *Idempotents lift modulo JR* if whenever e is an idempotent in R/JR, then there is an idempotent $f \in R$ with $f + JR = e$. If JR is nilpotent, then idempotents lift modulo JR. Finally, a ring R is *semiperfect* if R/JR is left Artinian and idempotents lift modulo JR. For example, a left Artinian ring is semiperfect. If M is a finitely generated left R-module and $(JR)M = 0$, then $M = 0$, a result commonly referred to as *Nakayama's lemma* [Anderson, Fuller 74].

Lemma 1.5.1 *Assume $U = (U_0, U_i : i \in S) \in \text{rep}(S, k)$ is a generic representation. Then $D = \text{End}\, U / J\text{End}\, U$ is a division k-algebra, $(J\text{End}\, U)^m = 0$ for some positive integer m, and U_0 has the structure of a right D-module with D-dimension $U_0 = $ endolength U.*

PROOF. Assume that U is a generic representation. Then

$$(J\text{End}\, U)U_0 \supseteq (J\text{End}\, U)^2 U_0 \supseteq \cdots$$

is a descending chain of End U-submodules of U_0. Since U_0 has finite length as an End U-module, $(J\text{End}\, U)^m U_0 = (J\text{End}\, U)^{m+1} U_0$ for some positive integer m. Also, U_0 is finitely generated as an End U-module. By Nakayama's lemma, $(J\text{End}\, U)^m U_0 = 0$. Hence, $(J\text{End}\, U)^m = 0$, since $(J\text{End}\, U)^m \subseteq \text{End}\, U$. Since U is indecomposable, the only idempotents of End $U/J\text{End}\, U$ are 0 and 1. But End $U/J\text{End}\, U$ is left Artinian, because U has finite endolength. It follows that the semisimple left Artinian ring End $U/J\text{End}\, U = D$ is a division k-algebra.

In fact, D is a unique simple End U module. This is because D is a division algebra and if M is a simple End U-module, then $(J\text{End}\, U)M = 0$ as an application of Nakayama's lemma. Consequently, M is isomorphic to D. Next, let

$M_0 = 0 \subset M_1 \subset \cdots \subset M_n = U_0$ be an End U composition series for U_0 with $n = $ endolength U. Each M_i / M_{i-1} is a simple module, hence isomorphic to D. Write $M_i / M_{i-1} = (x_i + M_{i-1})D$ for some $x_i \in M_i$. Then U_0 has the structure of a D-module with $U_0 = x_1 D \oplus \cdots \oplus x_n D$. In particular, the D-dimension of U_0 is n, the endolength of U.

Define $k(x) = \{ f(x)/g(x) : f(x), g(x) \in k[x], 0 \neq g(x) \}$, the field of quotients of the polynomial ring $k[x]$.

Example 1.5.2 *Given a field k, there is a generic k-representation U of S_5 with* End U *noncommutative*, JEnd$U \neq 0$, End U / JEnd$U = k(x)$, $(J$End $U)^2 = 0$, *and endolength* $U = 4$.

PROOF. Let R be the ring of 2×2 matrices of the form $\left(\begin{smallmatrix} \lambda & \sigma \\ 0 & \lambda \end{smallmatrix}\right)$ for $\lambda, \sigma \in k(x)$. Then $R/JR = k(x), JR \neq 0$, and $(JR)^2 = 0$, noting that JR consists of those matrices in R having all zeros on the diagonal. Moreover, R is a $k(x)$-vector space of dimension 2.
Define $U = (U_0, U_1, U_2, U_3, U_4, U_5)$, where

$$U_0 = R \oplus R, U_1 = R \oplus 0, U_2 = 0 \oplus R, U_3 = (1+1)R,$$
$$U_4 = (1+x)R, U_5 = (1+c)R,$$

and $c = \left(\begin{smallmatrix} 0 & 1 \\ 0 & 0 \end{smallmatrix}\right)$ is an element of R.
To see that End U is isomorphic to R, notice that R embeds in End U via $r \mapsto (\mu_r, \mu_r)$, where μ_r is left multiplication by r. Conversely, if $f \in$ End U, then $f = (g, g)$ with $g \in \mathrm{End}_k(R)$, since U_1, U_2, and U_3 are preserved by f. But $gx = xg$, since $f(U_4) \subseteq U_4$, and $gc = cg$, since $f(U_5) \subseteq U_5$. It follows that $g \in \mathrm{End}_R(R)$; hence g is multiplication by some r in R, as desired.
Finally, End U / JEnd $U = k(x)$, so that U is indecomposable. Moreover, U_0 has countably infinite k-dimension, and endolength $U = 4$ is the $k(x)$-dimension of $U_0 = R \oplus R$. This completes the proof.

The category rep(S, k) is *generically trivial* if there are no generic representations in rep(S, k).

Theorem 1.5.3 *Let S be a finite poset and k a field. Then* rep(S, k) *has finite representation type if and only if* rep(S, k) *is generically trivial.*

PROOF. Assume that rep(S, k) has finite representation type. If U is an indecomposable in Rep(S, k), then U_0 has finite k-dimension [Ringel, Tachikawa 75]. Thus, rep(S, k) has no generic representations.
Conversely, assume that rep(S, k) has infinite representation type. As a consequence of Theorem 1.3.6, it suffices to show that if S is one of the critical posets, namely $S_4, (2, 2, 2), (1, 3, 3), (1, 2, 5)$, or $(N, 4)$, then there is a generic representation in rep(S, k).
For each of the critical posets S, a generic representation U is listed below. Each representation is derived from the finite-dimensional representation of S

given in the proof of Theorem 1.3.6 with k^n replaced by the field $k(x)$. In each case, it can be shown, just as in the proof of Theorem 1.3.6, that End $U = k(x)$. In particular, U is an indecomposable k-representation with infinite k-dimension. Since U_0 has finite $k(x)$-dimension with $k(x) = $ End U, $U \in \mathrm{rep}(S, k)$ is a generic representation.

(i) $S = S_4$

$$U_0 = k(x) \oplus k(x), \qquad U_1 = k(x) \oplus 0,$$
$$U_2 = 0 \oplus k(x), \qquad U_3 = (1 + 1)k(x), \qquad U_4 = (1 + x)k(x).$$

(ii) $S = (2, 2, 2) = \begin{smallmatrix} 2 & 4 & 6 \\ | & | & | \\ 1 & 3 & 5 \end{smallmatrix}$

$$U_0 = k(x) \oplus k(x) \oplus k(x), \qquad U_1 = 0 \oplus 0 \oplus k(x),$$
$$U_2 = (1 + x)k(x) \oplus k(x), \qquad U_3 = (1 + 1)k(x) \oplus 0,$$
$$U_4 = k(x) \oplus k(x) \oplus 0, \qquad U_5 = 0 \oplus (1 + 1)k(x),$$
$$U_6 = k(x) \oplus (1 + 1)k(x).$$

(iii) $S = (1, 3, 3) = \begin{smallmatrix} & 4 & 7 \\ & | & | \\ 1 & 3 & 6 \\ & | & | \\ & 2 & 5 \end{smallmatrix}$

$$U_0 = k(x) \oplus k(x) \oplus k(x) \oplus k(x), \qquad U_1 = (1 + 1)k(x) \oplus (1 + 1)k(x),$$
$$U_2 = 0 \oplus (1 + 1)k(x) \oplus 0, \qquad U_3 = 0 \oplus k(x) \oplus k(x) \oplus 0,$$
$$U_4 = 0 \oplus k(x) \oplus k(x) \oplus k(x), \qquad U_5 = k(x) \oplus 0 \oplus 0 \oplus 0,$$
$$U_6 = k(x) \oplus 0 \oplus 0 \oplus k(x), \qquad U_7 = k(x) \oplus (1 + x)k(x) \oplus k(x).$$

(iv) $S = (N, 4) = \begin{smallmatrix} & & 8 \\ & & | \\ & & 7 \\ & & | \\ 1 & 3 & 6 \\ \diagdown | & | \\ 2 & 4 & 5 \end{smallmatrix}$

$$U_0 = k(x) \oplus k(x) \oplus k(x) \oplus k(x) \oplus k(x),$$
$$U_2 = (1 + 1)k(x) \oplus (1 + 1)k(x) \oplus 0,$$

$$U_1 = k(x) \oplus k(x) \oplus k(x) \oplus k(x) \oplus 0, \qquad U_4 = 0 \oplus (1 + 1)k(x) \oplus 0 \oplus 0,$$

$$U_3 = 0 \oplus k(x) \oplus k(x) \oplus (1 + 1)k(x), \qquad U_5 = 0 \oplus 0 \oplus 0 \oplus 0 \oplus k(x),$$

$$U_6 = k(x) \oplus 0 \oplus 0 \oplus 0 \oplus k(x), \qquad U_7 = k(x) \oplus 0 \oplus 0 \oplus k(x) \oplus k(x),$$

$$U_8 = k(x) \oplus (1 + x)k(x) \oplus k(x) \oplus k(x).$$

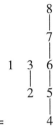

(v) $S = (1, 2, 5) =$

$$U_0 = k(x) \oplus k(x) \oplus k(x) \oplus k(x) \oplus k(x) \oplus k(x),$$

$$U_1 = (0 + 1 + 1 + 0 + 0 + 0)k(x) + (0 + 0 + 1 + 0 + 0 + 1)k(x)$$
$$\quad + (0 + 0 + 0 + 1 + 1 + 0)k(x),$$

$$U_2 = (1 + 1)k(x) \oplus (1 + 1)k(x) \oplus 0 \oplus 0,$$

$$U_3 = k(x) \oplus k(x) \oplus k(x) \oplus k(x) \oplus 0 \oplus 0,$$

$$U_4 = 0 \oplus 0 \oplus 0 \oplus 0 \oplus 0 \oplus k(x), \qquad U_5 = 0 \oplus 0 \oplus 0 \oplus 0 \oplus k(x) \oplus k(x),$$

$$U_6 = k(x) \oplus 0 \oplus 0 \oplus 0 \oplus k(x) \oplus k(x),$$

$$U_7 = k(x) \oplus 0 \oplus 0 \oplus k(x) \oplus k(x) \oplus k(x),$$

$$U_8 = k(x) \oplus (1 + x)k(x) \oplus k(x) \oplus k(x) \oplus k(x).$$

Call rep(S, k) *generically wild* if there is a generic representation U such that End U/JEnd U contains a copy of $k\langle x, y \rangle$.

Theorem 1.5.4 *Let S be a finite poset and k a field. For the following conditions,*
(a) $\Rightarrow$ (b) $\Rightarrow$ (c).

(a) rep(S, k) has wild representation type;
(b) rep(S, k) is generically wild;
(c) rep(S, k) is endowild.

Moreover, if $S = S_n$, then (a), (b), and (c) are all equivalent to the condition that $n \geq 5$.

PROOF. (a) $\Rightarrow$ (b) Let $N = (N_0, N_i : i \in S) \in$ rep$(S, k\langle x, y \rangle)$ be as given in the definition of wild representation type. There is a division k-algebra D with countable k-dimension containing $k\langle x,y\rangle$ by Lemma 3.5.7. Define $U = N \otimes_{k\langle x,y\rangle} D$ $= (U_0, U_i : i \in S) = (N_0 \otimes_{k\langle x,y\rangle} D, N_i \otimes_{k\langle x,y\rangle} D : i \in S) \in \mathrm{rep}(S, D)$. Then U_0 has countably infinite k-dimension and finite D-dimension, since N_0 is a finitely generated $k\langle x,y\rangle$-module. Moreover, U is indecomposable in $\mathrm{Rep}(S, k)$, because $\mathrm{rep}(S, k)$ has strongly wild representation type by Proposition 1.4.5. Since D is contained in End U and U_0 is a finite-dimensional D-space, it follows that U has finite endolength. Hence, U is a generic representation with $k\langle x,y\rangle \subseteq$ End U/JEnd U. This proves that $\mathrm{rep}(S, k)$ is generically wild.

(b) $\Rightarrow$ (c) Let $U = (U_0, U_i : i \in S)$ be a generic representation for $\mathrm{rep}(S, k)$ with End U/JEnd $U = D$, a division k-algebra containing $k\langle x,y\rangle$. As a consequence of Lemma 1.5.1, $U = WD$ for some $W = (W_0, W_i : i \in S) \in \mathrm{rep}(S, k\langle x,y\rangle)$.

Suppose A and B are two nonzero $m \times m$ k-matrices with minimal polynomials $f(x)^i$ and $g(y)^j$, respectively, and $f(x)$ and $g(y)$ irreducible polynomials. Let $R = k\langle x,y\rangle/K$, where $K = \langle xy, f(x)^i, g(y)^j\rangle$ is an ideal of $k\langle x,y\rangle$. Then $V = W \otimes_{k\langle x,y\rangle} R = (W_0/KW_0, (W_i + KW_0)/KW_0) \in \mathrm{rep}(S, k)$. Moreover, $C(A, B)$ is contained in $\mathrm{End}_{k\langle x,y\rangle} V$, a k-subalgebra of End V, observing that x acts on W_0/KW_0 by A and y on W_0/KW_0 by B. In particular, $C(A, B)$ is a k-subalgebra of End V. As noted in Section 1.1, for each finite-dimensional k-algebra Λ, there are matrices A and B with Λ isomorphic to $C(A, B)$. This shows that $\mathrm{rep}(S, k)$ is endowild.

As for the final statement, suppose that $\mathrm{rep}(S_n, k)$ does not have wild representation type. Then $n \leq 4$ by Theorem 1.4.6. If U is an indecomposable in $\mathrm{rep}(S_4, k)$, then End U is a factor k-algebra of $k[x]$ (Example 6.2.8). Thus, $\mathrm{rep}(S, k)$ is not endowild. This proves (c) $\Rightarrow$ (a) for the case that $S = S_n$ is an antichain.

Notice that (c) $\Rightarrow$ (a) of Theorem 1.5.4 is the open question given in Section 1.4. The final example of this section demonstrates that generic representations in $\mathrm{rep}(S, k)$, necessarily of infinite dimension, can be used to construct indecomposable finite-dimensional representations in $\mathrm{rep}(S, k)$. The proof is a routine computation, just as in Theorem 1.5.3.

Example 1.5.5 *Let U be one of the generic representations for $\mathrm{rep}(S, k)$ listed in the proof of Theorem 1.5.3 and $g(x)$ an irreducible polynomial in $k[x]$. Define $U/g(x)^e U = (U_0/g(x)^e U_0, (U_i + g(x)^e U_0)/g(x)^e U_0 : i \in S) \in \mathrm{rep}(S, k)$. Then $U/g(x)^e U$ is indecomposable with endomorphism ring $k[x]/\langle g(x)^e\rangle$.*

The following questions are motivated by analogous results for finite-dimensional algebras over algebraically closed fields [Crawley-Boevey 91].

Open Questions:

1. *Is there an existence condition on generic representations that is equivalent to $\mathrm{rep}(S, k)$ having tame representation type?*

2. *Is there a condition on the endomorphism ring of generic representations that is equivalent to* rep(S, k) *having tame representation type?*

EXERCISE

1. Do the computations to complete the proof of Theorem 1.5.3 and confirm Example 1.5.5.

NOTES ON CHAPTER 1

The invention of a derivative of a finite partially ordered set by Nazarova and Roiter in the late 1960s or early 1970s was a seminal event in the subject of representations of finite partially ordered sets (see [Simson 92]). This derivative was used by Kleiner in his classification of posets of finite representation type and sincere posets (Theorems 1.3.6 and 1.3.7). Partitioned matrix representations of a field k over a partially ordered set, herein called M_U for a representation U, are used in [Nazarova, Roiter 75] for the investigation of modules over finite-dimensional k-algebras. The category of S-spaces over k, denoted in this manuscript by rep(S, k), is investigated in [Gabriel 72] as a tool for the computation of representation type of representations of quivers and categories of modules. Various relationships between representations of finite partially ordered sets and finitely generated modules over finite-dimensional k-algebras are given in [Ringel 84], [Ringel 91, Section 4], [Simson 92], and, from a historical perspective, [Gustafson 82].

In the literature on modules over k-algebras, it is commonly assumed that k is an algebraically closed field. Applications to abelian group theory frequently involve either the field of rational numbers or finite fields, which are not algebraically closed. Hence, from the perspective of abelian group theory, results that do not depend on the field k, such as the representation type of rep(S, k), are most useful.

The elementary introductory material in Sections 1.1, 1.2, and the beginning of Section 1.3 is folklore. Included in [Simson 92] is a comprehensive treatment of I-spaces for a finite partially ordered set I. There are a number of proofs of Kleiner's theorems, Theorems 1.3.6 and 1.3.7. The proofs of these theorems in [Ringel 84] employ the theory of integral quadratic forms, while those in [Simson 92] use the Zavadiskij derivative, a refined version of the original Nazarova–Roiter derivative. An alternative proof of Theorem 1.3.6 is given in [Kerner 81].

Matrix problems for representations of partially ordered sets have been put in more general contexts such as vector space categories and bocses; see, for example, [Crawley-Boevey 88], [Ringel 84], [Rojter 80], and [Simson 92]. This is, in part, an attempt to find categorical settings for the combinatorial classification, via the Nazarova–Roiter or Zavadskij derivative, of indecomposable representations for finite posets with finite or tame representation type.

The notion of wild representation type for group representations, in the sense of Proposition 1.4.2(b), is observed in [Krugljak 64]. Tame representation type, in the sense that the representation type is infinite but the indecomposables are classifiable, is demonstrated for group representations in [Heller, Reiner 61]. Tame representation type for rep(S, k) has been subdivided into various kinds, e.g., domestic, finite growth, linear growth, polynomial growth, and one parameter. Characterizations for these subdivisions of the representation type of rep(S, k) in terms of the finite poset S are known; see [Nazarova 81], [Simson 92], and references therein.

Section 1.5 is motivated by more general results on generic modules in [Crawley-Boevey 91, 92]. The deepest of these results uses the theory of bocses, which is not available for fields that are not algebraically closed. Progress for the theory of generic representations over fields that are not algebraically closed awaits, among other things, the resolution of the problem of whether or not endowild is equivalent to wild representation type, as indicated in the open questions for Section 1.5.

2
Torsion-Free Abelian Groups

2.1 Quasi-isomorphism and Isomorphism at p

There are two equivalence relations on torsion-free abelian groups that are weaker than group isomorphism, namely quasi-isomorphism and isomorphism at a prime p. Properties of these equivalence relations are conveniently expressed in a categorical setting.

Abelian groups are written additively. An abelian group G is *torsion-free* if $na \neq 0$ for each nonzero integer n and nonzero $a \in G$. Hence, if n is a nonzero integer and a and b are elements of a torsion-free abelian group G with $na = nb$, then $a = b$. If G is a torsion-free abelian group, then G is isomorphic to a subgroup of a $\mathbb{Q}$-vector space $\mathbb{Q} \otimes_{\mathbb{Z}} G = \mathbb{Q}G$, where $\mathbb{Q}$ is the field of rational numbers. The group G is identified with its image in $\mathbb{Q}G$, so that $\mathbb{Q}G = \{qa : a \in G\}$.

Let $\mathbf{A}$ denote the category with torsion-free abelian groups as objects and group homomorphisms $\mathrm{Hom}(G, H)$ for G, H in $\mathbf{A}$ as morphisms. It is not difficult to verify that $\mathbf{A}$ is an additive category. A group homomorphism is a *monomorphism* if it is one-to-one and an *epimorphism* if it is onto. The group endomorphism ring of G is denoted by $\mathrm{End}\, G$. The *rank* of a torsion-free abelian group G is the cardinality of a maximal $\mathbb{Z}$-independent subset of G, equivalently the $\mathbb{Q}$-dimension of $\mathbb{Q}G$.

Define a category $\mathbf{A}_{\mathbb{Q}}$ with objects those of $\mathbf{A}$ but with morphism sets $\mathbb{Q}\mathrm{Hom}(G, H)$ for G, H in $\mathbf{A}$. Composition is defined by $(qf)(rg) = (qr)(fg)$ for $q, r \in \mathbb{Q}$ and $f, g \in \mathrm{Hom}(G, H)$. Composition is associative, and 1_G is an identity in the $\mathbf{A}_{\mathbb{Q}}$-endomorphism ring $\mathbb{Q}\mathrm{End}\, G$ of G.

Recall that $\mathbb{Q}\mathrm{End}\, G$ is semiperfect if $\mathbb{Q}\mathrm{End}\, G / J\mathbb{Q}\mathrm{End}\, G$ is left Artinian and idempotents lift modulo $J\mathbb{Q}\mathrm{End}\, G$.

Proposition 2.1.1

(a) *The category* $\mathbf{A}_{\mathbb{Q}}$ *is an additive category. If* $G = H \oplus L$ *in* $\mathbf{A}_{\mathbb{Q}}$, *then there is an idempotent* $e \in \mathbb{Q}\text{End } G$ *with* $H = e(G)$ *in* $\mathbf{A}_{\mathbb{Q}}$. *Conversely, if* G *is in* $\mathbf{A}_{\mathbb{Q}}$ *and* e *is an idempotent in* $\mathbb{Q}\text{End } G$, *then* $G = e(G) \oplus (1 - e)(G)$ *in* $\mathbf{A}_{\mathbb{Q}}$.

(b) *If* G *is indecomposable in* $\mathbf{A}_{\mathbb{Q}}$ *and* $\mathbb{Q}\text{End } G$ *is semiperfect, then* $\mathbb{Q}\text{End}(G)$ *is a local ring. In particular,* $\mathbb{Q}\text{End } G$ *is a local ring if* G *has finite rank.*

PROOF. (a) Notice that $\mathbb{Q}\text{Hom}(G, H)$ is an abelian group for each G, H in $\mathbf{A}_{\mathbb{Q}}$ with $(f + g)(x) = f(x) + g(x)$ for each $x \in G$ and that composition is distributive with respect to $+$. As for finite direct sums, let G_i be in $\mathbf{A}_{\mathbb{Q}}$ and $K = G_1 \oplus G_2 \oplus \cdots \oplus G_n$ the group direct sum with injections $i_j \in \text{Hom}(G_j, K)$ and projections $p_j : K \to G_j$. Then K is a direct sum in $\mathbf{A}_{\mathbb{Q}}$ with the same injections and projections. To see this, let $f_j \in \mathbb{Q}\text{Hom}(G_j, H)$. Choose a nonzero integer m with $mf_j \in \text{Hom}(G_j, H)$ for each j. Since K is a direct sum in $\mathbf{A}$, there is a unique $f \in \text{Hom}(K, H)$ with $fi_j = mf_j$ for each j. Then $(1/m)f$ is a unique element of $\mathbb{Q}\text{Hom}(K, H)$ with $(1/m)fi_j = f_j$ for each j, as required. Notice that $e_j = i_j p_j$ is an idempotent in $\mathbb{Q}\text{End } K$ with $1 = e_1 + \cdots + e_n$ and $e_j e_m = 0$ if $j \neq m$.

Conversely, let $e = (m/n)f$ be an idempotent in $\mathbb{Q}\text{End } G$ for some m/n in $\mathbb{Q}$ and $f \in \text{End } G$. Then G is the direct sum of $e(G)$ and $(1 - e)(G)$ in $\mathbf{A}_{\mathbb{Q}}$ with inclusions as injections.

(b) If G is indecomposable in $\mathbf{A}_{\mathbb{Q}}$, then $\mathbb{Q}\text{End } G$ has no idempotents other than 0 or 1 by (a). Hence, $\mathbb{Q}\text{End } G/J\mathbb{Q}\text{End } G$ has no nontrivial idempotents, since idempotents lift modulo $J\mathbb{Q}\text{End } G$. Moreover, $\mathbb{Q}\text{End } G/J\mathbb{Q}\text{End } G$ is semisimple Artinian, whence $\mathbb{Q}\text{End } G/J\mathbb{Q}\text{End } G$ is a division ring. Consequently, $\mathbb{Q}\text{End } G$ is a local ring. Finally, if G has finite rank, then $\mathbb{Q}\text{End } G$ is a finite-dimensional $\mathbb{Q}$-algebra. In this case, $\mathbb{Q}\text{End } \mathbb{Q}$ is Artinian, hence semiperfect.

Categorical notions of isomorphism and direct sums in $\mathbf{A}_{\mathbb{Q}}$ have group-theoretic interpretations that in fact, precede the categorical concepts [Jónsson 57, 59]. An abelian group G is *bounded* if there is a nonzero integer n with $nG = 0$.

Corollary 2.1.2 *Let* G *and* H *be objects of* $\mathbf{A}_{\mathbb{Q}}$.

(a) *The group* G *is isomorphic to* H *in* $\mathbf{A}_{\mathbb{Q}}$ *if and only if there is* $f \in \text{Hom}(G, H)$, $g \in \text{Hom}(H, G)$, *and a nonzero integer* n *with* $fg = n1_H$ *and* $gf = n1_G$.

(b) *If* $f \in \text{Hom}(G, H)$, *then* f *is an isomorphism in* $\mathbf{A}_{\mathbb{Q}}$ *if and only if* f *is a monomorphism and* $H/f(G)$ *is bounded. Moreover, if* $f \in \text{End } G$ *is a monomorphism and* G *has finite rank, then* f *is an isomorphism in* $\mathbf{A}_{\mathbb{Q}}$.

(c) *The group* G *is a direct sum* $G_1 \oplus \cdots \oplus G_n$ *in* $\mathbf{A}_{\mathbb{Q}}$ *if and only if there is a monomorphism* f *from the group direct sum* $G_1 \oplus \cdots \oplus G_n$ *into* G *such that* $G/\text{image } f$ *is bounded.*

(d) *The group* G *is indecomposable in* $\mathbf{A}_{\mathbb{Q}}$ *if and only if whenever* $G/(H \oplus K)$ *is bounded, then* $H = 0$ *or* $K = 0$.

(e) *The group* H *is a summand of* G *in* $\mathbf{A}_{\mathbb{Q}}$ *if and only if there is* $f \in \text{Hom}(G, H)$, $g \in \text{Hom}(H, G)$, *and a nonzero integer* n *with* $fg = n1_H$.

(f) *If* QEnd G *is semiperfect, then* G *is a finite direct sum of indecomposables in* $\mathbf{A}_\mathbb{Q}$. *Such a decomposition is unique up to isomorphism in* $\mathbf{A}_\mathbb{Q}$ *and the order of the summands. If* G *has finite rank, then* QEnd G *is semiperfect.*

PROOF.   (a) First suppose G and H are isomorphic in $\mathbf{A}_\mathbb{Q}$. Then there are $r/s, m/n \in \mathbb{Q}$, $f \in \mathrm{Hom}(G, H)$, and $g \in \mathrm{Hom}(H, G)$ with $(r/s)f(m/n)g = 1_H$ and $(m/n)g$ $(r/s)f = 1_G$. Hence, $(rf)(mg) = ns1_H$ and $(mg)(rf) = ns1_G$, with $rf \in \mathrm{Hom}(G, H)$ and $mg \in \mathrm{Hom}(H, G)$, as desired. Conversely, if $f \in \mathrm{Hom}(G, H)$, $g \in \mathrm{Hom}(H, G)$, and n is a nonzero integer with $fg = n1_H$ and $gf = n1_G$, then $(1/n)f \in \mathrm{QHom}(G, H)$ with $(1/n)fg = 1_H$ and $g(1/n)f = 1_G$. This shows that $(1/n)f$ is an isomorphism in $\mathbf{A}_\mathbb{Q}$.

(b) Assume that $f : G \to H$ is an isomorphism in $\mathbf{A}_\mathbb{Q}$. By (a), there is $g \in \mathrm{Hom}(H, G)$ and a nonzero integer n with $fg = n1_H$ and $gf = n1_G$. Thus, f is a monomorphism and $H/f(G)$ is bounded, since $nH = fg(H) \subseteq f(G) \subseteq H$.

Conversely, suppose f is a monomorphism and n is a nonzero integer with $nH \subseteq f(G) \subseteq H$. Then $g = (f^{-1})n \in \mathrm{Hom}(H, G)$ with $fg = n1_H$ and $gf = n1_G$. Now apply (a).

Next, suppose rank G is finite and $f \in \mathrm{End}\, G$ is a monomorphism. Then QEnd $G \supseteq f(\mathrm{QEnd}\, G) \supseteq \cdots$ is a descending chain of right ideals of QEnd$G$. Since G has finite rank, QEnd G is right Artinian. Choose n with $f^n(\mathrm{QEnd}\ G) = f^{n+1}(\mathrm{QEnd}\ G)$. Then there is $g \in \mathrm{End}\ G$ and $a/b \in \mathbb{Q}$ with $f^n = f^{n+1}(a/b)g$. So $f^n(b - afg) = 0$ and $b = afg$, since f is a monomorphism. Hence, $bG = afg(G) \subseteq f(G)$, $G/f(G)$ is bounded, and f is an isomorphism in $\mathbf{A}_\mathbb{Q}$ by the first part of the proof.

Assertions (c), (d), and (e) follow from (a), (b), and the definitions of direct sums and indecomposability in $\mathbf{A}_\mathbb{Q}$.

(f) Since QEnd $G/J\mathrm{QEnd}\ G$ is Artinian and idempotents lift modulo $J\mathrm{QEnd}\ G$, it follows that G can be written as the direct sum of finitely many groups $G_1 \oplus \cdots \oplus G_n$ with each G_i indecomposable in $\mathbf{A}_\mathbb{Q}$ [Lambek 66]. This uses the fact that for each such sum, $1 = e_1 + \cdots + e_n$ with each e_i idempotent and $e_i e_j = 0$ if $i \neq j$, as noted in the proof of Proposition 2.1.1(a). Each QEnd G_i is a local ring by Proposition 2.1.1(b).

Now suppose $G = H_1 \oplus \cdots \oplus H_m$ in $\mathbf{A}_\mathbb{Q}$ with each H_j indecomposable in $\mathbf{A}_\mathbb{Q}$. Then $m = n$, and there is a permutation σ of $\{1, \ldots, n\}$ such that G_i is isomorphic to $H_{\sigma(i)}$ for each $1 \leq i \leq n$ (Exercise 1.2.6). The proof is essentially the same as that of Theorem 1.2.2, using the assumption that each G_i has a local endomorphism ring in $\mathbf{A}_\mathbb{Q}$. The last statement follows from Proposition 2.1.1(b).

Standard terminology in abelian group theory for each of the conditions of Corollary 2.1.2 is as follows:

(a) G is *quasi-isomorphic* to H.

(b) f is a *quasi-isomorphism* from G to H or a *quasi-automorphism* of G if $H = G$.

(c) G is a *quasi-direct sum* of $G_1, \ldots, G_n$.

(d) G is *strongly indecomposable.*
(e) H is a *quasi-summand* of G.
(f) G has the *Krull–Schmidt* property in $\mathbf{A}_{\mathbb{Q}}$. The full subcategory of $\mathbf{A}_{\mathbb{Q}}$ consisting of finite rank groups is a *Krull–Schmidt* category.

Much of the structure of a group in $\mathbf{A}$ is lost by passing to the category $\mathbf{A}_{\mathbb{Q}}$. There are many examples of indecomposable groups in $\mathbf{A}$ that are not indecomposable in $\mathbf{A}_{\mathbb{Q}}$, for instance any indecomposable almost completely decomposable group (Chapter 5). This problem can be remedied to some extent by localizing a homomorphism group at a prime p instead of tensoring the homomorphism group with $\mathbb{Q}$, as is the case with $\mathbf{A}_{\mathbb{Q}}$.

Let p be a prime and define a category $\mathbf{A}_p$ with objects those of $\mathbf{A}$ but with morphism sets $\mathrm{Hom}(G, H)_p = \mathbb{Z}_p \otimes_{\mathbb{Z}} \mathrm{Hom}(G, H)$ for G, H in $\mathbf{A}$ where $\mathbb{Z}_p = \{m/n \in \mathbb{Q} : \gcd\{n, p\} = 1\}$ is a subring of $\mathbb{Q}$ called the *localization of $\mathbb{Z}$ at a prime p*. Composition is defined by $(q \otimes f)(r \otimes g) = qr \otimes fg$. It is routine to verify that composition is associative and 1_G is an identity in $(\mathrm{End}\ G)_p$, the $\mathbf{A}_p$-endomorphism ring of G.

It is important to observe that the homomorphism groups, not the groups themselves, are localized for the definition of the category $\mathbf{A}_p$. Localization commutes with homomorphisms for finitely generated torsion-free groups but not for finite rank torsion-free groups in general. For example, if H is the subgroup of $\mathbb{Q}$ generated by $\{1/p : p \text{ a prime}\}$, then $\mathbb{Z}_p$ is isomorphic to H_p for each prime p. However, $\mathrm{Hom}(H, \mathbb{Z}) = 0$, so that H is not isomorphic to $\mathbb{Z}$ in $\mathbf{A}_p$ for any p.

Proposition 2.1.3
(a) *The category $\mathbf{A}_p$ is an additive category. If G is in $\mathbf{A}_p$ and e is an idempotent in $Q\mathrm{End}\ G$, then $G = e(G) \oplus (1 - e)(G)$ in $\mathbf{A}_p$.*
(b) *If G is indecomposable in $\mathbf{A}_p$ and $(\mathrm{End}\ G)_p$ is semiperfect, then $(\mathrm{End}\ G)_p$ is a local ring.*

PROOF. The proof is analogous to that of Proposition 2.1.1, the only difference being that elements of $\mathbb{Z}_p$ are those elements m/n of $\mathbb{Q}$ with $m, n \in \mathbb{Z}$ subject to the additional condition that n is prime to p.

Corollary 2.1.4 *Let G and H be objects of $\mathbf{A}_p$.*

(a) *The group G is isomorphic to H in $\mathbf{A}_p$ if and only if there is $f \in \mathrm{Hom}(G, H)$, $g \in \mathrm{Hom}(H, G)$, and a nonzero integer n prime to p with $fg = n1_H$ and $gf = n1_G$.*
(b) *If $f \in \mathrm{Hom}(G, H)$, then f is an isomorphism in $\mathbf{A}_p$ if and only if f is a monomorphism and $H/f(G)$ is bounded by an integer prime to p.*
(c) *The group G is a direct sum $G_1 \oplus \cdots \oplus G_n$ in $\mathbf{A}_p$ if and only if there is a monomorphism f from the group direct sum $G_1 \oplus \cdots \oplus G_n$ into G such that $G/\mathrm{image}\ f$ is bounded by an integer prime to p.*

(d) *The group G is indecomposable in $\mathbf{A}_p$ if and only if whenever $G/(H \oplus K)$ is bounded by an integer prime to p, then $H = 0$ or $K = 0$.*

(e) *The group H is a summand of G in $\mathbf{A}_p$ if and only if there is $f \in \mathrm{Hom}(G, H)$, $g \in \mathrm{Hom}(H, G)$, and a nonzero integer n prime to p with $fg = n1_H$.*

(f) *If $(\mathrm{End}\, G)_p$ is semiperfect, then G is a finite direct sum of indecomposables in $\mathbf{A}_p$. Moreover, such a decomposition is unique up to isomorphism in $\mathbf{A}_p$ and the order of the summands.*

PROOF. Just as in Corollary 2.1.2.

Terminology for each of the conditions of Corollary 2.1.4 is as follows:

(a) G is *isomorphic at p to H.*

(b) f is an *isomorphism at p.*

(c) G is a *direct sum of $G_1, \ldots, G_n$ at p.*

(d) G is *indecomposable at p.*

(e) H is a *summand of G at p.*

(f) G has the *Krull–Schmidt property in $\mathbf{A}_p$.*

Despite the obvious similarities between the categories $\mathbf{A}_{\mathbb{Q}}$ and $\mathbf{A}_p$, there is one important distinction, namely that $(\mathrm{End}\, G)_p$ need not be a local ring for each finite rank G that is indecomposable in $\mathbf{A}_p$. Moreover, the subcategory of $\mathbf{A}_p$ with objects those torsion-free abelian groups of finite rank need not be a Krull–Schmidt category by Example 2.1.11.

Example 2.1.5 *There is a finite rank torsion-free abelian group G and a prime p such that G is indecomposable in $\mathbf{A}_p$ but $(\mathrm{End}\, G)p$ is not a local ring.*

PROOF. Let $\mathbb{Z}[i]$ denote the ring of Gaussian integers with $i^2 = -1$. Then $5\mathbb{Z}[i] = I_1 I_2$, where $I_1 = (1 + 2i)\mathbb{Z}[i]$ and $I_2 = (1 - 2i)\mathbb{Z}[i]$ are distinct maximal ideals of $\mathbb{Z}[i]$. Now, $\mathbb{Z}[i]_5$ is a subring of $\mathbb{Q}[i]$, $5\mathbb{Z}[i]_5$ is the Jacobson radical of $\mathbb{Z}[i]_5$, and $\mathbb{Z}[i]_5/5\mathbb{Z}[i]_5 = \mathbb{Z}[i]/5\mathbb{Z}[i]$ is isomorphic to $\mathbb{Z}[i]/I_1 \oplus \mathbb{Z}[i]/I_2$. In particular, $\mathbb{Z}[i]_5$ is not a local ring. By Theorem 2.4.6, there is a finite rank torsion-free abelian group G with $\mathrm{End}\, G$ isomorphic to $\mathbb{Z}[i]$. Then $(\mathrm{End}\, G)_5 = \mathbb{Z}[i]_5$ is not a local ring, but G is indecomposable in $\mathbf{A}_5$, since $\mathbb{Z}[i]_5$ has no idempotents other than 0 or 1.

The next example demonstrates that isomorphism is stronger than isomorphism at p, in fact stronger than isomorphism at p for each prime p. A torsion-free abelian group G is *completely decomposable* if G is a finite direct sum of rank-1 groups and *almost completely decomposable* if G is quasi-isomorphic to a completely decomposable group. Equivalently, G contains a completely decomposable group C as a subgroup with G/C bounded.

Example 2.1.6 *There are rank-2 indecomposable, almost completely decomposable groups G and H such that G and H are isomorphic at p for each* *prime p but G and H are not isomorphic as groups. Moreover, G and H are decomposable in* $\mathbf{A}_p$ *for all but one prime p.*

PROOF. Let $X = \mathbb{Z}[\frac{1}{11}]$, the subring of Q generated by $\frac{1}{11}$ and $Y = \mathbb{Z}[\frac{1}{31}]$. Define $G = X \oplus Y + \mathbb{Z}(\frac{1}{5}, \frac{1}{5})$ and $H = X \oplus Y + \mathbb{Z}(\frac{1}{5}, \frac{2}{5})$, subgroups of $\mathbb{Q} \oplus \mathbb{Q}$. Then both G and H are almost completely decomposable groups of rank 2. Since $11X = X$ and $31Y = Y$, it follows that $\text{Hom}(X, Y) = \text{Hom}(Y, X) = 0$. Consequently, if $f \in$ End G, then $f : X \to X$ and $f : Y \to Y$. Since X is a subring of $\mathbb{Q}$, End $X = X$, where an element x of X is regarded as an endomorphism of X by multiplication. Similarly, End $Y = Y$.

Restriction induces an isomorphism from End G to the ring $R_G = \{(x, y) \in X \times Y : x \equiv y \pmod 5\}$. To see this, let $f \in$ End G. Then $f = (x, y) : X \oplus Y \to X \oplus Y$ for some (x, y) in $X \oplus Y$. Moreover, $(x, y) = 5f(1, 1) = z(1, 1) + 5(x', y')$ for some integer z, x' in X, and y' in Y. This shows that $x \equiv y \pmod 5$. Conversely, if $(x, y) \in X \times Y$ with $x \equiv y \pmod 5$, then (x, y) induces, a well-defined endomorphism of G.

It follows that G is indecomposable as a group, since $0 = (0, 0)$ and $1 = (1, 1)$ are the only idempotents of R_G. A similar argument shows that End H is isomorphic to $R_H = \{(x, y) \in X \times Y : x \equiv 2y \pmod 5\}$, whence H is also indecomposable.

Since $5G \subseteq X \oplus Y \subseteq G$ and $5H \subseteq X \oplus Y \subseteq H$, both G and H are isomorphic at p to $X \oplus Y$ for each prime $p \neq 5$. As for $p = 5$, define $f : G \to H$ by $f(x, y) = (x, 2y)$ and $f(\frac{z}{5}, \frac{z}{5}) = (\frac{z}{5}, \frac{2z}{5})$ for x in X, y in Y, and z in $\mathbb{Z}$. Then f is a well-defined group homomorphism with $2H \subseteq f(G) \subseteq H$, and so G is isomorphic to H at $p = 5$ as well.

Finally, assume that $f : G \to H$ is a group isomorphism. Then $f_X = f : X \to X$ and $f_Y = f : Y \to Y$ are automorphisms, hence units of X and Y, respectively. In particular, $f_X = 11^i$ or -11^i for some integer i and $f_Y = 31^j$ or -31^j for some integer j. It follows that f_X and f_Y are congruent mod 5 to either 1 or -1, since $11 \equiv 31 \equiv 1 \pmod 5$. But f is onto, so that $(\frac{1}{5}, \frac{2}{5}) = f(x + \frac{z}{5}, y + \frac{z}{5})$ for some x in X, y in Y, and z in $\mathbb{Z}$. Hence, $5x + z \in X$ and $5y + z \in Y$ with $f_X(5x + z) = 5f_X(x) + f_X(z) = 1$ and $f_Y(5y + z) = 5f_Y(y) + f_Y(z) = 2$. Therefore, $zf_X(1) \equiv 1 \pmod 5$ and $zf_Y(1) \equiv 2 \pmod 5$. However, there is no such integer z, since both f_X and f_Y are congruent to either 1 or -1 mod 5. The conclusion is that G and H are not isomorphic.

A subgroup H of a torsion-free abelian group G is a *pure subgroup* of G if $nG \cap H = nH$ for each nonzero integer n. Equivalently, G/H is a torsion-free group. It is easy to confirm that a summand of a torsion-free abelian group is a pure subgroup. A subgroup H of a torsion-free abelian group G is a *p-pure* subgroup of G for a prime p if $pG \cap H = pH$, and G is *p-reduced* if $p^\omega(G) = \cap \{p^i G : i \geq 0\} = 0$.

Lemma 2.1.7 *Let G and H be torsion-free abelian groups of finite rank.*

(a) *If G and H are isomorphic in* $\mathbf{A}_p$, *then there are* $f \in \text{Hom}(G, H)$ *and* $g \in \text{Hom}(H, G)$ *such that* $fg - 1_H \in p\text{End } H$ *and* $gf - 1_G \in p\text{End } G$.

(b) *If G and H are p-reduced and there is $f \in \mathrm{Hom}(G, H)$ and $g \in \mathrm{Hom}(H, G)$ such that $fg - 1_H \in p\mathrm{End}\,H$ and $gf - 1_G \in p\mathrm{End}\,G$, then G and H are isomorphic in $\mathbf{A}_p$.*

PROOF. (a) Assume that G and H are isomorphic in $\mathbf{A}_p$. By Corollary 2.1.4(a), there are $f \in \mathrm{Hom}(G, H)$, $g \in \mathrm{Hom}(H, G)$, and a nonzero integer n prime to p with $fg = n1_H$ and $gf = n1_G$. Write $1 = rn + sp$ for some r, s in Z. Then $(rg)f = rn1_G = (1 - sp)1_G$ and $f(rg) = rn1_H = (1 - sp)1_H$. In particular, $rg \in \mathrm{Hom}(H, G)$ with $f(rg) - 1_H \in p\mathrm{End}\,H$ and $(rg)f - 1_G \in p\mathrm{End}\,G$, as desired.

(b) Suppose $x \in K = \ker fg \subseteq H$. Then $-x = fg(x) - x = (fg - 1)(x) = ph(x)$ for some $h \in \mathrm{End}\,H$. Thus, $K \subseteq pH \cap K = pK$, since K is a pure subgroup of H. But H is p-reduced, so that $K = \ker fg = 0$. Hence, fg is a monomorphism.

Moreover, image fg is p-pure in H. To see this, let $x \in H$ with $px = fg(y) = y + ph(y)$ for some $h \in \mathrm{End}\,H$ and $y \in H$. Then $y = pa \in pH$, for $a = x - h(y) \in H$, and so $px = fg(pa) = pfg(a)$. Consequently, $x = fg(a) \in$ image fg, because H is torsion-free.

Since $fg \in \mathrm{End}\,H$ is a monomorphism with p-pure image and H has finite rank, there is some integer n prime to p with $nH \subseteq fg(H) \subseteq H$, as an application of Corollary 2.1.2(b). Consequently, $nH \subseteq f(G) \subseteq H$. A similar argument shows that gf, hence f, is a monomorphism. By Corollary 2.1.4(b), G is isomorphic to H in $\mathbf{A}_p$.

Let frA denote the subcategory of $\mathbf{A}$ consisting of torsion-free abelian groups of finite rank and frA_p the full subcategory of $\mathbf{A}_p$ with objects those of frA. The criterion for isomorphism in fr $\mathbf{A}_p$ given in Lemma 2.1.7 can be placed in a categorical setting.

Define a category frA/p with objects p-reduced torsion-free groups of finite rank and morphism sets $\mathrm{Hom}(G, H)/p\mathrm{Hom}(G, H)$. It can be readily verified that frA/p is an additive category with direct sums in frA/p induced by direct sums in frA. In view of Lemma 2.1.7, if G and H are p-reduced, then G and H are isomorphic in frA_p if and only if they are isomorphic in frA/p. Notice that if G is in frA/p, then the endomorphism ring of G in frA/p, $\mathrm{End}\,G/p\mathrm{End}\,G$, is a finite-dimensional algebra over the field $\mathbb{Z}/p\mathbb{Z}$. Thus, in contrast to frA_p (Example 2.1.5), G is indecomposable in frA/p if and only if its endomorphism ring in frA/p is a local ring.

In view of the preceding observations, frA/p would be a Krull-Schmidt category if idempotents split in frA/p, i.e., if e is an idempotent in $\mathrm{End}\,G/p\mathrm{End}\,G$, then there exist some $f \in \mathrm{Hom}(G, H)/p\mathrm{Hom}(G, H)$ and $g \in \mathrm{Hom}(H, G)/p\mathrm{Hom}(H, G)$ with $gf = e$ and $fg = 1 \in \mathrm{End}\,H/p\mathrm{End}\,H$ (see Exercise 1.2.6). In this case, H is a summand of G in frA/p that can be identified with $e(G)$. However, idempotents need not split in frA/p, as demonstrated in Example 2.1.5. In that example, G is 5-reduced, $(\mathrm{End}\,G)_5/5(\mathrm{End}\,G)_5$ has nontrivial idempotents, but G is indecomposable in frA_5, hence in $\mathrm{frA}/5$ by Lemma 2.1.7.

The following theorem shows how to embed frA/p in a Krull–Schmidt category obtained by inserting the "missing" summands resulting from the fact that idempotents need not split.

Theorem 2.1.8 [Lady 75] *There is a Krull–Schmidt category* $(\mathrm{fr}A/p)^*$ *and a fully faithful functor* $F : \mathrm{fr}A/p \to (\mathrm{fr}A/p)^*$.

PROOF. The proof is outlined; details are left as an exercise. Define $(\mathrm{fr}A/p)^*$ to be the category with objects (e, G) such that e is an idempotent in $\mathrm{End}\, G/p\mathrm{End}\, G$. Morphisms are $\alpha : (e, G) \to (f, H)$, where $\alpha \in \mathrm{Hom}(G, H)$ with $\alpha = f\alpha e$. Routine arguments show that $(\mathrm{fr}A/p)^*$ is an additive category with finite direct sums given by $(e, G) \oplus (f, H) = (e \oplus f, G \oplus H)$. Moreover, idempotents split in $(\mathrm{fr}A/p)^*$, so that (e, G) is indecomposable if and only if the endomorphism ring of (e, G) in $(\mathrm{fr}A/p)^*$ is a local ring. Then $(\mathrm{fr}A/p)^*$ is a Krull–Schmidt category.

Finally, define $F : \mathrm{fr}A/p \to (\mathrm{fr}A/p)^*$ by $F(G) = (1, G)$ and conclude that F is fully faithful. In particular, $F(G)$ is isomorphic to $F(H)$ if and only if G is isomorphic to H.

Even though $\mathrm{fr}A/p$ is not a Krull–Schmidt category, as a consequence of Example 2.1.11 and Corollary 2.1.9 there are some uniqueness properties with regard to direct sums derived from the embedding of Theorem 2.1.8.

Corollary 2.1.9 [Lady 75] *Assume that* G, H, *and* K *are finite rank, p-reduced, torsion-free abelian groups.*

(a) *If* $G \oplus K$ *is isomorphic at* p *to* $H \oplus K$, *then* G *is isomorphic to* H *at* p.
(b) *If* G^n *is isomorphic at* p *to* H^n *for some positive integer n, then* G *is isomorphic to* H *at* p.

PROOF. (a) The group $G \oplus K$ is isomorphic to $H \oplus K$ in $\mathrm{fr}A/p$ by Lemma 2.1.7, hence in $(\mathrm{fr}A/p)^*$ by Theorem 2.1.8. Since the latter category is a Krull–Schmidt category, G is isomorphic to H in $(\mathrm{fr}A/p)^*$, hence in $\mathrm{fr}A/p$ and $\mathrm{fr}A_p$.

Assertion (b) is proved by an argument similar to that of (a), using the fact that $(\mathrm{fr}A/p)^*$ is a Krull–Schmidt category.

There are some conditions on the endomorphism ring of a group G in $\mathrm{fr}A$ sufficient to construct nonunique direct sum decompositions in $\mathrm{fr}A_p$.

Proposition 2.1.10 *Suppose* G *is a strongly indecomposable torsion-free abelian group of finite rank, p is a prime, and there are right ideals* $I_1, \ldots, I_n$ *of* $\mathrm{End}\, G$ *with* $(\mathrm{End}\, G)_p = (I_1)_p + \cdots + (I_n)_p$. *If for each i there is a subgroup* G_i *of* G *with* $I_i G \subseteq G_i$ *such that* G *is not isomorphic at* p *to* G_i, *then* $H = G_1 \oplus \cdots \oplus G_n$ *has a nonunique direct sum decomposition in* $\mathrm{fr}A_p$.

PROOF. To see that G is a summand of H at p, choose $r_i \in I_i$ and $m_i/k_i \in \mathbb{Z}_p$ with $m_i, k_i \in \mathbb{Z}$ such that $1 = (m_1/k_1)r_1 + \cdots + (m_n/k_n)r_n \in (\mathrm{End}\, G)_p$. Let $k = k_1 \cdots k_n$, an integer relatively prime to p. Then $k = m'_1 r_1 + \cdots + m'_n r_n$, where

$m'_i = km_i/k_i \in \mathbb{Z}$. Define $f \in \text{Hom}(G, H)$ by $f(x) = (m'_1 r_1 x, \ldots, m'_n r_n x)$ and $g \in \text{Hom}(H, G)$ by $g(x_1, \ldots, x_n) = x_1 + \cdots + x_n$, observing that each $m'_i r_i x \in I_i G \subseteq G_i$. Then $gf = k1_G$, and so G is a summand of H at p.

The group G is strongly indecomposable, hence indecomposable at p. Moreover, G is not isomorphic to a summand at p of any G_i, since rank $G_i \leq$ rank G is finite and G is not isomorphic at p to G_i. This shows that H has a nonunique direct sum decomposition in frA_p.

Example 2.1.11

(a) Let $p \geq 5$ be a prime and $R = \mathbb{Z}[x]/\langle h(x)\rangle$, where $h(x) = x(x + 1)(x + 2)(x + 3) + p(x^2 + 2)$. Then R is a subring of an algebraic number field, and there are ideals I_1 and I_2 of R with $R_p = (I_1)_p + (I_2)_p$.

(b) Choose a finite rank torsion-free abelian group G with $\text{End } G = R$. There are subgroups G_i of G with $I_i G \subseteq G_i$ such that G is not isomorphic at p to G_i for each i.

(c) The group $H = G_1 \oplus G_2$ has a nonunique decomposition into indecomposables in frA_p.

PROOF. (a) Multiplying polynomials yields $h(x) = x^4 + 6x^3 + (11 + p)x^2 + 6x + 2p$, an irreducible polynomial in $\mathbb{Q}[x]$ by Eisenstein's criterion [Hungerford 74]. Thus, R is a subring of the algebraic number field $\mathbb{Q}R = \mathbb{Q}[x]/\langle h(x)\rangle$ with R/pR a 4-dimensional $\mathbb{Z}/p\mathbb{Z}$-vector space. Let $\alpha = x + \langle h(x)\rangle \in R$. Now, $M_1 = \alpha R + pR$, $M_2 = (\alpha + 1)R + pR$, $M_3 = (\alpha + 2)R + pR$, and $M_4 = (\alpha + 3)R + pR$ are distinct ideals of R with each $(M_i)_p$ a maximal ideal of R_p containing pR_p. This is so because each $R_p/(M_i)_p = \mathbb{Z}_p/p\mathbb{Z}_p = \mathbb{Z}/p\mathbb{Z}$ is a field. Define $I_1 = M_1 M_2$ and $I_2 = M_3 M_4$. Then $(I_1)_p + (I_2)_p = R_p$, as a consequence of the maximality of the $(M_i)_p$'s.

(b) By Theorem 2.4.6, there is a finite rank torsion-free group G with $\text{End } G = R$. Let $a = a_1 + a_2$, where $a_1 \in M_1 G \backslash (M_2 G)_p$ and $a_2 \in M_2 G \backslash (M_1 G)_p$. Such a choice is possible, since $(M_1)_p$ and $(M_2)_p$ are distinct maximal ideals in R_p. Consequently, $na \notin M_1 G \cup M_2 G$ for each integer n prime to p. Recall that $I_1 = M_1 M_2$ and $I_2 = M_3 M_4$.

Define $G_1 = I_1 G + \mathbb{Z}a$, so that $I_1 G \subseteq G_1 \subseteq G$. Assume, by way of contradiction, that G is isomorphic at p to G_1, say $f \in \text{End } G = R$ with $nG_1 \subseteq f(G) \subseteq G_1$ for some integer n relatively prime to p. Then $f \notin M_1 \cup M_2$; otherwise, $na \in M_1 G \cup M_2 G$, which is impossible by the choice of a. On the other hand, $(I_1)_p \subseteq R_p f$. This is so because $f^{-1}n I_1(G) \subseteq G$, whence $f^{-1}n I_1 \subseteq \text{End } G = R$ and $n I_1 \subseteq Rf$ with $\gcd\{n, p\} = 1$. Since $(M_1)_p$ and $(M_2)_p$ are the only maximal ideals of R_p containing $(I_1)_p$, f is a unit in R_p and $R_p f = R_p$. Thus, $(G_1)_p = G_p$. But $\mathbb{Z}/p\mathbb{Z}$ is isomorphic to $(G_1)_p/(I_1 G)_p = G_p/(I_1 G)_p$ and $G_p/(I_1 G)_p$ is isomorphic to $G_p/(M_1 G)p \oplus G_p/(M_2 G)_p$, a contradiction. Similarly, using the ideals M_3 and M_4, there is a subgroup G_2 of G containing $I_2 G$ such that G_2 is not isomorphic at p to G.

Assertion (c) is a consequence of Proposition 2.1.10.

EXERCISES

1. Assume that G is a torsion-free abelian group of finite rank. Prove that if H is a subgroup with G/H bounded, then G/H is finite.

2. Assume that G and H are torsion-free abelian groups of finite rank and p is a prime. Prove that G is isomorphic at p to H if and only if G is a summand of H at p and H is a summand of G at p.

3. A torsion-free abelian group G is p-*local* for a prime p if $qG = G$ for each prime $q \neq p$.
   (a) Show that if G is a torsion-free abelian group, then G_p is a p-local group. Prove that G is p-local if and only if G is a torsion-free Z_p-module.
      Assume that G and H are p-reduced p-local groups of finite rank.
   (b) Show that if rank $G = 1$, then G is isomorphic to $\mathbb{Q}$ or $\mathbb{Z}_p$.
   (c) Prove that G and H are isomorphic if and only if they are isomorphic at p.
   (d) Prove that G is isomorphic to H if either G^n is isomorphic to H^n for some $n \geq 1$ or else there is a finite rank torsion-free abelian group K with $G \oplus K$ isomorphic to $H \oplus K$.
   (e) Find a p-local group of finite rank that does not have the Krull-Schmidt property.

4. Complete the details of the proof of Theorem 2.1.8.

5. Let $G = \mathbb{Z}[\frac{1}{11}] \oplus \mathbb{Z}[\frac{1}{31}] + \mathbb{Z}(\frac{1}{5}, \frac{1}{5})$. Use the techniques of the proof of Example 2.1.6 to find all groups H that are isomorphic at 5 to G and all groups H isomorphic to G that have a completely decomposable subgroup $\mathbb{Z}[\frac{1}{11}] \oplus \mathbb{Z}[\frac{1}{31}]$.

2.2 Near-isomorphism of Finite Rank Groups

The category frA$_p$ is still not completely satisfactory from a group-theoretic point of view, since indecomposables in frA need not be indecomposable in frA$_p$ (Example 2.1.6). Also, in order to fully exploit the category frA$/p$ there is an annoying assumption that groups in frA$_p$ be p-reduced. Both of these difficulties can be overcome by introducing an equivalence relation called near-isomorphism, for which isomorphism at p is a local version. In fact, a torsion-free abelian group G of finite rank is indecomposable if and only if it is indecomposable with respect to near-isomorphism (Corollary 2.2.11). Consequently, G is indecomposable if and only if it is indecomposable at p for each prime p.

Two torsion-free abelian groups G and H are *nearly isomorphic* if G and H are isomorphic at p for each prime p. Moreover, G is a *near summand* of H if G is a summand of H at p for each prime p.

A torsion-free abelian group G is *divisible* if $nG = G$ for each nonzero integer n. A divisible group is a $\mathbb{Q}$-vector space, hence, as a group, a direct sum of copies of $\mathbb{Q}$. A torsion-free abelian group G has a unique maximal divisible subgroup $d(G) = \cap\{nG : 1 \leq n \in \mathbb{Z}\}$. Then $d(G)$ is a summand of G, being injective as an abelian group. If $G = d(G) \oplus H$, then H is *reduced*, that is, $d(H) = 0$.

Lemma 2.2.1 *If G is a reduced torsion-free abelian group of finite rank, then there is a nonzero square-free integer k such that $kY \neq Y$ for each nonzero subgroup Y of G.*

PROOF. Let $\Pi = \{p_1, \ldots, p_n, \ldots\}$ be an indexing of the set of primes. Then $p_1^\omega(G) \supseteq p_1^\omega(p_2^\omega G) \supseteq \cdots$ is a descending chain of pure subgroups of G. Since G is reduced with finite rank, there is some i with $p_1^\omega p_2^\omega G \cdots p_i^\omega G = 0$. Let $k = p_1 \cdots p_i$. If $kY = Y$, then $Y \subseteq p_1^\omega p_2^\omega G \cdots p_i^\omega G = 0$.

The next theorem is a globalization of the definition of near-isomorphism.

Theorem 2.2.2 *Suppose G and H are reduced torsion-free abelian groups of finite rank.*

(a) *The groups G and H are nearly isomorphic if and only if for each nonzero integer n, there is $f \in \mathrm{Hom}(G, H)$, $g \in \mathrm{Hom}(H, G)$, and an integer m relatively prime to n with $fg = m1_H$ and $gf = m1_G$.*

(b) *The group G is a near summand of H if and only if for each nonzero integer n', there exist some $f \in \mathrm{Hom}(G, H)$, $g \in \mathrm{Hom}(G, H)$, and integer m relatively prime to n' with $gf = m1_G$.*

PROOF. It is sufficient to prove (b). In fact, if G and H are nearly isomorphic, then G is a near summand of H and H is a near summand of G. Conversely, if G and H have finite rank, G is a near summand of H, and H is a near summand of G, then G and H are nearly isomorphic. This follows from the observation that if G is a summand of H at p and H is a summand of G at p, then G and H are isomorphic at p (Exercise 2.1.2).

($\Leftarrow$) To see that G is a near summand of H, let p be a prime. Choosing $n' = p$ shows that G is a summand of H at p, whence G is a near summand of H.

($\Rightarrow$) Assume that G is a near summand of H and let p be a prime. Since G is a summand of H at p, there exist $f \in \mathrm{Hom}(G, H)$, $g \in \mathrm{Hom}(H, G)$, and an integer m prime to p with $gf = m1_G$ by Corollary 2.1.4(e). As in Lemma 2.1.7(a), there are $f_p \in \mathrm{Hom}(G, H)$ and $g_p \in \mathrm{Hom}(H, G)$ with $g_p f_p - 1_G \in p\mathrm{End}\, G$.

More generally, for each square-free integer n, there exist $f_n \in \mathrm{Hom}(G, H)$ and $g_n \in \mathrm{Hom}(G, H)$ with $g_n f_n - 1_G \in n\mathrm{End}\, G$. The proof is by induction on n. The case that n is prime is confirmed in the previous paragraph. Now assume that n is not prime and write $n = mk$ for relatively prime square-free integers $m < n$ and $k < n$. By the induction hypothesis, there are $f_k, f_m \in \mathrm{Hom}(G, H)$ and $g_k, g_m \in \mathrm{Hom}(H, G)$ with $g_m f_m - 1_G \in m\mathrm{End}\, G$ and $g_k f_k - 1_G \in k\mathrm{End}\, G$. Write $1 = rm + sk$ for integers r and s and define $f_n = skf_m + rmf_k$ and $g_n = skg_m + rmg_k$. A straightforward computation shows that $g_n f_n - 1_G \in n\mathrm{End}\, G$, as desired.

By Lemma 2.2.1, there is a square-free integer k such that if Y is a subgroup of H with $kY = Y$, then $k = 0$. To complete the proof, let n' be an integer and n the square-free integer consisting of the product of the distinct prime divisors of

$n'k$. Notice that k, being square-free, divides n. By the preceding paragraph, there exist some $f_n \in \mathrm{Hom}(G, H)$ and $g_n \in \mathrm{Hom}(H, G)$ with $g_n f_n - 1_G \in n\mathrm{End}\, G$. Then $n(\ker g_n f_n) = \ker g_n f_n$, since $(g_n f_n - 1)(x) = g_n f_n(x) - x = nh(x) \in nG$ for some $h \in \mathrm{End}\, G$ and $\ker g_n f_n$ is a pure subgroup of G. Since k divides n, $k(\ker g_n f_n) = \ker g_n f_n$, so that $\ker g_n f_n = 0$ by the choice of k. Thus, $g_n f_n$ is a monomorphism. Also, image $g_n f_n$ is q-pure for each prime q dividing n, again since $g_n f_n(x) - x \in nG$.

Since $g_n f_n$ is a monomorphism and G has finite rank, there is some integer m prime to n with $mG \subseteq g_n f_n(G) \subseteq G$ by Corollary 2.1.2. Finally, $h_n = (g_n f_n)^{-1}m \in \mathrm{Hom}(G, G)$, $h_n g_n \in \mathrm{Hom}(H, G)$, $f_n \in \mathrm{Hom}(G, H)$, and $(h_n g_n)f_n = m1_G$. But m is also relatively prime to n', by the choice of n, so that the proof is complete.

Corollary 2.2.3 *Let G and H be reduced torsion-free abelian groups of finite rank. Then G is nearly isomorphic to H if and only if G is isomorphic to H in* frA/p *for each p.*

PROOF. Apply Lemma 2.1.7(a) and the proof of Theorem 2.2.2, wherein it is deduced that G is nearly isomorphic to H from the assumption that for each prime p, there are $f_p \in \mathrm{Hom}(G, H)$ and $g_p \in \mathrm{Hom}(H, G)$ with $g_p f_p - 1_G \in p\mathrm{End}\, G$ and $f_p g_p - 1_H \in p\mathrm{End}\, H$.

Corollary 2.2.4 *Suppose G, H, and K are finite rank torsion-free abelian groups.*

(a) *If $G \oplus K$ is nearly isomorphic to $H \oplus K$, then G is nearly isomorphic to H.*
(b) *If G^n is nearly isomorphic to H^n for some positive integer n, then G is nearly isomorphic to H.*

PROOF. It is sufficient to assume that G, H, and K are reduced groups, since if A and B are torsion-free abelian groups of finite rank, then A and B are nearly isomorphic if and only if $A/d(A)$ and $B/d(B)$ are nearly isomorphic (Exercise 2.2.5). Now apply Corollary 2.2.3 and Theorem 2.1.8.

A torsion-free abelian group is *semilocal* if $pG = G$ for all but a finite number of primes p. For semilocal groups, isomorphism and near-isomorphism are equivalent:

Corollary 2.2.5 *Let G, H, and K be finite rank torsion-free abelian groups. If G and H are semilocal, then:*

(a) *G is nearly isomorphic to H if and only if G is isomorphic to H.*
(b) *G is a near summand of H if and only if G is a summand of H.*
(c) *G has no nontrivial near summands if and only if G is indecomposable as a group.*

(d) *If $G \oplus K$ is isomorphic to $H \oplus K$, then G is isomorphic to H.*

(e) *If G^n is isomorphic to H^n for some positive integer n, then G is isomorphic to H.*

PROOF. It is sufficient to prove (b). Then (a) follows, since if G is a summand of H and H is a summand of G, then G and H are isomorphic. Moreover, (c) is an immediate consequence of (b), and (d) and (e) follow from (a) and Corollary 2.2.4.

(b) Assume that G is a near summand of H and let n be the product of those primes p such that $pG \neq G$. Since G is semilocal, n is an integer. By Theorem 2.2.2(b), there is an integer m prime to n, $f \in \operatorname{Hom}(G, H)$, and $g \in \operatorname{Hom}(H, G)$ with $gf = m1_G$. Then m is a unit of the endomorphism ring of G by the choice of n. Hence, $m^{-1}g \in \operatorname{Hom}(H, G)$ with $(m^{-1}g)f = 1_G$. This shows that G is a summand of H. Conversely, a summand is clearly a near summand.

The next corollary provides some justification for the term near summand.

Corollary 2.2.6 *Suppose G, H, and K are torsion-free abelian groups of finite rank. If G is a near summand of both H and K, then G is a group summand of $H \oplus K$.*

PROOF. Let p be a prime and choose, via Theorem 2.2.2(b), an integer m prime to p and $f_p \in \operatorname{Hom}(G, H)$, $g_p \in \operatorname{Hom}(H, G)$ with $g_p f_p = m1_G$. Another application of Theorem 2.2.2(b) yields an integer n prime to m and $f_m \in \operatorname{Hom}(G, K)$, $g_m \in \operatorname{Hom}(K, G)$ with $g_m f_m = n1_G$. Write $1 = rm + sn$ for some integers r and s. Then $f \in \operatorname{Hom}(G, H \oplus K)$, where $f(x) = (f_p(x), f_m(x))$, and $g \in \operatorname{Hom}(H \oplus K, G)$ with $g(x, y) = rg_p(x) + sg_m(y)$. A simple computation shows that $gf = 1_G$, as desired.

Let G be a torsion-free abelian group of finite rank. As noted in Section 2.1, $\mathbb{Q}\operatorname{End} G$ is a finite-dimensional $\mathbb{Q}$-algebra, $J\mathbb{Q}\operatorname{End} G$ is nilpotent, and $\mathbb{Q}\operatorname{End} G/J\mathbb{Q}\operatorname{End} G$ is a semisimple Artinian $\mathbb{Q}$-algebra. Define $N\operatorname{End} G = (J\mathbb{Q}\operatorname{End} G) \cap \operatorname{End} G$. Then $N\operatorname{End} G$ is a nilpotent ideal of $\operatorname{End} G$, and $\operatorname{End} G/N\operatorname{End} G$ is contained in $\mathbb{Q}\operatorname{End} G/J\mathbb{Q}\operatorname{End} G = K_1 \times \cdots \times K_n$, where each $K_i = \operatorname{Mat}_n(D_i)$ is a full matrix ring over a division $\mathbb{Q}$-algebra D_i.

Following is a brief summary of basic definitions and properties of orders. Further details and proofs can be found in [Arnold 82] or [Reiner 75]. Let $K = \operatorname{Mat}_n(D)$, D a finite-dimensional division algebra over $\mathbb{Q}$, and let F be the center of K. Then F is the center of D, hence an algebraic number field. Let S be a subring of F. A subring R of K with $\mathbb{Q}R = K$ is called an *S-order in K* if R is an S-algebra that is finitely generated as an S-module. Then R is a *maximal S-order* if $R = R'$ for each S-order R' with $R \subseteq R' \subset K$. For example, if S is a principal ideal domain with quotient field F, then $\operatorname{Mat}_n(S)$ is a maximal S-order in $\operatorname{Mat}_n(F)$ [Arnold 82, Corollary 10.3].

Lemma 2.2.7 *Assume that G is a torsion-free abelian group of finite rank and $\mathbb{Q}\operatorname{End} H/J\mathbb{Q}\operatorname{End} H = K_1 \times \cdots \times K_n$ with each $K_i = \operatorname{Mat}_n(D_i)$ a full matrix* ring over a division $\mathbb{Q}$-algebra D_i. There is a smallest positive integer $n(G)$ with $n(G)R^* \subseteq \mathrm{End}\, G/N\mathrm{End}\, G \subseteq R^* = R_1 \times \cdots \times R_n$ and each R_i a maximal S_i-order in K_i for some Dedekind subring S_i of the center of D_i.

PROOF. [Arnold 82, Corollary 10.14]. ∎

For some groups G, the integer $n(G)$ is easy to compute. For example, if X is a rank-1 group and $G = X^n$ is the direct sum of n copies of X, then $\mathrm{End}\, G = \mathrm{Mat}_n(\mathrm{End}\, X)$ with $\mathrm{End}\, X$ a subring of $\mathbb{Q}$, $\mathbb{Q}\mathrm{End}\, G = \mathrm{Mat}_n(\mathbb{Q})$, and $J\mathbb{Q}\mathrm{End}\, G = N\mathrm{End}\, G = 0$. In this case, $n(G) = 1$, since $S = \mathrm{End}\, X$ is a principal ideal domain by Exercise 2.2.6. More generally, it is proved in Example 5.1.7 that if G is a torsion-free abelian group of finite rank and C is a fully invariant completely decomposable subgroup of G with $nG \subseteq C \subseteq G$ for some least positive integer n, then $n = n(G)$.

With the notation of Lemma 2.2.7, $R = \mathrm{End}\, G/N\mathrm{End}\, G$ is a subring of R^* and $n(G)R^* \subseteq R \subseteq R^*$. An R-module M is an R-lattice if M is a finitely generated R-module with additive group a torsion-free abelian group, necessarily of finite rank. Examples of R-lattices include finitely generated projective R-modules. Two R-lattices M and N are in the same genus class if M_p and N_p are isomorphic R_p-modules for each prime p of $\mathbb{Z}$, recalling that $M_p = \mathbb{Z}_p \otimes_{\mathbb{Z}} M$.

Lemma 2.2.8 *Assume that G is a torsion-free abelian group of finite rank and $n(G)R^* \subseteq R = \mathrm{End}\, G/N\mathrm{End}\, G \subseteq R^*$ as given in Lemma 2.2.7.*

(a) *If M is an R-lattice, then R^*M is projective as an R^*-module.*
(b) *Two R-lattices M and N are in the same genus class if and only if either $n(G) = 1$ and $\mathbb{Q}M$ is isomorphic to $\mathbb{Q}N$ as $\mathbb{Q}R$-modules or else $n(G) > 1$ and M_p is isomorphic to N_p as R_p-modules for each prime divisor p of $n(G)$.*

PROOF. Assertion (a) follows from the observation that R^* is a finite product of maximal orders R_i, R_iM is an R_i-lattice, and lattices over maximal orders are projective [Arnold 82, Theorem 11.3].
(b) [Arnold 82, Corollary 12.4]. ∎

The integer $n(G)$ provides a single integer to test for near isomorphism. Proofs of the next corollary and proposition use the functorial correspondence $K \mapsto \mathrm{Hom}(G, K)/\mathrm{Hom}(G, K)N\mathrm{End}\, G$ from the category of summands of finite direct sums of copies of G to the category of finitely generated projective modules over $\mathrm{End}\, G/N\mathrm{End}\, G$ given in Theorem 2.4.3. Briefly, this correspondence identifies near isomorphism classes of groups with genus classes of R-lattices.

Corollary 2.2.9 *Let G be a torsion-free group of finite rank. A torsion-free abelian group K of finite rank is nearly isomorphic to G if and only if K is a summand of $G \oplus G$ and either $n(G) = 1$ and G is quasi-isomorphic to K or else $n(G) > 1$ and G is isomorphic at p to K for each prime p dividing $n(G)$.*

PROOF. Assume that K is nearly isomorphic to G. Then K is quasi-isomorphic to G and isomorphic at p to G for each p. Moreover, K is a summand of $G \oplus G$ by Corollary 2.2.6.

Conversely, suppose either $n(G) = 1$ or else $n(G) > 1$ and G is isomorphic to K at p for each prime p dividing n. Let $R = $ End G/N End G. Then R and $M = $ Hom$(G, K)/$Hom$(G, K)N$End G are finitely generated projective R-modules. By Theorem 2.4.3(a) and (c), either $n(G) = 1$ and $\mathbb{Q}M$ is $\mathbb{Q}R$-isomorphic to $\mathbb{Q}R$ or else $n(G) > 1$ and M_p is R_p-isomorphic to R_p for each prime divisor p of $n(G)$. Hence, R and M are in the same genus class as an application of Lemma 2.2.8(b). By Theorem 2.4.3(d), G and K are nearly isomorphic.

The next technical proposition is the key to confirming in Corollary 2.2.11(b) that a finite rank group H is indecomposable if and only if it is indecomposable at p for each prime p.

Proposition 2.2.10 *Let G be a finite rank torsion-free group and $n(G)$ as given in Lemma 2.2.7.*

(a) *If K is a near summand of G and $g \in$ Hom(G, K), then there is $h \in$ Hom(image g, G) with $gh = n(G)1_{\text{image } g}$.*

(b) *If X and K are near summands of G and there is an integer m prime to $n(G)$ with $mK \subseteq X \subseteq K$, then X is nearly isomorphic to K.*

PROOF. (a) Suppose K is a near summand of G, a summand of $G \oplus G$ by Corollary 2.2.6, and that $g \in$ Hom(G, K). Let $R = $ End G/NEnd G and $M = $ Hom$(G, K)/$Hom$(G, K)N$End G. Then R and M are finitely generated projective R- modules. Since $K \mapsto M$ is a functor, there is an R-homomorphism $g' : R \to M$ induced by $g'(f) = fg$ for each $f \in$ Hom(G,K) (Theorem 2.4.3). But $n(G)R^* \subseteq R \subseteq R^*$ and R^* is torsion-free, so that g' extends to an R^*- homomorphism $g^* = (1/n(G))g' : R^* \to R^*M$. Now, image g^* is an R^*-lattice, hence a projective R^*-module by Lemma 2.2.8(a). Thus, there is an R^*-homomorphism $h^* :$ image $g^* \to R^*$ with $g^*h^* = 1_{\text{image } g^*}$. Moreover, $h' = n(G)h^* :$ image $g' \to R$ is an R-homomorphism with $g'h' = n(G)1_{\text{image } g'}$. It follows from Theorem 2.4.3 that there is a homomorphism $h :$ image $g \to G$ with $gh = n(G)1_{\text{image } g}$.

(b) In view of Corollary 2.2.6, both X and K are summands of $G \oplus G$. Then $M = $ Hom$(G, X)/$Hom$(G, X)N$End G and $N = $ Hom$(G, K)/$Hom$(G, K)N$ End G are finitely generated projective R-modules, where $R = $ End G/NEnd G. As a consequence of the hypothesis, there is an integer m prime to $n(G)$ with $mN \subseteq M \subseteq N$. Hence, $mN_p = M_p = N_p$ for each prime p dividing $n(G)$. By Lemma 2.2.8(b), M and N are in the same genus class. Consequently, X and K are nearly isomorphic by Theorem 2.4.3(d).

Corollary 2.2.11 *Assume that G and H are torsion-free abelian groups of finite rank.*

(a) *The group H is a near summand of G if and only if $G = X \oplus X'$ for some groups X and X' with X nearly isomorphic to H.*

(b) *The group G is indecomposable if and only if 0 and G are the only near summands of G.*

PROOF. (a) Assume that H is a near summand of G and let $n(G)$ be as given in Lemma 2.2.7. There are $f \in \text{Hom}(H, G)$, $g \in \text{Hom}(G, H)$, and an integer m prime to $n(G)$ with $gf = m1_H$. Write $1 = rm + sn(G)$ for integers r and s and let $X = \text{image } g \subseteq H$. By Proposition 2.2.10(a), there is $h \in \text{Hom}(X, G)$ with $gh = n(G)1_X$. Define $\phi \in \text{Hom}(X, G)$ by $\phi(x) = rf(x) + sh(x)$. Then $g\phi = rgf + sgh = rm + sn(G) = 1_X$, whence X is a summand of G. Furthermore, $mH = gf(H) \subseteq g(G) = X \subseteq H$, so that X is nearly isomorphic to H by Proposition 2.2.10(b).

Assertion (b) is a consequence of (a).

EXERCISES

The groups in Exercises 1–5 are assumed to be torsion-free abelian of finite rank.

1. Prove that if G is nearly isomorphic to $H_1 \oplus H_2$, then $G = G_1 \oplus G_2$ with G_i nearly isomorphic to H_i.

2. Show that if G is nearly isomorphic to H, then G is indecomposable if and only if H is indecomposable.

3. Prove that if G is nearly isomorphic to H, then $G \oplus G'$ is isomorphic to $H \oplus H$ for some G' nearly isomorphic to G.

4. Prove that if G is nearly isomorphic to H^n, then G is isomorphic to $H^{n-1} \oplus G'$ for some G' nearly isomorphic to H.

5. (a) Prove that $d(G \oplus H) = d(G) \oplus d(H)$.
   (b) Prove that G is nearly isomorphic to H if and only if $G/d(G)$ is nearly isomorphic to $H/d(H)$.

6. Show that if R is a subring of the rationals $\mathbb{Q}$, then R is a principal ideal domain.

2.3   Stable Range Conditions for Finite Rank Groups

The endomorphism ring of a torsion-free abelian group of finite rank has 2 in the stable range (Theorem 2.3.6). As a consequence, two torsion-free abelian groups G and H of finite rank are nearly isomorphic if and only if for some positive integer n, the direct sum of n copies of G is isomorphic to the direct sum of n copies of H (Theorem 2.3.12).

There is a deceptively easy, but false, argument that if A, H, and K are abelian groups with $A \oplus H$ isomorphic to $A \oplus K$, then H is isomorphic to K. The argument is that H is isomorphic to $(A \oplus H)/A$, $(A \oplus K)/A$ is isomorphic to K, $(A \oplus H)/A$ is isomorphic to $(A \oplus K)/A$, and so H is isomorphic to K. The flaw in this argument, which is correct if $A \oplus H$ is actually equal to $A \oplus K$, is a confusion between equality and isomorphism. This distinction motivates the notion of substitution.

An abelian group A has the *substitution property* if given $G = A_1 \oplus H = A_2 \oplus K$ with A isomorphic to both A_1 and A_2, then there is a subgroup C of G with $G = C \oplus H = C \oplus K$. In this case, H and K are isomorphic. The group A has *1-substitution* if whenever $G = A \oplus B$ and $f : G \to A$, $g : A \to G$ with $fg = 1_A$, then there is $\phi : A \to G$ with $f\phi = 1_A$ and $G = \phi(A) \oplus B$. A ring R has 1 *in the stable range* if for $f_1 g_1 + f_2 g_2 = 1$ with $f_i, g_i \in R$, there is some $h \in R$ with $f_1 + f_2 h$ a unit of R.

Proposition 2.3.1 [Warfield 80], [Fuchs 70B] *The following statements are equivalent for an abelian group A:*

(a) *A has 1-substitution;*
(b) *End A has 1 in the stable range;*
(c) *A has the substitution property.*

PROOF. (a) $\Rightarrow$ (b) Assume $f_1 g_1 + f_2 g_2 = 1$ with $f_i, g_i \in \mathrm{End}\, A$. Define $G = A \oplus A$, $f = (f_1, f_2) : G \to A$, and $g = (g_1, g_2) : A \to G$, where f and g are defined by $f(a_1, a_2) = f_1(a_1) + f_2(a_2)$ and $g(a) = (g_1(a), g_2(a))$ for $a, a_i \in A$. By (a), there is $\phi = (\phi_1, \phi_2) : A \to G$ with $\phi_i \in \mathrm{End}\, A$ such that $f\phi = f_1\phi_1 + f_2\phi_2 = 1_A$ and $G = \phi(A) \oplus A$. Since $G = A \oplus A$, there is an isomorphism $h_1 : A \to A$ with $1_A = (h_1, 0)(\phi_1, \phi_2) = h_1 \phi_1$. Hence, ϕ_1 is an isomorphism and $f_1 + f_2(\phi_2 \phi_1^{-1}) = \phi_1^{-1}$ is a unit of End A, as desired.

(b) $\Rightarrow$ (c) Suppose $G = A \oplus H = A_2 \oplus K$ with $\gamma : A_2 \to A$ an isomorphism. Let $f : G \to A$ be a projection of G onto A_2 (with kernel K) followed by γ and write $f = (f_1, \alpha)$. Also, let $g = \gamma^{-1} : A \to A_2 \subseteq G = A \oplus H$ and write $g = (g_1, \beta)$. Then $1_A = \gamma\gamma^{-1} = fg = f_1 g_1 + (\alpha\beta)1_A$. By (b), there is a unit u of End A and $h \in \mathrm{End}\, A$ with $f_1 u + (\alpha\beta)h = 1_A$ and $\theta = (u, \beta h) : A \to G = A \oplus H$ with $1_A = f_1 u + \alpha\beta h = f\theta$. Therefore, $G = \theta(A) \oplus \ker f = \theta(A) \oplus K$, recalling that $\ker f = K$. Moreover, $\phi = (u^{-1}, 0) : G = A \oplus H \to A$ with $\ker\phi = H$ and $1_A = \phi\theta = (u^{-1}, 0)(u, \beta h)$. Hence, if $\theta(A) = C$, then $G = C \oplus \ker\phi = C \oplus H$. Since $G = C \oplus K$, the proof is complete.

(c) $\Rightarrow$ (a) Given $G = A \oplus B$, $f : G \to A$, and $g : A \to G$ with $fg = 1_A$, then $G = g(A) \oplus \ker f$ with $g(A)$ isomorphic to A. By (c), $G = C \oplus B = C \oplus \ker f$ for some subgroup C of G. Hence, there is an isomorphism $\phi : A \to C$. Now, $f\phi(A) = f(C) = f(C \oplus \ker f) = f(G) = A$, so that $f\phi$ is a unit of End A. Then, $\phi' = \phi(f\phi)^{-1} : A \to G$ with $f\phi' = 1_A$ and $G = C \oplus B = \phi(A) \oplus B = \phi'(A) \oplus B$, as desired.

Example 2.3.2
(a) [Bass 68] *If R is a ring and R/JR is Artinian, then 1 is in the stable range of R.*
(b) *1 is not in the stable range of* $\mathbb{Z}$.

PROOF. (a) It is sufficient to assume $JR = 0$. To see this, let $f_1 g_1 + f_2 g_2 = 1$ with $f_i, g_i \in R$. Then $f_1 g_1 + f_2 g_2 \equiv 1 \pmod{JR}$, i.e., the coset $f_1 g_1 + f_2 g_2 + JR$ is equal to $1 + JR$, and $J(R/J(R)) = 0$. If there are $h, u \in R$ with $f_1 + f_2 h \equiv u \pmod{JR}$ and u a unit modulo JR, then $f_1 + f_2 h = u + x$ for some x in JR, and there is $v \in R$ with $1 - (u + x)v \in JR$. Hence, $(u + x)v = 1 - (1 - (u + x)v)$ is a unit of R. Thus, $u + x$ is a unit of R with $f_1 + f_2 h = u + x$, as desired.

Now assume that R is Artinian with $JR = 0$, i.e., R is a semisimple Artinian ring. Then the right ideal $f_1 R$ is a summand of R [Hungerford 74], and so $R = f_1 R \oplus I$ for some right ideal $I \subseteq f_2 R$. Define $\phi : R \to f_1 R$ by $\phi(r) = f_1 r$. Since R is semisimple Artinian, $\ker \phi$ is a summand of R, say $\theta : R \to \ker \phi$ with $\theta(x) = x$ for each $x \in \ker \phi$. Hence, $R = f_1 R \oplus I$ is isomorphic to $f_1 R \oplus \ker \phi$. By the classical Krull–Schmidt theorem for decompositions of finitely generated modules over semisimple Artinian rings [Anderson, Fuller 74], there is an isomorphism $\sigma : \ker \phi \to I$.

Now, $(\phi, \theta) : R \to f_1 R \oplus \ker \phi$, $(1, \sigma) : f_1 R \oplus \ker \phi \to R = f_1 R \oplus I$, and the composite $u = (1, \sigma)(\phi, \theta)$ is an R-automorphism of R as a right R-module. Thus, u is multiplication by some unit $u(1) = (f_1, x) = f_1 + x$ of R with $x \in I \subseteq f_2 R$. Write $x = f_2 h$ for some $h \in R$ so that $f_1 + f_2 h = u(1)$ is a unit of R. This shows that 1 is in the stable range of R.

(b) Simply observe that $3.5 + 7(-2) = 1$ but $3 + 7h$ is not a unit of $\mathbb{Z}$ for any h in $\mathbb{Z}$.

Lemma 2.3.3 *If A is a torsion-free abelian group of finite rank and $f_1 g_1 + f_2 g_2$ is a monomorphism for $f_i, g_i \in \text{End } A$, then there is some $h \in \text{End } A$ with $f_1 + f_2 h$ a monomorphism.*

PROOF. As a consequence of Corollary 2.1.2(b), $f \in \text{End } A$ is a monomorphism if and only if f is a unit in $\mathbb{Q}\text{End } A$. Since $\mathbb{Q}\text{End } A$ is Artinian, this is equivalent to f being a unit modulo $J\mathbb{Q}\text{End } A$. Moreover, f is a unit modulo $J\mathbb{Q}\text{End } A$ if and only if f is not a left zero divisor modulo $J\mathbb{Q}\text{End } A$, since $\mathbb{Q}\text{End } A/J\mathbb{Q}\text{End } A$ is semisimple Artinian.

It is now sufficient to assume that $J\mathbb{Q}\text{End } A = 0$ and $f_1 g_1 + f_2 g_2$ is a unit of $\mathbb{Q}\text{End } A$ and prove that $f_1 + f_2 h$ is not a left zero divisor in $\mathbb{Q}\text{End } A$. As in Example 2.3.2, $\mathbb{Q}\text{End } A = f_1 \mathbb{Q}\text{End } A \oplus I$ for some right ideal $I \subseteq f_2 \mathbb{Q}\text{End } A$. Write $1 = f_1 + f_2 h'$ for some $h' = (1/m)h \in \mathbb{Q}\text{End} A$ with $0 \neq m \in \mathbb{Z}$ and $h \in \text{End } A$. Then $m = m f_1 + f_2 h$. However, $f_1 + f_2 h$ is not a left zero divisor of $\mathbb{Q}\text{End } A$. This is so because if $(f_1 + f_2 h)r = 0$, then $0 = f_1 r + f_2 h r \in f_1 \mathbb{Q}\text{End} A \oplus I = \mathbb{Q}\text{End} A$. Hence, $f_1 r = f_2 h r = 0$, and so $mr = (m f_1 + f_2 h)r = 0$. Since $0 \neq m, r = 0$. Consequently, $f_1 + f_2 h$ is not a left zero divisor.

Theorem 2.3.1 provides conditions under which a group A can be canceled from a direct sum. There is a weaker condition that allows a partial cancellation. A group A has the *2-substitution property* if given $G = A \oplus A \oplus B$ and $f : G \to A$, $g : A \to G$ with $fg = 1_A$, then there is $\phi : A \to G$ with $f\phi = 1_A$ and $G = \phi(A) \oplus C \oplus B$ for some $C \subseteq A \oplus A$. In this case, $A \oplus A$ is isomorphic to $\phi(A) \oplus C$ with $\phi(A)$ isomorphic to A.

Lemma 2.3.4 *If an abelian group A has the 2-substitution property and $A \oplus A \oplus X$ is isomorphic to $A \oplus Y$ for abelian groups X and Y, then Y is isomorphic to $A' \oplus X$ for some A' with $A \oplus A'$ isomorphic to $A \oplus A$.*

PROOF. Let $G = A_1 \oplus A_2 \oplus X = A \oplus Y$ with A isomorphic to A_i for each i and let $f : G \to A$ be a projection with ker $f = Y$. Since A has the 2-substitution property, there is $\phi : A \to G$ with $f\phi = 1_A$ and $G = A_1 \oplus A_2 \oplus X = \phi(A) \oplus A' \oplus X$ for some $A' \subseteq A \oplus A$. Also, $G = \phi(A) \oplus \ker f = \phi(A) \oplus Y$ as $f\phi = 1_A$. Thus, Y can be shown to be isomorphic to $A' \oplus X$, by canceling $\phi(A)$. Furthermore, $G = A_1 \oplus A_2 \oplus X = \phi(A) \oplus A' \oplus X$, so that $A_1 \oplus A_2$ is isomorphic to $\phi(A) \oplus A'$, shown by canceling X. But ϕ is a monomorphism, so that $\phi(A)$ is isomorphic to A. Since A is isomorphic to A_i for $i = 1, 2$, $A \oplus A$ is isomorphic to $A \oplus A'$ with Y isomorphic to $A' \oplus X$, as desired.

Given a ring R, 2 *is in the stable range of R* if $f_1g_1 + f_2g_2 + f_3g_3 = 1$ with $f_i, g_i \in R$ implies there is $h_i, k_i \in R$ with $1 = (f_1 + f_3h_1)k_1 + (f_2 + f_3h_2)k_2$.

Theorem 2.3.5 [Warfield 80] *An abelian group A has the 2-substitution property if and only if 2 is in the stable range of* End A.

PROOF. ($\Rightarrow$) Assume that A has the 2-substitution property and let $f_1g_1 + f_2g_2 + f_3g_3 = 1_A$ with $f_i, g_i \in$ End A. Define $G = A \oplus A \oplus A$, $f = (f_1, f_2, f_3) : G \to A$, and $g = (g_1, g_2, g_3) : A \to G$, so that $fg = f_1g_1 + f_2g_2 + f_3g_3 = 1_A$. Since A has the 2-substitution property, there is $\phi = (h_1, h_2, h_3) : A \to G$ with $1_A = f\phi = f_1h_1 + f_2h_2 + f_3h_3$ and $G = \phi(A) \oplus C \oplus A$ for some $C \subseteq A \oplus A \oplus 0$. Canceling A from G gives $\theta = (k_1, k_2, 0) : G \to A$ with $\theta\phi = k_1h_1 + k_2h_2 = 1_A$. Therefore,

$$(f_1 + f_3(h_3k_1))h_1 + (f_2 + f_3(h_3k_2))h_2 = f_1h_1 + f_2h_2 + f_3h_3(k_1h_1 + k_2h_2)$$
$$= f_1h_1 + f_2h_2 + f_3h_3 = 1_A.$$

This shows that 2 is in the stable range of End A.

($\Leftarrow$) Let $G = A \oplus A \oplus B$, $f = (f_1, f_2, \alpha) : G \to A$, and $g = (g_1, g_2, \beta) : A \to G$ with $1_A = fg = f_1g_1 + f_2g_2 + (\alpha\beta)1_A$. Since 2 is in the stable range of End A, $1_A = (f_1 + (\alpha\beta)h_1)k_1 + (f_2 + (\alpha\beta)h_2)k_2$ for some $h_i, k_i \in$ End A. Define $\phi = (k_1, k_2, \beta(h_1k_1 + h_2k_2)) : A \to G$, so that $f\phi = f_1k_1 + f_2k_2 + \alpha\beta(h_1k_1 + h_2k_2) = (f_1 + (\alpha\beta)h_1)k_1 + (f_2 + (\alpha\beta)h_2)k_2 = 1_A$.

Then $\theta = (f_1 + (\alpha\beta)h_1, f_2 + (\alpha\beta)h_2, 0) : G = A \oplus A \oplus B \to A$ with $\theta\phi = (f_1 + (\alpha\beta)h_1)k_1 + (f_2 + (\alpha\beta)h_2)k_2 = 1_A$. In particular, $G = \phi(A) \oplus \ker\theta$. But $G = A \oplus A \oplus B$ and $B \subseteq \ker\theta$, so that $\ker\theta = B \oplus (\ker\theta) \cap (A \oplus A)$ with $C = (\ker\theta) \cap (A \oplus A) \subseteq A \oplus A$. Thus, $G = \phi(A) \oplus B \oplus C$ with $C \subseteq A \oplus A$.

Theorem 2.3.6 [Warfield 80] *If G is a torsion-free abelian group of finite rank, then 2 is in the stable range of* End G.

PROOF. Assume $f_1 g_1 + f_2 g_2 + f_3 g_3 = 1$ with $f_i, g_i \in \text{End } G$. It is sufficient to assume that f_1 is a unit of $\mathbb{Q}\text{End } G$. To see this, there is, by Lemma 2.3.3, an h in End G such that $f_1' = f_1 + (f_2 g_2 + f_3 g_3)h$ is a unit of $\mathbb{Q}\text{End } G$. Now,

$$f_1' g_1 + f_2(g_2 - g_2 h g_1) + f_3(g_3 - g_3 h g_1)$$
$$= (f_1 + (f_2 g_2 + f_3 g_3)h)g_1 + f_2(g_2 - g_2 h g_1) + f_3(g_3 - g_3 h g_1)$$
$$= f_1 g_1 + f_2 g_2 + f_3 g_3 = 1.$$

Hence, if $(f_1' + f_3 h_1)k_1 + (f_2 + f_3 h_2)k_2 = 1$ for some $h_i, k_i \in \text{End } G$, then substituting for f_1' gives an equation

$$1_G = (f_1 + (f_2 g_2 + f_3 g_3)h + f_3 h_1)k_1 + (f_2 + f_3 h_2)k_2$$
$$= (f_1 + f_3(g_3 h + h_1 - h_2 g_2 h))k_1 + (f_2 + f_3 h_2)(g_2 h k_1 + k_2)$$

of the desired form to show that 2 is in the stable range of End G.

So, assume that f_1 is a unit in $\mathbb{Q}\text{End } G$. There is a positive integer n with n End $G \subseteq f_1 \text{End } G \subseteq \text{End } G$. By Example 2.3.2, 1 is in the stable range of End $G/n\text{End } G$, since End $G/n\text{End } G$ is finite. Thus, $f_2 g_2 + (f_1 g_1 + f_3 g_3)1 \equiv 1 (\text{mod } n\text{End } G)$ yields an $h \in \text{End } G$ such that $f_2 + (f_1 g_1 + f_3 g_3)h \equiv u (\text{mod } n\text{End } G)$ with u a unit modulo $n\text{End } G$. Since $n\text{End } G \subseteq f_1 \text{End } G$, $f_2 + f_3 g_3 h \equiv u (\text{mod } f_1 \text{End } G)$. Because u is a unit modulo $n\text{End } G$, there is $v \in \text{End } G$ with $uv - 1 \in n\text{End } G$. Then $(f_2 + f_3 g_3 h)v \equiv 1 (\text{mod } f_1 \text{End } G)$, since $n\text{End } G \subseteq f_1 \text{End } G$. Write $1 = f_1 k_1 + (f_2 + f_3 g_3 h)v$ for some $k_1 \in \text{End } G$. This yields $(f_1 + f_3 \cdot 0)k_1 + (f_2 + f_3(g_3 h))v = 1$, whence 2 is in the stable range of End G.

Corollary 2.3.7 *Suppose G is a torsion-free abelian group of finite rank.*

(a) *The group G has the 2-substitution property.*
(b) *If $G \oplus G \oplus X$ is isomorphic to $G \oplus Y$ for finite rank torsion-free abelian groups X and Y, then Y is isomorphic to $G' \oplus X$ for some G' with $G \oplus G'$ isomorphic to $G \oplus G$.*

PROOF. Apply Lemma 2.3.4 and Theorems 2.3.5 and 2.3.6.

It is known that if $G \oplus G'$ is isomorphic to $G \oplus G$, then G is isomorphic to G' for "most" finite rank torsion-free abelian groups G. A finite-dimensional simple $\mathbb{Q}$-algebra K is a *totally definite quaternion algebra* if K is a division algebra with center F, K has F-dimension 4, and for each archimedean valuation v of F, the v-completion F_v of F is the reals and $F_v \otimes_F K$ is the ring of Hamiltonian quaternions.

Proposition 2.3.8 *Suppose G is a torsion-free abelian group of finite rank and write $\mathbb{Q}\text{End } G/J\mathbb{Q}\text{End } G = K_1 \times \cdots \times K_n$, with each K_i a matrix ring over a division algebra D_i. If no K_i is a totally definite quaternion algebra and $G \oplus G'$ is isomorphic to $G \oplus G$, then G is isomorphic to G'.*

PROOF. [Arnold 82].

It follows from Theorem 3.1.8 that if a torsion-free abelian group G is quasi-isomorphic to a finite direct sum of rank-1 groups, then G satisfies the hypotheses of the next example.

Example 2.3.9 *Let G be a finite rank torsion-free abelian group and write $Q\text{End }G/J Q\text{End }G = \text{Mat}_{n(1)}(D_1) \times \cdots \times \text{Mat}_{n(m)}(D_m)$ with each D_i a division algebra and, for each i, either $n(i) > 1$ or else $n(i) = 1$ and D_i is a field. If $G \oplus G \oplus X$ is isomorphic to $G \oplus Y$ for finite rank torsion-free abelian groups X and Y, then Y is isomorphic to $G \oplus X$.*

PROOF. A consequence of Corollary 2.3.7(b) and Proposition 2.3.8.

Lemma 2.3.10 *If G is a torsion-free abelian group of finite rank, then there are only finitely many isomorphism classes of summands of G.*

PROOF. A consequence of the Jordan–Zassenhaus theorem for $\mathbb{Z}$-orders ([Lady 74A] or see [Arnold 82]).

Lemma 2.3.11 *Assume that G is a torsion-free abelian group of finite rank. There is a positive integer n such that if G is nearly isomorphic to H, then $G^n \oplus C$ is isomorphic to $H^n \oplus C$ with C a summand of G^m for some $m \geq 1$.*

PROOF. Define the *Grothendieck group* of G to be $K_0(G) = F/R$, where F is the free abelian group with a basis consisting of isomorphism classes (A) of groups A such that A is a summand of G^m for some $m \geq 1$ and R the subgroup of F generated by elements of the form $(A \oplus B) - (A) - (B)$. Define $[A]$ to be the coset $(A) + R$. Then $[A \oplus B] = [A] \oplus [B]$ and $n[A] = [A^n]$ for a positive integer n.

It follows that each element of $K_0(G)$ is of the form $x = [A] - [B]$ and that $x = 0$ if and only if there is some C with $A \oplus C$ isomorphic to $B \oplus C$ and C a summand of G^m for some m. The latter condition can be seen by writing $(A) - (B)$ as a linear combination of the generators of R, rearranging this equation so that only positive coefficients are on each side of the equation, and using the fact that F is a free abelian group.

Let U be the set of elements of $K_0(G)$ of the form $[H] - [G]$ such that H is nearly isomorphic to G. To see that U is a subgroup, it suffices to show that U is closed under sums and additive inverses. If $[H] - [G]$ and $[K] - [G]$ are elements of U, then by Corollary 2.2.6, $G \oplus L = H \oplus K$ for some group L. But L is nearly isomorphic to G by Corollary 2.2.4, since G is nearly isomorphic to both H and K. Hence, $([H] - [G]) + ([K] - [G]) = [L] - [G] \in U$, since $[H] + [K] = [H \oplus K] = [G \oplus L] = [G] + [L]$ and $K_0(G)$ is a group. By Corollaries 2.2.6 and 2.2.4, $H \oplus M = G \oplus G$ for some group M nearly isomorphic to G. Thus, $[M] - [G] = -([H] - [G]) \in U$.

If $[H] - [G] \in U$, then H is a summand of $G \oplus G$. By Lemma 2.3.10, there are only finitely many isomorphism classes of summands of $G \oplus G$. In particular, U must be finite. Choose a positive integer n with $nU = 0$. If H is nearly isomorphic to G, then $[H] - [G] \in U$, and so $n([H] - [G]) = 0$. Thus, $n[G] = [G^n] = [H^n] = nH \in K_0(G)$ and $G^n \oplus C$ is isomorphic to $H^n \oplus C$ with C a summand of G^m for some $m \geq 1$.

Theorem 2.3.12 [Warfield 80] *Two torsion-free abelian groups G and H of finite rank are nearly isomorphic if and only if there is some integer n with G^n isomorphic to H^n.*

PROOF. ($\Leftarrow$) is Corollary 2.2.4(b).

($\Rightarrow$) By Lemma 2.3.11, there is some positive integer m with $G^m = K \oplus K'$ and positive integer n with $G^n \oplus K$ isomorphic to $H^n \oplus K$. Then, G^{n+m} is isomorphic to $G^n \oplus K \oplus K'$, which is isomorphic to $H^n \oplus K \oplus K' = H^n \oplus G^m$. Hence, G^{n+m} is isomorphic to $H^n \oplus G^m$.

First assume $m \geq 2$. By Corollary 2.3.7(b), $H^n \oplus G^{m-1}$ is isomorphic to $G' \oplus G^{m+n-2}$ with $G' \oplus G$ isomorphic to G^2. Thus, $H^n \oplus G^{m-1}$ is isomorphic to $G^{n+(m-1)}$. This reduces m by 1. It now suffices to assume that $m = 1$, i.e., G^{n+1} is isomorphic to $H^n \oplus G$.

Next, suppose $n \geq 2$. Then $G \oplus G \oplus G^{n-1}$ is isomorphic to $G \oplus H^n$. Again, by Corollary 2.3.7(b), H^n is isomorphic to $G' \oplus G^{n-1}$ for some G' with $G \oplus G'$ isomorphic to G^2. Hence, H^n is isomorphic to $(G' \oplus G) \oplus G^{n-2}$ and so to G^n, as desired.

The only remaining case is $m = n = 1$, i.e., G^2 is isomorphic to $G \oplus H$. Since G is nearly isomorphic to H, $G \oplus G'$ is isomorphic to H^2 for some G'. Then $H^3 = H^2 \oplus H$ is isomorphic to, sequentially, $G \oplus G' \oplus H$, $G \oplus H \oplus G'$, $G \oplus G \oplus G'$, $G \oplus H \oplus H$, and $G^2 \oplus H$. By Corollary 2.3.7, H has the 2-substitution property and G^2 is isomorphic to $H \oplus H'$ for some H' with $H \oplus H'$ isomorphic to H^2. Thus, G^2 is isomorphic to H^2, as desired.

EXERCISES

1. [Goodearl 76] Suppose S is a subring of R with $nR \subseteq S$ for some nonzero integer n and R/nR Artinian. Prove that if 1 is in the stable range of R, then 1 is in the stable range of S.

2. Suppose G is a reduced, semilocal, torsion-free abelian group of finite rank. Let n be the product of the primes p with $pG \neq G$. Show that $n\text{End } G \subseteq J\text{End } G$ and conclude that 1 is in the stable range of End G.

3. A torsion-free abelian group G of finite rank has *self-cancellation* if $G \oplus G$ isomorphic to $G \oplus G'$ implies G is isomorphic to G'. Show G has self-cancellation in the following cases:
   (a) If End $G \oplus M$ is isomorphic to End $G \oplus$ End G as right End G-modules, then M is isomorphic to End G.

(b) G is isomorphic to H^2 for some H.

(c) $G = H \oplus K$ and both H and K have self-cancellation.

4. Assume that G, H, and K are torsion-free abelian groups of finite rank with $G \oplus K$ isomorphic to $H \oplus K$. Prove that G is isomorphic to H in the following cases:

(a) Either G or K is semilocal.

(b) K is isomorphic to a summand of G and K has self-cancellation.

(c) For some $n > 1$, G has self cancellation and $G^n = Y \oplus X$ for some X nearly isomorphic to K.

5. [O'Meara, Vinsonhaler 99] Let G and H be torsion-free abelian groups of finite rank. Prove that G^n is isomorphic to H^n for each integer $n \geq 2$ if and only if $G \oplus G$, $G \oplus H$, and $H \oplus H$ are all isomorphic.

2.4 Self-Small Groups and Endomorphism Rings

There is a category equivalence from the category of summands of direct sums of a self-small abelian group A to the category of projective End A-modules (Theorem 2.4.1). In particular, finite rank torsion-free abelian groups can be investigated in the context of finitely generated projective modules over subrings of semisimple finite-dimensional $\mathbb{Q}$-algebras (Theorem 2.4.3). Moreover, each reduced subring of a finite-dimensional $\mathbb{Q}$-algebra may be realized as the endomorphism ring of a finite rank torsion-free abelian group (Theorem 2.4.6).

Write $\oplus_I A$ for the direct sum of copies of an abelian group A indexed by the set I. An abelian group A is *self-small* if for each $f \in \mathrm{Hom}(A, \oplus_I A)$, there is a finite subset J of I with $f(A) \subseteq \oplus_J A$. For example, if A is a torsion-free abelian group with finite rank, then A is self-small. To see this, let $f \in \mathrm{Hom}(A, \oplus_I A)$. Since A has finite rank, there is a finitely generated free subgroup B of A with rank $B = \mathrm{rank} A$ and A/B a torsion group. Then $f(B) \subseteq \oplus_J A$ for some finite subset J of I. But $\oplus_J A$ is a pure subgroup of $\oplus_I A$, being a summand. Hence, $f(A) \subseteq C \subseteq \oplus_J A$, where C is the pure subgroup of $\oplus_J A$ generated by $f(B)$.

For an abelian group A, $P(A)$ is the category with summands of direct sums of copies of A as objects and group homomorphisms as morphism sets. For a ring R, $P(R)$ denotes the category of projective right R-modules with right R-module homomorphisms as morphism sets. Both $P(A)$ and $P(R)$ are additive categories.

Suppose $F, G : \mathbf{X} \to \mathbf{X}'$ are functors with $\mathbf{X}$ and $\mathbf{X}'$ additive categories. A *natural equivalence* of F and G is an isomorphism $\phi_M : F(M) \to G(M)$ for each object M of $\mathbf{X}$ such that if $f : M \to N$ is a morphism in $\mathbf{X}$, then $G(f)\phi_M = \phi_N F(f)$. A functor $F : \mathbf{X} \to \mathbf{X}'$ is a *category equivalence* if F is both fully faithful and *dense*, for each object V of $\mathbf{X}'$, there is an object U of $\mathbf{X}$ with $F(U)$ isomorphic to V, equivalently, if there is a functor $G : \mathbf{X}' \to \mathbf{X}$ with FG naturally equivalent to $I_{\mathbf{X}'}$, the identity functor on $\mathbf{X}'$, and GF naturally equivalent to $I_{\mathbf{X}}$ (Exercise 2.4.1).

Theorem 2.4.1 *If A is a self-small abelian group, then there is a category equivalence $H_A : P(A) \to P(\mathrm{End} A)$ given by $H_A(G) = \mathrm{Hom}(A, G)$.*

PROOF. The group A is a left End A-module with $fa = f(a)$ for $f \in$ End A and $a \in A$. Then Hom(A, G) is a right End A-module, where fr $\in$ Hom(A, G) is the composite of $f \in$ Hom(A, G) and $r \in$ End A. If $f \in$ Hom(G, K), then $H_A(f)$: Hom$(A, G) \to$ Hom(A, K) is defined by $H_A(f)(g) = fg$ for each $g \in$ Hom(A, G). Thus, H_A is a functor from $P(A)$ to the category of right End A-modules, observing that $H_A(ff') = H_A(f)H_A(f')$ and $H_A(1_G) = 1$. Moreover, $H_A(G \oplus K) = H_A(G) \oplus H_A(K)$, noticing that if $G \oplus K$ has injections i_G and i_K, then $H_A(i_G)$, $H_A(i_K)$ are injections for $H_A(G) \oplus H_A(K)$.

Since A is self-small, $H_A(\oplus_I A) = \oplus_I H_A(A)$. To see this, for each $j \in I$, let i_j be an injection and π_j a projection for the direct sum $\oplus_I A$. If $f \in$ Hom$(A, \oplus_I A)$, then $f = (\pi_j f i_j)_I \in \oplus_IHom(A, A)$, since $\pi_j f i_j = 0$ for $j \in I \backslash J$, where J is a finite subset of I with $f(A) \subseteq \oplus_J A$. It now follows that $H_A : P(A) \to P($End $A)$ is a functor.

Define a functor $T_A : P($End$A) \to P(A)$ by $T_A(M) = M \otimes_{\text{End } A} A$ and $T_A(f) = f \otimes 1 : T_A(M) \to T_A(N)$ for $f \in$ Hom$_R(M, N)$. There is a natural homomorphism $\theta_M : T_A H_A(M) \to M$ given by $\theta_M(f \otimes a) = f(a)$ for $f \in$ Hom(A, M) and $a \in A$. Since $\theta_A : T_A H_A(A) \to A$ is an isomorphism, $\theta : T_A H_A(\oplus_I A) \to \oplus_I A$ is an isomorphism. Hence, if $G \oplus H = \oplus_I A$, then $\theta_G : T_A H_A(G) \to G$ is an isomorphism. This shows that $\theta : T_A H_A \to I_{P(A)}$ is a natural equivalence from the functor $T_A H_A$ to the identity functor on $P(A)$.

A similar argument shows that $\phi : I_{P(\text{End}A)} \to H_A T_A$ is a natural equivalence of functors, where $\phi_M : M \to H_A T_A(M)$ is defined by $\phi_M(m)(a) = m \otimes a$ for $M \in P($End $A)$, $m \in M$, and $a \in A$. This follows from the observation that $\phi_{\text{End}A} :$ End $A \to H_A T_A($End $A)$ is an isomorphism.

Corollary 2.4.2 *If A is a self-small abelian group, then each group in $P(A)$ is isomorphic to a direct sum of copies of A if and only if each projective right End A-module is free.*

PROOF. A consequence of Theorem 2.4.1.

For an abelian group A, $P_f(A)$ is the category with summands of finite direct sums of copies of A as objects and group homomorphisms as morphism sets. For a ring R, $P_f(R)$ denotes the category of finitely generated projective right R-modules with right R-module homomorphisms as morphism sets. Modules in $P_f(R)$ are precisely summands of finitely generated free R-modules. If $R =$ End A/NEnd A with A a torsion-free abelian group of finite rank, then each M in $P_f(R)$ is an R-lattice as defined in Section 2.2. This follows from the observation that the additive group of R is torsion-free with finite rank, since NEnd $A =$ End $A \cap J\mathbb{Q}$End A is a pure subgroup of the finite rank torsion-free group End A. Consequently, a summand of a finite direct sum of copies of R is a finitely generated R-module with finite rank torsion-free additive group.

The next proposition demonstrates the intimate connection between torsion-free abelian groups of finite rank and lattices over subrings of finite-dimensional semisimple $\mathbb{Q}$-algebras.

Theorem 2.4.3 *Assume that A is a torsion-free abelian group of finite rank and* N*End* $A =$ End $A \cap J\mathbb{Q}$End A. *There is a functor* $H : P_f(A) \to P_f($End $A/$ NEnd $A)$ *given by* $H(G) = H_A(G)/H_A(G)N$End A *such that if G and K are in P(A), then G and K are:*

(a) *quasi-isomorphic if and only if* $\mathbb{Q}H(G)$ *and* $\mathbb{Q}H(K)$ *are isomorphic as right* $\mathbb{Q}$*End* $A/J\mathbb{Q}$*End* A*-modules;*
(b) *isomorphic if and only if* $H(G)$ *is isomorphic to* $H(K)$*;*
(c) *isomorphic at a prime p if and only if* $H(G)_p$ *is isomorphic to* $H(K)_p$ *as a right* (End A/NEnd $A)_p$*-module;*
(d) *nearly isomorphic if and only if the* End A/NEnd A*-lattices* $H(G)$ *and* $H(K)$ *are in the same genus class.*

PROOF. Notice that H is a functor, being the composite of the functor H_A of Theorem 2.4.2 and the functor $P($End $A) \to P($End A/NEnd $A)$ given by $M \mapsto M/MN$EndA. In particular, if $f \in$ Hom(G, K), then $H_A(f) : H_A(G) \to H_A(K)$ induces $H(f) : H(G) = H_A(G)/H_A(G)N$End $A \to H(K) = H_A(K)/H_A(K)$ NEnd A.

Assume that $G, K \in P(A)$, $f \in$ Hom(G, K), $g \in$ Hom(K, G), and m is a nonzero integer with $fg = m1_K$ and $gf = m1_G$. Then $H(f)H(g) = m1_{H(K)}$ and $H(g)H(f) = m1_{H(G)}$, since H is a functor.

Conversely, assume that $\alpha' : H(G) \to H(K)$ and $\beta' : H(K) \to H(G)$ are End A/NEnd A-homomorphisms and m is a nonzero integer with $\alpha'\beta' = m1_{H(K)}$ and $\beta'\alpha' = m1_{H(G)}$. Since $H_A(G)$ and $H_A(K)$ are projective End A-modules, there exist $\alpha : H_A(G) \to H_A(K)$ and $\beta : H_A(K) \to H_A(G)$ with $\alpha\beta(H_A(K)) + H_A(K)N$End $A = mH_A(K) + H_A(K)N$End A. Now, NEnd $A \subseteq J$End A and $H_A(G)$ is finitely generated as an End A-module, so, by Nakayama's lemma, $\alpha\beta(H_A(K)) = mH_A(K)$. Therefore, $\gamma = m^{-1}\alpha\beta : H_A(K) \to H_A(K)$ must be an automorphism. By Theorem 2.4.2, there are $f : G \to K$ and $g : K \to G$ with $H_A(f) = \alpha$ and $H_A(g) = \beta\gamma^{-1}$. Hence, $H_A(gf) = H_A(g)H_A(f) = (\beta\gamma^{-1})\alpha = \beta\beta^{-1}\alpha^{-1}$ $m\alpha = m = H_A(m1_K)$, and $H_A(fg) = H_A(f)H_A(g) = \alpha(\beta\gamma^{-1}) = \alpha\beta\beta^{-1}\alpha^{-1}$ $m = H_A(m1_G)$. But H_A is a faithful functor, whence $fg = m1_K$ and $gf = m1_G$.

The proof is completed by choosing various values of m. For (a), any nonzero m will do, noticing that $\mathbb{Q}($End A/NEnd $A) = \mathbb{Q}$End $A/J\mathbb{Q}$End A. For (b), let $m = 1$, and for (c), let m be an integer prime to p. Finally, (d) is a consequence of (c) and the definitions of near-isomorphism and genus class of R-lattices given in Section 2.2.

Let R be an integral domain and M a *torsion-free R-module*, that is, if $rm = 0$ with $r \in R$ and $m \in M$, then either $r = 0$ or $m = 0$. If M is a torsion-free R-module and N is a submodule of M, then N is a *pure submodule* if $rM \cap N = rN$ for each $r \in R$; equivalently, M/N is a torsion-free R-module. If N is a submodule of M, then $\{m \in M : rm \in N$ for some $0 \neq r \in R\}$ is the *pure submodule of M generated by N*. The *rank of M* is the cardinality of a maximal R-independent subset of M.

Theorem 2.4.4 *Suppose R is a principal ideal domain and M is a countably generated torsion-free R-module. Then M is a free R-module if and only if each pure finite rank submodule of M is a finitely generated free module.*

PROOF. ($\Rightarrow$) Assume that M is a free R-module and N is a finite rank pure R-submodule of M. Then N is contained in, hence a pure submodule of, a finitely generated free summand of M. Thus, it is sufficient to assume that M is a finitely generated free R-module. In this case, M/N is a finitely generated torsion-free R-module, hence a free R-module (Exercise 2.4.2). Because free modules are projective, N is a summand of M. Since M is a finitely generated free module and R is a principal ideal domain, N is a finitely generated free R-module.

($\Leftarrow$) Let $\{x_1, \ldots, x_n, \ldots\}$ be a countable set of generators for M. Define M_n to be the pure submodule of M generated by $\{x_1, \ldots, x_n\}$. Then $M_1 \subseteq \cdots \subseteq M_n \subseteq \cdots \subseteq M$ is an ascending chain of pure finite rank submodules of M with $M = \cup\{M_n : n \geq 1\}$. By assumption, each M_n is a free R-module. For each n, M_{n+1}/M_n is a finitely generated torsion-free R-module, hence free by Exercise 2.4.2. Since free modules are projective, $M_{n+1} = N_{n+1} \oplus M_n$ for some free R-module N_n and $n \geq 1$. Define $N_1 = M_1$ and observe that for each n, $M_n = N_1 \oplus \cdots \oplus N_{n-1}$. Then $M = \oplus\{N_n : n \geq 1\}$ is a free R-module, since $M = \cup\{M_n : n \geq 1\}$ and each N_n is a free R-module.

Corollary 2.4.5 [Pontryagin 34] *Assume that G is a countable torsion-free abelian group of finite rank. Then G is a free group if and only if each pure finite rank subgroup is a free group.*

PROOF. Apply Theorem 2.4.4 with $R = \mathbb{Z}$.

Theorem 2.4.6 [Corner 63] *Suppose R is a ring with identity and that the additive group of R is countable, reduced, and torsion-free. There is a countable torsion-free abelian group G with* End *G isomorphic to R. Moreover, if the additive group of R has rank n, then G may be chosen with* rank $G = 2n$.

PROOF. Only an outline of this proof is given. See, for example, [Fuchs 73] for more detailed arguments. Suppose that the additive group of R is torsion-free with rank n. The $\mathbb{Z}$-*adic topology* is the topology on R with open sets $r + nR$ such that $r \in R$ and $0 \neq n \in \mathbb{Z}$. Since R is reduced, R is Hausdorff in this topology. Let R^* be the completion of R in the $\mathbb{Z}$-adic topology. Then R^* is a commutative $\mathbb{Z}^*$-algebra with additive group uncountable and torsion-free and $R \subseteq R^*$. In fact, $\mathbb{Z}^* = \Pi_p \mathbb{Z}_p^*$ with $\mathbb{Z}_p^*$ the p-adic integers, an uncountable integral domain, and the transcendence degree of the quotient field of $\mathbb{Z}_p^*$ over the algebraic closure of $\mathbb{Q}$ is uncountable.

First, assume that the additive group of R has finite rank and let $\{1 = r_1, \ldots, r_n\}$ be a maximal $\mathbb{Z}$-linearly independent subset of R. Since $\mathbb{Q}R$ is countable, there is a subset $\{\alpha_1, \ldots, \alpha_n\}$ of $\mathbb{Z}^*$ that is *algebraically independent over* $\mathbb{Q}R$, i.e., for each nonzero polynomial $h(x_1, \ldots, x_n)$ with coefficients in $\mathbb{Q}R$, $h(\alpha_1, \ldots, \alpha_n) \neq 0$.

Define G to be the pure subgroup of R^* generated by the group $R + R\beta$, where $\beta = r_1\alpha_1 + \cdots + r_n\alpha_n \in R^*$. Then rank $G = 2n$, and R is isomorphic to a subring of End G, via $r \mapsto$ left multiplication by r. On the other hand, let $f \in$ End G. Then f extends uniquely to a $\mathbb{Z}^*$-endomorphism g of R^*. For some nonzero integer m, $mf(\beta) = a + b\beta \in R + R\beta$ and $mf(r_i) = a_i + b_i\beta \in R + R\beta$ for each i. Hence, $mf(\beta) = mg(\beta) =$

$$mg(r_1)\alpha_1 + \cdots + mg(r_n)\alpha_n = mf(r_1)\alpha_1 + \cdots + mf(r_n)\alpha_n = a + b\beta.$$

Substitution gives

$$(a_1 + b_1\beta)\alpha_1 + \cdots + (a_n + b_n\beta)\alpha_n = a + br_1\alpha_1 + \cdots + br_n\alpha_n.$$

Since the α_i's are algebraically independent over $\mathbb{Q}R$, each b_i equals 0, $a = 0$, and $br_i = a_i$ for each i. Hence, $mf(r_i) = br_i$ for each i. In fact, $mf(1) = b$, since $r_1 = 1$. This shows that $mf(x) = bx$ for each $x \in R + R\beta$ with $b \in R$. Since R is pure in R^* and $mf(1) = b \in R$, it follows that f is left multiplication by $f(1) \in R$. Consequently, R is isomorphic to End G.

For the general case, assume that the additive group of R has countably infinite rank. Define G to be the pure subgroup of R^* generated by $\{R, R\beta_r : r \in R\}$, where $\beta_r = \rho_r 1_R + \sigma_r r$ and $\{\rho_r, \sigma_r : r \in R\}$ is a countable subset of $\mathbb{Z}^*$ that is algebraically independent over $\mathbb{Q}R$. Then G is a countable group, and calculations like those above show that End G is isomorphic to R.

Theorem 2.4.7 [Beaumont, Pierce 61A] *If R is a subring of a finite-dimensional $\mathbb{Q}$-algebra, then the additive group of R is quasi-isomorphic to the group $NR \oplus R/NR$.*

PROOF. See, for example, [Arnold 82].

EXERCISES

1. Prove that a functor $F : \mathbf{X} \to \mathbf{X}'$ is a category equivalence of additive categories if and only if there is a functor $G : \mathbf{X}' \to \mathbf{X}$ with FG naturally equivalent to the identity functor $I_{\mathbf{X}'}$ and GF naturally equivalent to the identity functor $I_{\mathbf{X}}$.

2. Show that if R is a principal ideal domain and M is a finitely generated torsion-free R-module, then M is a free R-module.

3. Assume that A is a torsion-free abelian group of finite rank and End A is a principal ideal domain.
   (a) Prove that if G is a subgroup of $\oplus_I A$ with $\mathrm{Hom}(A, G)A = G$, then G is isomorphic to $\oplus_J A$ for some subset J of I.
   (b) Prove that if G is a pure subgroup of $\oplus_I A$ with I finite, then G is a summand of $\oplus_I A$. Conclude that G is a finite direct sum of copies of A.

NOTES ON CHAPTER 2

Chapter 2 is a brief introduction to some fundamental techniques for countable torsion-free abelian groups. Further results and historical background, especially for groups of infinite rank, may be found in [Fuchs 73], a comprehensive exposition of the status of the subject up to 1973. Much of Chapter 2 is a revised and extended version of portions of [Arnold 82], lecture notes focusing on torsion-free abelian groups of finite rank and their relationship to modules over subrings of finite-dimensional $\mathbb{Q}$-algebras. Included in [Arnold 82] are complete arguments for properties of orders and lattices over orders referenced in Chapter 2.

The notion of quasi-isomorphism of torsion-free abelian groups of finite rank is due to [Jónsson 57, 59]. The categorical version here is as given in [Walker 64]. In [Jónsson 59], it is proved directly that $\text{frA}_{\mathbb{Q}}$ is a Krull–Schmidt category. The approach in Section 2.1, establishing that G is strongly indecomposable if and only if $\mathbb{Q}\text{End}\,G$ is a local ring, follows that of [Reid 63] for the finite rank case. Examples of torsion-free abelian groups of finite rank with nonunique direct sum decompositions are given in [Levi 17], a paper that did not receive wide distribution; see [Mader 93]. The pathology of direct sum decompositions for torsion-free abelian groups of finite rank is aptly demonstrated in [Fuchs 73, Sections 90 and 91], beginning with examples taken from [Jónsson 57]. Some positive results for direct sum decompositions are given in [Arnold 82].

Near-isomorphism for torsion-free abelian groups of finite rank is introduced in [Lady 75] and motivated by genus class for lattices over $\mathbb{Z}$-orders [Jacobinski 68]. The equivalent definition given in Section 2.1 more closely imitates the concept of quasi-isomorphism. Fundamental properties of genus class for finite rank torsion-free modules over a Dedekind subring R of an algebraic number field are given in [Arnold 77]. Genus class for these modules simultaneously generalizes near-isomorphism for torsion-free abelian groups of finite rank and genus class for lattices over R-orders.

The results of Section 2.3, like many others herein, hold in a more general context [Warfield 80]. The condition that $G^n \approx H^n$ implies $G \approx H$ is known as power cancellation; see [Goodearl 76]. Examples of nonisomorphic torsion-free abelian groups G and H of finite rank with $G^n \approx H^n$ for all integers $n \geq 2$ are generated in [O'Meara, Vinsonhaler, 99] (see Exercise 2.3.5 for an equivalent condition).

Self-cancellation, $G \oplus G \approx G \oplus G'$ implies $G \approx G'$, is also known as the strongly separative property [Ara, Goodearl, O'Meara, Pardo 98]. A sufficient condition, called Eichler's criterion, for self-cancellation of torsion-free abelian groups of finite rank is given in Proposition 2.3.8. An example for which self-cancellation fails is given in [Arnold 82], based on an example of [Swan 62] for $\mathbb{Z}$-orders. Other examples are given in [O'Meara, Vinsonhaler 99]. A necessary and sufficient condition for a finite rank torsion-free module over a Dedekind subring of an algebraic number field to have self-cancellation is given in [Blazhenov 96].

A characterization of rank-1 torsion-free abelian groups G of finite rank with the cancellation property, $G \oplus H \approx G \oplus K$ implies that $H \approx K$, is given in [Fuchs–Loonstra 71]. This was generalized in [Stelzer 85, 87] to give a sufficient condition for torsion-free abelian groups of finite rank to satisfy the cancellation property. This program is completed by a necessary and sufficient condition for a torsion-free module of finite rank over a Dedekind subring of an algebraic number field to have the cancellation property in [Blazhenov 96]. Specifically, for torsion-free abelian groups of finite rank, G has the cancellation property if and only if $G = F \oplus H$, where F is a free group, for each nonzero integer n each unit of $\text{End}\,H/n\text{End}\,H$ lifts to a unit of $\text{End}\,H$, and each quasi-summand of G satisfies a weak Eichler criterion.

Theorem 2.4.3 is the crucial result for an interpretation of torsion-free abelian groups of finite rank as finitely generated projective modules over subrings of finite-dimensional $\mathbb{Q}$-algebras. Such an interpretation provides justification for the complexity of these groups as well as for positive results for groups with restricted endomorphism rings. This point of view, as well as some far-reaching generalizations of Theorem 2.4.1, is exploited in [Albrecht 83, 89, 91], [Albrecht, Goeters 89, 94], and references therein.

Corner's theorem, Theorem 2.4.6, can be used to construct pathological direct sum decompositions [Corner 69]. This theorem has also provided motivation for a large number of papers on realization of endomorphism rings; see, for example, [Brenner 67], [Dugas, Göbel 82, 97], [Corner, Göbel 85], [Göbel, Shelah 95], and references therein.

3

Butler Groups

3.1 Types and Completely Decomposable Groups

Finite direct sums of torsion-free abelian groups of rank 1, called completely decomposable groups, can be classified in terms of types. Included in this section is a compilation of some of the fundamental properties of types and fully invariant subgroups determined by types.

The isomorphism class of a rank-1 torsion-free abelian group A is denoted by type A. Types of rank-1 groups are represented by isomorphism classes of subgroups of $\mathbb{Q}$. If X and Y are subgroups of $\mathbb{Q}$ and $f : X \to Y$ is a nonzero homomorphism, then f extends to a unique $\mathbb{Q}$-endomorphism g of $\mathbb{Q}$, with $g(qx) = qf(x)$ for each $q \in \mathbb{Q}$ and $x \in X$. Since $\mathbb{Q}$-endomorphisms of $\mathbb{Q}$ are multiplication by elements of $\mathbb{Q}$, $\mathrm{Hom}(X, Y)$ is isomorphic to a subgroup of $\mathbb{Q}$, each $0 \neq f \in \mathrm{Hom}(X, Y)$ is a monomorphism, and $\mathrm{End}\, X$ is isomorphic to a subring of $\mathbb{Q}$.

Lemma 3.1.1
(a) *If X and Y are torsion-free abelian groups of rank 1, then the following statements are equivalent:*
   (i) $\mathrm{Hom}(X, Y)$ *and* $\mathrm{Hom}(Y, X)$ *are both nonzero;*
   (ii) *X and Y are quasi-isomorphic;*
   (iii) *X and Y are isomorphic.*
(b) *The set of types form a distributive lattice, where for subgroups X and Y of $\mathbb{Q}$,*
   (i) *type $X \leq$ type Y if and only if $\mathrm{Hom}(X, Y) \neq 0$;*
   (ii) *the meet, type $X \cap$ type Y, of type X and type Y is type $X \cap Y$; and*
   (iii) *the join, type $X \cup$ type Y, of type X and type Y is type $X + Y$.*

PROOF. Exercise 3.1.1.

A consequence of Lemma 3.1.1(a) is that if X and Y are nonzero subgroups of $\mathbb{Q}$, then type $X = $ type Y if and only if $(X + Y)/X$ and $(X + Y)/Y$ are finite.

Let G be a torsion-free abelian group. Recall from Section 2.1 that a subgroup K of G is a pure subgroup of G if $K \cap nG = nK$ for each nonzero integer n, equivalently, G/K is torsion-free. It is routine to confirm that if K is a pure subgroup of H and H is a pure subgroup of G, then K is a pure subgroup of G. Let H be a subset of G and define $\langle H \rangle$ to be the subgroup of G generated by H and define $\langle H \rangle_* = \{x \in G : nx \in \langle H \rangle\}$ for some $0 \neq n \in Z\}$, the pure subgroup of G generated by H. In particular, $\langle H \rangle_*$ is a pure subgroup of G.

If x is a nonzero element of a torsion-free abelian group G and X is the pure rank-1 subgroup of G generated by x, then type x is the isomorphism class of X. Define typeset $G = \{$type $x : 0 \neq x \in G\}$. If K is a pure subgroup of G and x is a nonzero element of K, then x has the same type in K as it does in G. Pure rank-1 subgroups of a torsion-free group G are like simple modules in the sense that if X and Y are pure rank-1 subgroups of G, then either $X = Y$ or $X \cap Y = 0$.

Types can be viewed as equivalence classes of *height sequences* $h = (h_p)_\Pi$, indexed by the set of primes Π, where h_p is a nonnegative integer or ∞ for each prime p. The terminology arises from the notion of p-heights. Let G be a torsion-free abelian group and x a nonzero element of G. Define $h_p(x)$, the *p-height of x* in G, to be ∞ if $x \in p^\omega G$ and n if $x \in p^n G \backslash p^{n+1} G$. Then $h(x) = (h_p(x))_\Pi$ is a height sequence. Let X be the pure subgroup of G generated by x. Since X is a pure subgroup of G, the p-height of x in G is exactly the same as the p-height of x in X for each prime. Hence, rank-1 groups give rise to height sequences.

There is an equivalence relation on height sequences that determines precisely when associated rank-1 groups are isomorphic. Two height sequences h and h' are *equivalent* if they differ in at most finitely many finite entries. For example, $(1, 1, \ldots, 1, \ldots)$ is equivalent to $(0, 0, 0, 1, 1, \ldots)$ but not to $(0, \infty, 1, 1, \ldots)$. A routine verification shows that this is an equivalence relation. Observe that if m and n are nonzero integers and $0 \neq x \in G$, then $h(mx)$ is equivalent to $h(nx)$. It follows that if G is a torsion-free group, $0 \neq x \in G$, and $X = \langle x \rangle_*$, then $h(y)$ is equivalent to $h(x)$ for each $0 \neq y \in X$.

Theorem 3.1.2 *There is a one-to-one correspondence from the set of types to the set of equivalence classes of height sequences.*

PROOF. The correspondence is given by $\phi($type $X) = [h(x)]$, where $[h(x)]$ is the equivalence class of the height sequence of $0 \neq x \in X$. To see that ϕ is well-defined, assume that X and Y are rank-1 groups, $0 \neq x \in X, 0 \neq y \in Y$, and $f : X \to Y$ is an isomorphism. Then $h(x) = h(f(x))$. Since $Y = \langle f(x) \rangle_* = \langle y \rangle_*$, $h(f(x))$ is equivalent to $h(y)$, and so $h(x)$ is equivalent to $h(y)$.

As for one-to-one, suppose that $h(x)$ and $h(y)$ are equivalent. Then there are nonzero integers m and n with $h(mx) = h(ny)$. Define $f : \mathbb{Q}mx \to \mathbb{Q}ny$ by

$f(qmx) = qny$ for each $q \in \mathbb{Q}$. Then $f : X \to Y$ is a well-defined isomorphism, since $h(mx) = h(ny)$, $X = \langle mx \rangle_*$, and $Y = \langle ny \rangle_*$. Finally, let $h = (h_p)_{\Pi}$ be a height sequence and define X to be the subgroup of $\mathbb{Q}$ generated by $\{1/p^i : i \leq h_p, p \in \Pi\}$. Then $1_{\mathbb{Q}} \in X$, $h(1_{\mathbb{Q}}) = h$, and $\phi(\text{type } X) = [h]$. This completes the proof.

The type of an element of a direct sum can be computed from the types of its components. Suppose $G = A \oplus B$ is a direct sum of torsion-free abelian groups and $x = (a, b)$ is a nonzero element of G. Then type $x = (\text{type } a) \cap (\text{type } b)$, the meet of type a and type b. This is because $h_p(x) = \min\{h_p(a), h_p(b)\}$ for each prime p and the type of an element is determined by the equivalence class of its height sequence. More generally, if $G = A + B$, not necessarily a direct sum, and $x = a + b$ with a in A and b in B, then type $x \geq (\text{type } a) \cap (\text{type } b)$, since $h_p(x) \geq \min\{h_p(a), h_p(b)\}$ for each prime p.

Lemma 3.1.3 *If G is a torsion-free abelian group with finite typeset, then the typeset of G is closed under meets.*

PROOF. If rank $G = 1$, then each nonzero element of G has the same type. Assume that rank $G \geq 2$ and let x and y be $\mathbb{Z}$-independent elements of G. Then $\{x + my : m \text{ nonzero integer}\}$ is an infinite subset of G. Since typeset G is finite, there are nonzero integers m and m' with $m \neq m'$ and $\tau = \text{type } x + my = \text{type } x + m'y \in \text{typeset } G$. Now, $\tau \geq \text{type } x \cap \text{type } my = \text{type } x \cap \text{type } y$. On the other hand, $(m - m')y = (x + my) - (x + m'y)$, so that type $y = \text{type }(m - m')y \geq (\text{type } x + my) \cap (\text{type } x + m'y) = \tau$. Similarly, type $x \geq \tau$, since $(m' - m)x = m'(x + my) - m(x + m'y)$. Hence, type $x \cap \text{type } y \geq \tau$, and so type $x \cap \text{type } y = \tau \in \text{typeset } G$.

Let τ be a type and G a torsion-free abelian group. Define

$$G(\tau) = \{x : \text{type } x \geq \tau\},$$

called the τ-*socle* of G, a pure subgroup of G. Define

$$G^*(\tau) = \langle \{x \in G : \text{type } x > \tau\} \rangle$$

to be the subgroup of G generated by $\{x \in G : \text{type } x > \tau\}$ and

$$G^{\#}(\tau) = \langle G^*(\tau) \rangle_*,$$

the pure subgroup of G generated by $G^*(\tau)$.

For example, if X is a rank-1 group, then $X(\tau) = X$ if type $X \geq \tau$, and $X(\tau) = 0$ otherwise; $X^*(\tau) = X$ if type $X > \tau$, and $X^*(\tau) = 0$ otherwise; and $X^{\#}(\tau) = X^*(\tau)$ for each τ.

Lemma 3.1.4 *Suppose G and H are torsion-free abelian groups and σ and τ are types.*

(a) *If $\tau \le \sigma$, then $G(\sigma) \subseteq G(\tau)$.*
(b) *$G(\tau)(\sigma) = G(\tau \cup \sigma) = G(\tau) \cap G(\sigma)$.*
(c) *If H is a pure subgroup of G, then $H(\tau) = H \cap G(\tau)$.*
(d) *If $f \in \text{Hom}(G, H)$ then $f(G(\tau)) \subseteq H(\tau)$, $f(G^*(\tau)) \subseteq H^*(\tau)$, and $f(G^\#(\tau)) \subseteq H^\#(\tau)$.*
(e) *$(G \oplus H)(\tau) = G(\tau) \oplus H(\tau)$, $(G \oplus H)^*(\tau) = G^*(\tau) \oplus H^*(\tau)$, and $(G \oplus H)^\#(\tau) = G^\#(\tau) \oplus H^\#(\tau)$.*

PROOF. Assertions (a), (b), and (c) are consequences of the definitions, while (d) follows from the fact that if X is a pure rank-1 subgroup of G with $f(X) \ne 0$, then type $X \le$ type $f(X)$, since $h_p(x) \le h_p f(x)$ for each $x \in X$ and prime p. To confirm (e), apply (d) to injections and projections for the direct sum $G \oplus H$.

Recall that a completely decomposable group is a finite direct sum of rank-1 torsion-free abelian groups. The τ-socle of a completely decomposable group C is completely determined by the rank-1 summands of C.

Example 3.1.5 *Suppose $C = X_1 \oplus \cdots \oplus X_n$ is a completely decomposable group with rank $X_i = 1$ for each i.*

(a) *If τ is a type, then $C(\tau) = \oplus\{X_i : \text{type } X_i \ge \tau\}$, $C^*(\tau) = \oplus\{X_i : \text{type } X_i > \tau\} = C^\#(\tau)$, and $C(\tau)/C^\#(\tau)$ is isomorphic to $\oplus\{X_i : \text{type } X_i = \tau\}$.*
(b) *The typeset of C is the meet closure of $\{\text{type } X_i : 1 \le i \le n\}$. In particular, typeset C is finite.*

PROOF. If $x = x_1 \oplus \cdots \oplus x_n$ is a nonzero element of C with x_i in X_i, then type $x = \cap\{\text{type } x_i : x_i \ne 0\}$ and type $X_i =$ type x_i if x_i is nonzero. This observation can be used to confirm both (a) and (b).

The *critical typeset* of a torsion-free abelian group G is $T_{cr}(G) = \{\tau : G(\tau)/G^\#(\tau) \ne 0\}$. In view of Example 3.1.5(a), the critical typeset of a completely decomposable group records the types of rank-1 summands of G. Specifically, if $C = X_1 \oplus \cdots \oplus X_n$ with rank $X_i = 1$ for each i, then $T_{cr}(C) = \{\text{type }(X_i) : 1 \le i \le n\}$.

A torsion-free abelian group G is *homogeneous* if there is a type τ with type $x = \tau$ for each nonzero element x of G. In this case, G is called τ-*homogeneous*. A rank-1 group X is homogeneous. It follows from Example 3.1.5 that if G is homogeneous completely decomposable, then G is isomorphic to X^n for some rank-1 group X with type $X = \tau$.

Homogeneous completely decomposable groups have some special properties.

Lemma 3.1.6 [Baer 37] *Assume that C is a τ-homogeneous completely decomposable group.*

(a) *If $C = G/K$ for some torsion-free abelian group G and $G(\tau) + K = G$, then K is a summand of G.*

(b) *Each pure subgroup of C is a summand.*

(c) *Each summand of C is τ-homogeneous completely decomposable.*

PROOF. (a) It is sufficient to assume that C has rank 1 and type τ. For the general case project onto a rank-1 summand of C and induct on the number of rank-1 summands of C. Let $\pi : G \to C = G/K$ denote the canonical epimorphism with $\ker \pi = K$. Define $I = \{\pi h : h \in \text{Hom}(C, G)\}$, a right ideal of End C. Then $IC = C$. To see this, let $0 \neq x \in C$. Since $G(\tau) + K = G$ and π is onto with $\ker \pi = K$, there is $y \in G(\tau)$ with $x = \pi(y)$. Because type $y \geq \tau = \text{type } x$ and $x \in C$ with rank $C = 1$, there is $h \in \text{Hom}(C, G)$ with $y \in h(C)$. Thus, $x = \pi(y) \in \pi h(C) \subseteq IC$.

Since End C is a principal ideal domain, $I = f \text{End } C$ for some $f \in \text{End } C$. Then $fC = IC = C$, whence $f \in I$ is an automorphism of C, and so $I = \text{End } C$. But $1_C \in I$, so that there is $h \in \text{Hom}(C, G)$ with $\pi h = 1_C$. This shows that $K = \ker \pi$ is a summand of G.

Both (b) and (c) are consequences of Exercise 2.4.3. Following is a more direct argument.

(b) It is sufficient to assume that $C = X^n$ for some subgroup X of $\mathbb{Q}$ of type τ with $\mathbb{Z} \subseteq X$, since a homogeneous completely decomposable group C is a direct sum $X_1 \oplus \cdots \oplus X_n$ with each X_i a rank-1 group isomorphic to a common subgroup X of $\mathbb{Q}$ containing $\mathbb{Z}$.

Write $C = X^n = X(\mathbb{Z}^n)$ with $C/\mathbb{Z}^n$ a torsion group. If H is a pure subgroup of C, then $H \cap \mathbb{Z}^n$ is a pure subgroup, hence a summand, of $\mathbb{Z}^n$. This is because $\mathbb{Z}^n / H \cap \mathbb{Z}^n$ is free, being finitely generated and torsion-free. Write $\mathbb{Z}^n = H \cap \mathbb{Z}^n \oplus D$. Then $H = X(H \cap \mathbb{Z}^n)$, since $H/(H \cap \mathbb{Z}^n)$ is torsion and $X(H \cap \mathbb{Z}^n)$ is a pure subgroup of C. Thus, $C = X(\mathbb{Z}^n) = X(H \cap \mathbb{Z}^n) \oplus XD = H \oplus XD$, and H is a summand of C.

(c) Assume that H is a summand of C. As in (b), $C = H \oplus H'$, where $\mathbb{Z}^n = (H \cap \mathbb{Z}^n) \oplus D$ with $X(H \cap \mathbb{Z}^n) = H$ and $H' = XD$. Then H is τ-homogeneous completely decomposable, since $H \cap \mathbb{Z}^n$ is a finitely generated free group, being a summand of the free group $\mathbb{Z}^n$, and X is a rank-1 group of type τ.

Theorem 3.1.7 [Baer 37]

(a) *Summands of completely decomposable groups are completely decomposable.*

(b) *Let $C = X_1 \oplus \cdots \oplus X_m$ and $D = Y_1 \oplus \cdots \oplus Y_n$ be two completely decomposable groups such that X_i and Y_j are rank-1 groups. The following statements are equivalent:*

  (i) *C and D are isomorphic;*

  (ii) *C and D are quasi-isomorphic;*

  (iii) *$m = n$ and there is a permutation σ of $\{1, 2, \ldots, n\}$ with type $X_i = $ type $Y_{\sigma(i)}$ for each i;*

  (iv) *rank $C(\tau)/C^{\#}(\tau) = $ rank $D(\tau)/D^{\#}(\tau)$ for each type τ .*

PROOF. (a) Let G be a summand of a completely decomposable group C, say $C = G \oplus H$. If $\tau \in T_{cr}(C)$, then $C(\tau) = C_\tau \oplus C^*(\tau)$ with C_τ a τ-homogeneous completely decomposable group by Example 3.1.5(a). Moreover, $C = \oplus\{C_\tau : \tau \in T_{cr}(C)\}$.

The proof is an induction on n, the cardinality of $T_{cr}(C)$. If $n = 1$, then C is homogeneous and G is completely decomposable by Lemma 3.1.6(c). Assume that $n > 1$ and that τ is a maximal element of $T_{cr}(C)$. Then $C_\tau = C(\tau)$ is homogeneous completely decomposable and a summand of C. But $C(\tau) = G(\tau) \oplus H(\tau)$, by Lemma 3.1.4(e). Thus, $G(\tau)$ and $H(\tau)$ are homogeneous completely decomposable summands of C by Lemma 3.1.6, hence of G and H, respectively. Write $G = G(\tau) \oplus G'$ and $H = H(\tau) \oplus H'$ for some groups G' and H'. Then $\oplus\{C_\sigma : \sigma \in T_{cr}(C)\} = C = G \oplus H = C_\tau \oplus G' \oplus H'$, whence $G' \oplus H'$ is isomorphic to $\oplus\{C_\sigma : \sigma \in T_{cr}(C), \sigma \neq \tau\}$. By induction on n, G' is completely decomposable, as is $G = G(\tau) \oplus G'$.

(b) (ii) $\Rightarrow$ (iii) Assume that C and D are quasi-isomorphic and let τ be the type of some C_i or D_j. By Example 3.1.5(a), $C(\tau) = C_\tau \oplus C^*(\tau)$ and $D(\tau) = D_\tau \oplus D^*(\tau)$ with C_τ and D_τ τ-homogeneous completely decomposable groups. Then $C(\tau)$ is quasi-isomorphic to $D(\tau)$, $C^*(\tau)$ is quasi-isomorphic to $D^*(\tau)$, and so C_τ is quasi-isomorphic to D_τ. In particular, rank $C_\tau =$ rank D_τ for each $\tau \in$ critical typeset C. Thus, C_τ and D_τ are isomorphic, both being isomorphic to X^n for some rank-1 group X of type τ. It follows that $C = \oplus\{C_\tau : \tau \in T_{cr}(C)\}$ is isomorphic to $D = \oplus\{D_\tau : \tau \in T_{cr}(D)\}$. The implications (iii) $\Rightarrow$ (iv) $\Rightarrow$ (i) $\Rightarrow$ (ii) follow immediately from the definitions and Example 3.1.5(a).

Recall from Chapter 2.2 that if G is a torsion-free abelian group of finite rank, then $N\mathrm{End}\,G = (J\mathbb{Q}\mathrm{End}\,G) \cap \mathrm{End}\,G$ is a nilpotent ideal containing each nilpotent ideal of $\mathrm{End}\,G$.

Proposition 3.1.8 *If C is a completely decomposable group, then* $\mathrm{End}\,C/N\mathrm{End}\,C$ *is a finite product of matrix rings over subrings of $\mathbb{Q}$.*

PROOF. Let $C = X_1 \oplus \cdots \oplus X_n$ be a completely decomposable group with rank $X_i = 1$ for each i. By grouping together those X_i's that are isomorphic, write $C = A_1 \oplus \cdots \oplus A_m$, where each A_i is τ_i-homogeneous completely decomposable and the τ_i's are distinct types. Observe that $\mathrm{Hom}(A_i, A_j) \neq 0$ if and only $\tau_i \leq \tau_j$. If $\mathrm{Hom}(A_i, A_j) \neq 0$ and $\mathrm{Hom}(A_j, A_i) \neq 0$, then $i = j$, since the τ_i's are assumed to be distinct.

Since $\mathrm{End}\,C = \oplus\{\mathrm{Hom}(A_i, A_j) : 1 \leq i, j \leq m\}$, it is sufficient to prove that if $I = \oplus\{\mathrm{Hom}(A_i, A_j) : 1 \leq i \neq j \leq m\}$, then I is a nilpotent ideal of $\mathrm{End}\,C$. In this case, $I \subseteq N\mathrm{End}\,C$ and $(\mathrm{End}\,C)/I = \mathrm{End}\,A_1 \times \cdots \times \mathrm{End}\,A_n$ is isomorphic to $\mathrm{Mat}_{n(1)}(\mathrm{End}\,X_1) \times \cdots \times \mathrm{Mat}_{n(m)}(\mathrm{End}\,X_m)$ with each $\mathrm{End}\,X_i$ a subring of $\mathbb{Q}$ and rank $A_i = n(i)$ for each i. Consequently, $N((\mathrm{End}\,C)/I) = 0$ and $N\mathrm{End}\,C \subseteq I$, since $N\mathrm{End}\,C$ is a nilpotent ideal of $\mathrm{End}\,C$. Then $N\mathrm{End}\,C = I$ and $\mathrm{End}\,C/N\mathrm{End}\,C$ is a finite product of matrix rings over subrings of $\mathbb{Q}$.

To see that I is an ideal, let $0 \neq f \in \mathrm{Hom}(A_i, A_j)$ and $0 \neq g \in \mathrm{Hom}(A_r, A_s) \subseteq I$ with $r \neq s$. Then $\tau_r < \tau_s$. If $0 \neq fg \in \mathrm{Hom}(A_r, A_j)$, then $\tau_r < \tau_s \leq \tau_i \leq \tau_j$, whence $r \neq j$. Thus, $fg \in I$, and I is a left ideal of $\mathrm{End}\,C$. A similar argument shows that I is also a right ideal of $\mathrm{End}\,C$.

If n is a positive integer and $x \in I^n$, then x is a finite sum of a composition of homomorphisms $f_{i,j} \in \mathrm{Hom}(A_i, A_j) \subseteq I$ with $i \neq j$. Since m, the number of A_i's, is finite, there is a sufficiently large n such that any such composition has some subscript i repeated. Each such composition is 0; otherwise, there is some $j \neq i$ with both $\mathrm{Hom}(A_i, A_j)$ and $\mathrm{Hom}(A_j, A_i)$ nonzero, a contradiction. Consequently, $I^n = 0$.

The typeset of a torsion-free abelian group G records the isomorphism classes of pure rank-1 subgroups of G. There is also a set of types that records the isomorphism classes of rank-1 torsion-free homomorphic images of G. Up to isomorphism, a rank-1 torsion-free homomorphic image of G can be written as $f(G)$ for some $0 \neq f : G \to \mathbb{Q}$.

For a finite rank torsion-free abelian group G, define cotypeset G to be $\{\text{type } f(G) : 0 \neq f : G \to \mathbb{Q}\}$. Equivalently, cotypeset G is the set of all types of rank-1 homomorphic images of G. For a type τ, define

$$G[\tau] = \cap\{\text{kernel} f : f \in \mathrm{Hom}(G, \mathbb{Q}), \text{ type } f(G) \leq \tau\},$$

called the τ-radical of G and let

$$G^*[\tau] = \cap\{\text{kernel} f : f \in \mathrm{Hom}(G, \mathbb{Q}), \text{ type } f(G) < \tau\}.$$

Notice that $G^*[\tau] = \cap\{G[\sigma] : \sigma < \tau\}$, $G[\tau]$ is a subgroup of $G^*[\tau]$, and if $0 \neq x \in G$, then there is a pure subgroup K of G with rank $G/K = 1$ and $x + K \neq 0$. Hence, $G[\tau] = G$ if and only if $\sigma \not\leq \tau$ for each $\sigma \in$ cotypeset G, and $G[\tau] = 0$ if and only if $\tau \geq \sigma$ for each $\sigma \in$ cotypeset G.

For example, if X is a rank-1 group and τ is a type, then $X[\tau] = X$ if type $X \not\leq \tau$ and 0 if type $X \leq \tau$. Moreover, $X^*[\tau] = X$ if type $X \not< \tau$ and 0 if type $X < \tau$. In particular, if $\tau = $ type X, then $X[\tau] = 0$ and $X^*[\tau] = X$.

The next proposition shows that properties of τ-socles have analogues for τ-radicals.

Lemma 3.1.9 *Suppose G and H are torsion-free abelian groups of finite rank.*

(a) *If $\sigma \leq \tau$, then $G[\tau] \subseteq G[\sigma]$. Moreover, $(G/G[\tau])[\tau] = 0$, and if $H[\tau] = 0$, then each nonzero element of H has type less than or equal to τ.*

(b) *If $g \in \mathrm{Hom}(G, H)$ and τ is a type, then $g(G[\tau]) \subseteq H[\tau]$, and $g(G^*[\tau]) \subseteq H^*[\tau]$.*

(c) *If H is a pure subgroup of $G[\tau]$, then $(G/H)[\tau] = G[\tau]/H$.*

(d) *For each type τ, $(G \oplus H)[\tau] = G[\tau] \oplus H[\tau]$, and $(G \oplus H)^*[\tau] = G^*[\tau] \oplus H^*[\tau]$.*

(e) *If cotypeset G is finite, then it is closed under joins.*

PROOF. (a) Assume that $\sigma \leq \tau$. Then $G[\tau] = \cap\{\ker h : h : G \to \mathbb{Q}, \text{ type } h(G) \leq \tau\} \subseteq G[\sigma] = \cap\{\ker h : h : G \to \mathbb{Q}, \text{ type } h(G) \leq \sigma\}$, since if $h : G \to \mathbb{Q}$ with type $h(G) \leq \sigma$, then type $h(G) \leq \tau$.

Next, let $g : G \to G/G[\tau]$ be the canonical epimorphism defined by $g(a) = a + G[\tau]$. If $x + G[\tau] \in (G/G[\tau])[\tau] = \cap(\ker h : h : G/G[\tau] \to \mathbb{Q}, \text{type } h(G) \leq \tau\}$, then $x \in \cap(\ker hg : hg : G \to \mathbb{Q}, \text{type } h(G) \leq \tau\}$. On the other hand, if $f : G \to \mathbb{Q}$ with type $f(G) \leq \tau$, then $f(G[\tau]) = 0$ and $f = hg$ for some $h : G/G[\tau] \to \mathbb{Q}$ with type $h(G) \leq \tau$. Thus, $x \in G[\tau] = \cap\{\ker f : f : G \to \mathbb{Q}, \text{type } f(G) \leq \tau\}$ and so $(G/G[\tau])[\tau] = 0$.

Finally, let X be a pure rank-1 subgroup of H. If type $X \not\leq \tau$, then $f(X) = 0$ for each $f \in \text{Hom}(H, \mathbb{Q})$ with type $f(H) \leq \tau$. Hence, $X \subseteq H[\tau] = 0$, a contradiction.

(b) If $h : H \to \mathbb{Q}$ with type $h(H) \leq \tau$, then $hg : G \to \mathbb{Q}$ with type $hg(G) \leq$ type $h(H) \leq \tau$. Let $x \in G[\tau]$. Then $hg(x) = 0$ and $g(x) \in \ker h$ for each $h : H \to \mathbb{Q}$ with type $h(H) \leq \tau$, i.e., $g(x) \in H[\tau]$. A similar argument, using those $h : H \to \mathbb{Q}$ with type $h(H) < \tau$, shows that $g(G^*[\tau]) \subseteq H^*[\tau]$.

(c) Let K be the subgroup of G with $K/H = (G/H)[\tau]$. As a consequence of (b), $G[\tau]/H$ is contained in $(G/H)[\tau]$, and so $G[\tau]$ is contained in K. Conversely, if $h : G \to \mathbb{Q}$ with type $h(G) \leq \tau$, then $h(H) = 0$, since $H \subseteq G[\tau]$. Thus, $h : G/H \to \mathbb{Q}$ with type $h(G/H) \leq \tau$, and $h(K) = h(K/H) = h((G/H)[\tau]) = 0$. Therefore, $K \subseteq \cap\{\ker h : h : G \to \mathbb{Q}, \text{type } h(G) \leq \tau\} = G[\tau]$.

(d) This follows from (b) using projections and injections of $G \oplus H$.

(e) Let $0 \neq f, g \in \text{Hom}(G, \mathbb{Q})$ with type $f(G) \neq$ type $g(G)$. For each integer n, $h_n = f + ng : G \to \mathbb{Q}$ is a nonzero homomorphism with $h_n(G) \subseteq f(G) + g(G)$. In particular, type $h_n(G) \leq$ type $f(G) \cup$ type $g(G)$ for each n. Since cotypeset G is finite, there are integers $n \neq m$ with type $h_n(G) =$ type $h_m(G)$. Now, $h_n - h_m = (n - m)g$, and $-mh_n + nh_m = (n - m)f$ with $n - m \neq 0$. Hence, $(n - m)(f(G) + g(G)) \subseteq h_n(G) + h_m(G)$, and so type $f(G) \cup$ type $g(G) \leq$ type $h_n(G) \cup$ type $h_m(G) =$ type $h_n(G)$. Consequently, type $f(G) \cup$ type $g(G) =$ type $h_n(G) \in$ cotypeset G, as desired.

The *critical cotypeset* of a torsion-free abelian group G is $CT_{cr}(G) = \{\tau : G^*[\tau]/G[\tau] \neq 0\}$. The critical cotypeset is, in fact, a subset of the cotypeset of G, since if $G^*[\tau]/G[\tau] \neq 0$, there must be $h : G \to \mathbb{Q}$ with type $h(G) = \tau$. Just as for typesets, the critical cotypeset records the types of rank-1 summands of a completely decomposable group.

Example 3.1.10

(a) *If $C = X_1 \oplus \cdots \oplus X_n$ is a completely decomposable group with* rank $X_i = 1$ *for each i and τ is a type, then*

$$C[\tau] = \oplus\{X_i : \text{type } X_i \not\leq \tau\}, \quad C^*[\tau] = \oplus\{X_i : \text{type } X_i \not< \tau\},$$

$C/C[\tau]$ is isomorphic to $\oplus\{X_i : \text{type } X_i \leq \tau\}$, and $C^*[\tau]/C[\tau]$ is isomorphic to $\oplus\{X_i : \text{type } X_i = \tau\}$. Moreover, cotypeset C *is the join closure of* $\{\text{type } X_i : 1 \leq i \leq n\}$, *a finite set.*

(b) *If C and D are completely decomposable groups, then C and D are isomorphic if and only if* rank $C^*[\tau]/C[\tau] =$ rank $D^*[\tau]/D[\tau]$ *for each τ.*

PROOF. (a) The first statement follows from Lemma 3.1.9(d) and the computations of $X[\tau]$ and $X^*[\tau]$ for a rank-1 group X. As for the latter statement, let $\tau \in$ cotypeset C and $0 \neq f : C \to \mathbb{Q}$ with type $f(C) = \tau$. Then $f(C) = \Sigma\{f(X_i): f(X_i) \neq 0, \text{type } f(X_i) \leq \tau\}$, whence τ is the join of $\{\text{type } X_i : f(X_i) \neq 0\}$. Conversely, cotypeset C is closed under finite joins by Lemma 3.1.9(e).

Assertion (b) is a consequence of (a) and Lemma 3.1.9(b).

A *finite Boolean algebra* T is a finite complemented distributive lattice $(T, \leq)$ with a least element 0 and a greatest element 1. An *atom* of T is an element t of T with $0 < t$ such that if $s \in T$ and $0 < s \leq t$, then $s = t$. A finite Boolean algebra T is lattice isomorphic to the Boolean algebra $P(S)$ of all subsets of a set S with n elements. In $P(S)$, 0 is the empty set, 1 is the set S, and the atoms are precisely the 1-element subsets of S.

For an example of a Boolean algebra of types, let $S_3 = \{\text{type } \mathbb{Z}[1/p], \text{type } \mathbb{Z}[1/q], \text{type } \mathbb{Z}[1/r]\}$, where p, q, and r are distinct primes and $\mathbb{Z}[1/p] = \{m/p^j : m \in \mathbb{Z}, j \geq 0\}$, the subring of $\mathbb{Q}$ generated by $1/p$. The sublattice T_3 of the lattice of types generated by S_3 is a Boolean algebra with S_3 as the set of atoms.

Example 3.1.11 *Let T_3 be a finite Boolean algebra of types with exactly 3 atoms τ_1, τ_2, τ_3, and X_i a subgroup of $\mathbb{Q}$ with type $X_i = \tau_i$ for $i = 1, 2, 3$. Define $G = X_1 \oplus X_2 + X_3(1+1) \subseteq \mathbb{Q} \oplus \mathbb{Q}$, a torsion-free abelian group of rank 2. Then:*

(i) *typeset $G = \{\tau_0, \tau_1, \tau_2, \tau_3\}$, where τ_0 is the least element of T_3.*
(ii) *cotypeset $G = \{\tau_1 \cup \tau_2, \tau_2 \cup \tau_3, \tau_1 \cup \tau_3, \tau_1 \cup \tau_2 \cup \tau_3\}$.*
(iii) *$G(\tau_0) = G, G(\tau_i) = X_i$ for $i = 1, 2$, and $G(\tau_3) = X_3(1 + 1)$.*
(iv) *$G^*[\tau_1 \cup \tau_2 \cup \tau_3] = G[\tau_1 \cup \tau_2 \cup \tau_3] = 0, G[\tau_i \cup \tau_j] = X_k$ with $\{i, j, k\} = \{1, 2, 3\}$, and $G^*[\tau_i \cup \tau_j] = G$ whenever $i \neq j$.*

PROOF. Exercise 3.1.2.

As noted in Exercise 3.1.3, the lattice of types is not a complete lattice. Specifically, infinite meets and infinite joins need not exist. Nevertheless, there are types that bound all of the elements of the typeset and, respectively, the cotypeset of a finite rank torsion-free abelian group G even if the typeset of G, respectively the cotypeset of G, is infinite.

Let $S = \{x_1, \ldots, x_n\}$ be a maximal $\mathbb{Z}$-independent subset of G. Define the *inner type* of G to be $\text{IT}(G) = \cap\{\text{type } x_i : 1 \leq i \leq n\}$. The *outer type* of G is $\text{OT}(G) = \cup\{\text{type } G/Y_i : 1 \leq i \leq n\}$, where Y_i is the pure subgroup of G generated by $\{x_j : 1 \leq i \neq j \leq n\}$, noticing that rank $Y_i = n - 1$ and rank $G/Y_i = 1$ for each i.

Lemma 3.1.12 [Warfield 68] *Let G be a torsion-free abelian group of finite rank. Then*

(a) $\text{IT}(G)$ *and* $\text{OT}(G)$ *do not depend on the maximal $\mathbb{Z}$-independent set S;*

(b) $\mathrm{IT}(G) \leq \tau$ *for each* $\tau \in$ typeset G *and* $\mathrm{OT}(G) \geq \sigma$ *for each* $\sigma \in$ cotypeset G;

(c) *if* typeset G *is finite, then* $\mathrm{IT}(G) \in$ typeset G, *and if* cotypeset G *is finite, then* $\mathrm{OT}(G) \in$ cotypeset G; *and*

(d) $\mathrm{IT}(G) \leq \mathrm{OT}(G)$.

PROOF. (a) Let $S = \{x_1, \ldots, x_n\}$ and $S' = \{x_1', \ldots, x_n'\}$ be two maximal $\mathbb{Z}$-independent subsets of G. For each i, there is a nonzero integer m with $mx_i' = k_1x_1 + \cdots + k_nx_n$ for some $k_i \in \mathbb{Z}$. Then type $x_i' = $ type $mx_i' \geq \cap\{$type $k_ix_i : k_i \neq 0\} \geq \cap\{$type $x_i; 1 \leq i \leq n\}$. Hence, $\cap\{$type $x_i'; 1 \leq i \leq n\} \geq \cap\{$type $x_i : 1 \leq i \leq n\}$. The reverse inclusion is proved by a similar argument. The fact that $\mathrm{OT}(G)$ does not depend on S follows from the proof of (b).

(b) If $\tau \in$ typeset G, then $\tau = $ type x for some $0 \neq x \in G$. Extend x to a maximal independent subset S, so that type $x \geq \cap\{$type $s : s \in S\} = \mathrm{IT}(G)$.

Next, let X be a rank-1 group with $\mathrm{OT}(G) = \cup\{$type $G/Yi : 1 \leq i \leq n\} = $ type X. There is a monomorphism $h : G \to G/Y_1 \oplus \cdots \oplus G/Y_n$ defined by $h(x) = (x + Y_1, \ldots, x + Y_n)$. Since type $G/Y_i \leq$ type X, there is a monomorphism $g_i : G/Y_i \to X$ for each i. The composite $g = (g_1, \ldots, g_n)h : G \to X^n$ is a monomorphism with rank $G = n = $ rank X^n.

Let $\sigma \in$ cotypeset G and $f : G \to \mathbb{Q}$ with type $f(G) = \sigma$. Define C to be the pure subgroup of X^n generated by $g(\ker f)$. Then rank $C = n - 1$, and so X^n/C is a rank-1 group isomorphic to X by Lemma 3.1.6(b) and (c). Thus, g induces a nonzero homomorphism $G/\ker f \to X^n/C \to X$. By Lemma 3.1.1(b), $\sigma = $ type $f(G) \leq$ type $X = \mathrm{OT}(G)$, as desired.

Assertion (c) follows from (b) and Lemmas 3.1.3 and 3.1.9(e). As for (d), let $0 \neq x_1 \in G$, $S = \{x_1, x_2, \ldots, x_n\}$ a maximal $\mathbb{Z}$-independent subset of G, and Y_i the pure subgroup of G generated by $\{x_j : 1 \leq i \neq j \leq n\}$. Then type $x_i \leq$ type $G/Y_i \leq \mathrm{OT}(G)$, since $0 \neq x_i + Y_i \in G/Y_i$. It follows that $\mathrm{IT}(G) \leq \mathrm{OT}(G)$.

There are relationships between inner types and τ-socles and between outer types and τ-radicals.

Proposition 3.1.13 [Lady 79] *If G is a torsion-free abelian group of finite rank, then:*

(a) $G(\tau) = G$ *if and only if* $\mathrm{IT}(G) \geq \tau$;

(b) $G[\tau] = 0$ *if and only if* $\mathrm{OT}(G) \leq \tau$; *and*

(c) *the following statements are equivalent:*

    (i) $\tau = \mathrm{IT}(G) = \mathrm{OT}(G)$;

    (ii) G *is* τ-*homogeneous completely decomposable;*

    (iii) $G(\tau) = G$ *and* $G[\tau] = 0$.

PROOF. Statements (a) and (b) are consequences of the definitions.

(c) (i) $\Rightarrow$ (ii) Assume that $\mathrm{IT}(G) = \mathrm{OT}(G) = \tau$. Let $0 \neq f \in \mathrm{Hom}(G, \mathbb{Q})$. Then $G(\tau) = G$, since $\mathrm{IT}(G) = \tau$, and type $f(G) = \tau$, since $\mathrm{OT}(G) = \tau = \mathrm{IT}(G)$. By Lemma 3.1.6(a), $G = K \oplus Y$ with $K = \ker f$ and Y isomorphic to

$f(G)$. Then $\mathrm{IT}(K) = \tau = \mathrm{OT}(K)$ so, by induction on the rank of G, K is τ-homogeneous completely decomposable. Since Y is a rank-1 group with type τ, G is τ-homogeneous completely decomposable.

(ii) $\Rightarrow$ (iii) If G is τ-homogeneous completely decomposable, then $G(\tau) = G$ by Example 3.1.5(a) and $G[\tau] = 0$ by Example 3.1.10(a).

(iii) $\Rightarrow$ (i) By (a) and (b), $\mathrm{IT}(G) \geq \mathrm{OT}(G)$. Since $\mathrm{IT}(G) \leq \mathrm{OT}(G)$ by Lemma 3.1.12(d), the conclusion is that $\mathrm{IT}(G) = \mathrm{OT}(G)$.

Proposition 3.1.14 [Warfield 68] *Suppose G and H are torsion-free abelian groups of finite rank. Then* rank $\mathrm{Hom}(G, H) = $ (rank G)(rank H) *if and only if* $\mathrm{IT}(H) \geq \mathrm{OT}(G)$.

PROOF. Observe that $\mathrm{Hom}(G, H)$ is contained in $\mathrm{Hom}(\mathbb{Q}G, \mathbb{Q}H)$, a $\mathbb{Q}$-vector space with dimension equal to the product of the dimension of $\mathbb{Q}G$ and the dimension of $\mathbb{Q}H$. Then $\mathrm{Hom}(G, H)$ is a torsion-free abelian group with rank $\leq$ (rank G) (rank H).

Assume $\mathrm{IT}(H) \geq \mathrm{OT}(G)$. Let $S = \{x_1, \ldots, x_n\}$ be a maximal $\mathbb{Z}$-independent subset of H with $\mathrm{IT}(H) = \cap\{\text{type } X_i : 1 \leq i \leq n\}$, X_i the pure subgroup of H generated by x_i, and $T = \{y_1, \ldots, y_m\}$ a maximal $\mathbb{Z}$-independent subset of G with $\mathrm{OT}(G) = \cup\{\text{type } G/Y_j : 1 \leq j \leq n\}$, where Y_j is the pure subgroup of G generated by $\{y_i : 1 \leq i \neq j \leq n\}$. Then type $X_i \geq \mathrm{IT}(H) \geq \mathrm{OT}(G) \geq$ type G/Y_j. Hence, there are nonzero homomorphisms $f_{ji} : G \to G/Y_j \to X_i$. Now, $X_1 \oplus \cdots \oplus X_n \subseteq H$, so that $\{f_{ji} : 1 \leq j \leq m, 1 \leq i \leq n\}$ is a $\mathbb{Z}$-linearly independent subset of $\mathrm{Hom}(G, H)$. It now follows that rank $\mathrm{Hom}(G, H) = mn = $ (rank G)(rank H).

Conversely, assume rank $\mathrm{Hom}(G, H) = $ (rank G)(rank H). Then $\mathbb{Q}\mathrm{Hom}(G, H) = \mathrm{Hom}(\mathbb{Q}G, \mathbb{Q}H)$, since $\mathbb{Q}\mathrm{Hom}(G, H)$ is a subspace of $\mathrm{Hom}(\mathbb{Q}G, \mathbb{Q}H)$ with the same dimension. It suffices to prove that if G/K is a rank-1 quotient of G and C is a pure rank-1 subgroup of H, then $\mathrm{Hom}(G/K, C) \neq 0$. The composite $\mathbb{Q}G \to \mathbb{Q}(G/K) \to \mathbb{Q}C \subseteq \mathbb{Q}H$ is an element $f \in \mathrm{Hom}(\mathbb{Q}G, \mathbb{Q}H) = \mathbb{Q}\mathrm{Hom}(G, H)$ with $\mathbb{Q}K = \ker f$. Choose $0 \neq k \in \mathbb{Z}$ with $kf \in \mathrm{Hom}(G, H)$. Then $kf(G) \subseteq \mathbb{Q}C \cap H = C$, since C is a pure subgroup of H, and so kf induces a nonzero homomorphism $G/K \to C$, as desired.

Conditions on the τ-radical and τ-socle influence the structure of subgroups and torsion-free quotients of a finite rank torsion-free abelian group G. Notice that (c) of the next proposition is a partial "dual" to Lemma 3.1.6(a).

Proposition 3.1.15 [Lady 79] *Let G be a torsion-free abelian group of finite rank, τ a type, and K a pure subgroup of G.*

(a) *If $(G/K)(\tau) = G/K$ and $G[\tau] = 0$, then G/K is a τ-homogeneous completely decomposable group.*

(b) *If $K(\tau) = K$ and $G[\tau] = 0$, then K is a τ-homogeneous completely decomposable summand of G.*

(c) *If H is a τ-homogeneous completely decomposable subgroup of G with $H \oplus G[\tau]$ a pure subgroup of G, then $G = H \oplus L$ for some subgroup L of G with $G[\tau] \subseteq L$.*

PROOF. (a) By Proposition 3.1.13(a) and (b), $IT(G/K) \geq \tau \geq OT(G)$. Now, $OT(G) \geq OT(G/K) \geq IT(G/K)$, observing that a rank-1 quotient of G/K is also a rank-1 quotient of G. Then $OT(G/K) = \tau = IT(G/K)$. By Proposition 3.1.13(c), G/K is τ-homogeneous completely decomposable.

(b) Let p be a prime. By Proposition 3.1.13(a) and (b), $IT(K) \geq \tau \geq OT(G)$. Either K_p is divisible, or else G_p is a free $\mathbb{Z}_p$-module. To see this, let G/K be a torsion-free rank-1 quotient of G and X a pure rank-1 subgroup of K. Since type $G/K \leq OT(G) \leq IT(K) \leq$ type X, there is a nonzero $f : G/K \rightarrow X$, hence a nonzero homomorphism $f_p : (G/K)p \rightarrow X_p$. If $(G/K)_p$ is not a free $\mathbb{Z}_p$-module, then $(G/K)_p$ is isomorphic to $\mathbb{Q}$. This is because a rank-1 p-local group is isomorphic to either $\mathbb{Z}_p$ or $\mathbb{Q}$ by Exercise 2.1.3. Hence, X_p is isomorphic to $\mathbb{Q}$, and so either G_p is a free $\mathbb{Z}_p$-module or K_p is divisible.

Now, $\text{Hom}(G, K)_p$ is a p-pure subgroup of $\text{Hom}(G_p, K_p)$, and as a consequence of Proposition 3.1.14 and the assumption that $OT(G) \leq IT(K)$, rank $\text{Hom}(G, K) = \text{rank Hom}(G, K)_p = (\text{rank } G)(\text{rank } K) = (\text{rank } G_p)(\text{rank } K_p)$. Thus, $\text{Hom}(G, K)_p = \text{Hom}(G_p, K_p)$. Since K_p is divisible or G_p is a free $\mathbb{Z}_p$-module, the restriction homomorphism $\phi_p : \text{Hom}(G_p, K_p) \rightarrow \text{Hom}(K_p, K_p)$ is onto. But $\text{Hom}(G, K) = \cap\{\text{Hom}(G, K)_p : p \in \Pi\}$ and $\text{Hom}(K, K) = \cap\{\text{Hom}(K, K)_p : p \in \Pi\}$, by Exercise 3.1.9, so that the restriction homomorphism $\phi : \text{Hom}(G, K) \rightarrow \text{Hom}(K, K)$, is onto. Choosing $g \in \text{Hom}(G, K)$ with $\phi(g) = 1_K$ shows that K is a summand of G.

Finally, write $G = K \oplus H$. Then $0 = G[\tau] = K[\tau] \oplus H[\tau]$ by Lemma 3.1.9(d), whence $K[\tau] = 0$. Since $K(\tau) = K$, K is τ-homogeneous completely decomposable by Proposition 3.1.13(c).

(c) In this case, $K = (H \oplus G[\tau])/G[\tau]$ is a pure τ-homogeneous completely decomposable subgroup of $G/G[\tau]$ with $K(\tau) = K$ and $(G/G[\tau])[\tau] = 0$. By (b), K is a summand of $G/G[\tau]$, say $f : G/G[\tau] \rightarrow K$ is a projection. Define $g = \beta f \alpha : G \rightarrow H$, where $\alpha : G \rightarrow G/G[\tau]$ is the quotient map and $\beta : K \rightarrow H$ is the natural isomorphism. Then $G[\tau]$ is contained in ker g and $g(x) = x$ for each $x \in H$. Therefore, $G = H \oplus \ker g$ with $G[\tau] \subseteq \ker g$, as desired.

EXERCISES

1. [Fuchs 73] or [Arnold 82] Provide a proof for Lemma 3.1.1.

2. Verify Example 3.1.11.

3. (a) Show that the lattice of types is not a complete lattice.
   (b) Prove that there are infinite chains in the lattice of all types with no least element and no greatest element.
   (c) Prove that the lattice of types is uncountable.

4. Let T be a finite Boolean algebra of types with n atoms. Prove that if $\{p_1, \ldots, p_n\}$ is a set of n distinct primes, then T is lattice isomorphic to the lattice of types generated by $\{\text{type } \mathbb{Z}[1/p_1], \ldots, \text{type } \mathbb{Z}[1/p_n]\}$.

5. Let G be a torsion-free abelian group of finite rank and τ a type. Prove that if H is a pure subgroup of G with $H(\tau) = H$ and $H \oplus G[\tau]$ is a pure subgroup of G, then H is a summand of G and G/H is homogeneous completely decomposable. Give examples to show that the hypothesis that $H \oplus G[\tau]$ is a pure subgroup of G is necessary for this exercise and Proposition 3.1.15(c).

6. (a) Give an example of a rank-2 torsion-free abelian group G with infinite typeset.
   (b) Give an example of a rank-2 torsion-free abelian group G such that $\text{IT}(G) \notin$ typeset G.
   (c) Give an example of a rank-2 torsion-free abelian group G such that $\text{OT}(G) \notin$ cotypeset G.

7. Let X be a subgroup of $\mathbb{Q}$, $0 \neq x \in X$, and $h(x)$ the height sequence of x in X.
   (a) Give necessary and sufficient conditions on $h(x)$ for X to be a subring of $\mathbb{Q}$.
   (b) Let Y be another subgroup of $\mathbb{Q}$ and $0 \neq y \in Y$. Show that $\text{Hom}(X, Y)$ is a torsion-free abelian group of rank 1 and compute a height sequence for $\text{Hom}(X, Y)$ from $h(x)$ and $h(y)$.

8. Let G be a torsion-free abelian group of finite rank and F a free subgroup of G with rank $F = \text{rank } G$. Define the *Richman type* of G, $\text{RT}(G)$, to be the quasi-isomorphism class of the torsion group G/F, where two torsion groups T and T' are quasi-isomorphic if there is a homomorphism $f : T \to T'$ with ker f and $T'/\text{image } f$ finite groups.
   (a) Prove that $\text{RT}(G)$ does not depend on the choice of F.
   (b) Compute the Richman type of a rank-1 group G in terms of a height sequence $h(x)$ for G.
   (c) Write $G/F = \oplus\{T_p : p \in \Pi\}$, $T_p = \mathbb{Z}(p^{i(p,1)}) \oplus \cdots \oplus \mathbb{Z}(p^{i(p,n)})$ a direct sum of p-groups with $0 \leq i(p, 1) \leq \cdots \leq i(p, n) \leq \infty$. Prove that $\text{IT}(G)$ is the equivalence class of the height sequence $(i(p, 1))$ and $\text{OT}(G)$ is the equivalence class of the height sequence $(i(p, n))$.

9. Prove that if G is a torsion-free abelian group of finite rank and Π is the set of primes of $\mathbb{Z}$, then $G = \cap\{G_p : p \in \Pi\}$. Conclude that if H is another torsion-free abelian group of finite rank, then $\text{Hom}(G, H) = \cap\{\text{Hom}(G, H)_p : p \in \Pi\}$.

3.2   Characterizations of Finite Rank Groups

The class of finite rank Butler groups arises naturally. For instance, it is the smallest class of torsion-free abelian groups that contains all rank-1 groups and is closed under isomorphism, finite direct sums, pure subgroups, and torsion-free homomorphic images. Furthermore, the class of Butler groups is precisely the class of finite rank torsion-free abelian groups that arise as representatives of $\mathbb{Q}$-representations of finite posets, as demonstrated in Section 3.3.

The first example illustrates that pure subgroups or homomorphic images of completely decomposable groups need not be completely decomposable.

Example 3.2.1 *Let p_1, p_2, and p_3 be three distinct primes, X_i the localization of $\mathbb{Z}$ at p_i, and B the pure subgroup of the completely decomposable group $X_1 \oplus X_2 \oplus X_3$ generated by $(1, 1, 0)$ and $(0, 1, 1)$. Then B is a strongly indecomposable group of rank 2 and a homomorphic image of the completely decomposable group $(X_1 \cap X_2) \oplus (X_2 \cap X_3) \oplus (X_1 \cap X_3)$.*

PROOF. Clearly, rank $B = 2$. Now, type $(1, 1, 0) = $ type $X_1 \cap X_2$, type $(0, 1, 1) = $ type $X_2 \cap X_3$, and type $(1, 0, -1) = $ type $(X_1 \cap X_3)$. Since $(1, 0, -1) = (1, 1, 0) - (0, 1, 1) \in B$, the typeset of B contains three pairwise incomparable elements. Consequently, B is strongly indecomposable; otherwise, B is quasi-isomorphic to a completely decomposable group of rank 2, which can have at most two pairwise incomparable types by Example 3.1.5. A routine calculation shows that $B = \langle(1, 1, 0)\rangle_* + \langle(0, 1, 1)\rangle_* + \langle(1, 0, -1)\rangle_*$, whence B is a homomorphic image of $(X_1 \cap X_2) \oplus (X_2 \cap X_3) \oplus (X_1 \cap X_3)$.

Theorem 3.2.2 [Butler 65] *Suppose G is a torsion-free abelian group of finite rank. Then G is a pure subgroup of a completely decomposable group if and only if G is a homomorphic image of a completely decomposable group.*

PROOF. A consequence of the more general Lemma 3.2.3.

A group epimorphism $f : G \rightarrow H$ of torsion-free abelian groups is *balanced* if $f(G(\tau)) = H(\tau)$ for each type τ. In this case, typeset H is contained in typeset G, and f induces an onto homomorphism $G(\tau)/G^{\#}(\tau) \rightarrow H(\tau)/H^{\#}(\tau)$ for each type τ. As an example, the homomorphism in Example 3.2.1, $(X_1 \cap X_2) \oplus (X_2 \cap X_3) \oplus (X_1 \cap X_3) \rightarrow B$, is balanced.

A monomorphism $f : G \rightarrow H$ is *cobalanced* if $f(G)$ is a pure subgroup of H and f induces a monomorphism $f_\tau : G/G[\tau] \rightarrow H/H[\tau]$ with pure image for each type τ. In this case, cotypeset G is contained in cotypeset H. Notice that if $f : G \rightarrow H$ is a cobalanced monomorphism, then $f_\tau(G^*[\tau]/G[\tau])$ is a pure subgroup of $H^*[\tau]/H[\tau]$ for each τ.

The proof of Lemma 3.2.3 is constructive in the sense that if G is a pure subgroup of $X_1 \oplus \cdots \oplus X_n$ with rank $X_i = 1$, then there are rank-1 groups Y_j, constructed from the X_i's, such that G is a homomorphic image of $Y_1 \oplus \cdots \oplus Y_m$.

Lemma 3.2.3 *Let G be a torsion-free abelian group of finite rank.*

(a) *If G is a pure subgroup of a completely decomposable group, then there is a completely decomposable group D and a balanced epimorphism $f : D \rightarrow G$.*
(b) *If G is the homomorphic image of a completely decomposable group, then there is a completely decomposable group D and a cobalanced monomorphism $f : G \rightarrow D$.*

PROOF. (a) Assume that G is a pure subgroup of $C = X_1 \oplus \cdots \oplus X_n$ with each X_i a rank-1 group. Then typeset G is finite, as a consequence of Example 3.1.5(b), and closed under meets by Lemma 3.1.3. Given $x = x_1 \oplus \cdots \oplus x_n$ in C, let support $x = \{i : x_i \neq 0\}$, support $G = \{\text{support } x : 0 \neq x \in G\}$ ordered by inclusion, and S the set of minimal elements in support G. Then S is a finite set, and for each I in S, there is a nonzero x_I in G with support $x_I = I$. Let X_I be the pure rank-1 subgroup of G generated by x_I and notice that support $y = I$ for each nonzero y in X_I.

Define $D = \oplus\{X_I : I \in S\}$ and $f : D \to G$ by $f(\oplus x_I) = \Sigma x_I$ for $x_I \in X_I$. It is sufficient to show that $f(D(\tau)) = G(\tau)$ for each type τ, in which case $f(D) = f(D(\tau)) = G(\tau) = G$ for $\tau = \text{IT}(G)$, the inner type of G.

If τ is a type, then $D(\tau) = \oplus(X_I : \text{type } X_I \geq \tau)$, by Example 3.1.5(a), and $H = f(D(\tau)) = \Sigma\{X_I : \text{type } X_I \geq \tau\} \subseteq G(\tau)$. The proof is concluded by showing that $G(\tau)_p$ is contained in H_p for each prime p. In this case, via Exercise 3.1.9, $G(\tau) = \cap\{G(\tau)_p : p \text{ prime}\} \subseteq H = \cap\{H_p : p \text{ prime}\}$, and so $H = G(\tau)$.

Let $x = q_1 x_1 + \cdots + q_n x_n$ be a nonzero element of $G(\tau)_p$ with q_i in $\mathbb{Z}_p$ and x_i in X_i and choose I in S with $I \subseteq$ support x. Next choose a nonzero $x_I = k_1 x_1 + \cdots + k_n x_n$ in X_I with each k_i in $\mathbb{Z}$, support $x_I = I$, and k_j prime to p for some j in I, recalling that X_I is a pure subgroup of G. Then type $x_I = \cap\{\text{type } X_i : i \in I\} \geq \cap\{\text{type } X_i : q_i \neq 0\} \geq \tau$, since $I \subseteq$ support x and $x \in G(\tau)_p$. This shows that $x_I \in G(\tau)_p$. Furthermore, $y = x - (q_j/k_j)x_I \in G(\tau)_p$ with support $y \subseteq$ support $x \backslash \{j\} \subset$ support x. By induction on the cardinality of support x, $y \in H_p$, observing that if support x has one element, then $x = x_I \in H$. Therefore, $x = (q_j/k_j)x_I + y \in H_p$ and $G(\tau)_p \subseteq H_p$ as desired.

(b) Assume that $\pi : C \to G$ is an epimorphism for $C = X_1 \oplus \cdots \oplus X_n$ a completely decomposable group with rank $X_i = 1$ for each i. Given a nonzero $f : G \to \mathbb{Q}$, define cosupport $f = \{i : f\pi(X_i) = 0\}$, cosupport $G = \{\text{cosupport } f : 0 \neq f \in \text{Hom}(G, \mathbb{Q})\}$ ordered by inclusion, and S the set of maximal elements of cosupport G. For each $I \in$ cosupport G, choose $f_I \in \text{Hom}(G, \mathbb{Q})$ with cosupport $f_I = I$.

Define $D = \oplus\{f_I(G) : I \in S\}$ and $\phi : G \to D$ by $\phi(x) = \oplus\{f_I(x) : I \in S\}$. If τ is a type, then $D/D[\tau] = \oplus\{f_I(G) : \text{type } f_I(G) \leq \tau\}$ by Example 3.1.10(a). For each type τ, ϕ induces a homomorphism $\phi_\tau : G/G[\tau] \to D/D[\tau]$ with $\phi_\tau(x + G[\tau]) = \phi(x) + D[\tau]$. It suffices to prove that ϕ_τ is a monomorphism with pure image, in which case ϕ is a cobalanced monomorphism. In particular, cotypeset G has a largest element $\tau = \text{OT}(G)$ by Lemma 3.1.9(e), since cotypeset $G \subseteq$ cotypeset C is finite by Example 3.1.10(a). By Proposition 3.1.13(b), $G[\tau] = 0$. Hence, $G/G[\tau] = G$, and $\phi = \phi_\tau$ is a monomorphism with pure image.

To see that ϕ_τ is a monomorphism, it is sufficient to prove that if $0 \neq x + G[\tau] \in G/G[\tau]$, then $\phi_\tau(x) \neq 0$. Since $x \notin G[\tau]$, there is some $f : G \to \mathbb{Q}$ with type $f(G) \leq \tau$ and $f(x) \neq 0$. Choose $I \in S$ with cosupport $f \subseteq I$. Then type $f_I(G) = \cup\{\text{type } X_i : i \notin I\} \leq \cup\{\text{type } X_i : i \notin \text{cosupport} f\} = \text{type } f(G) \leq \tau$. If $f_I(x) \neq 0$, then $\phi_\tau(x + D[\tau]) \neq 0$ by the definition of ϕ_τ, and the proof is complete. Now suppose that $f_I(x) = 0$. Since $f_I \neq 0$, there is $1 \leq i \leq n$ with $i \notin I$; hence $i \notin$ cosupport f. Choose $0 \neq q \in \mathbb{Q}$ such that $0 \neq h = qf_I - f : G \to \mathbb{Q}$ with $i \in$ cosupport h, i.e., $h\pi(X_i) = 0$. Then type $h(G) \leq (\text{type } qf_I(G))\cup(\text{type } f(G)) \leq \tau$ with cosupport $h \subset$ cosupport f, since $i \in$ cosupport $h \backslash$ cosupport f. Furthermore,

$h(x) \neq 0$, since $f(x) \neq 0$ and $f_I(x) = 0$. By induction on the cardinality of cosupport f, there is some $I' \in S$ with $f_{I'}(x) \neq 0$. Hence, $\phi_\tau(x) \neq 0$, as desired.

It remains to prove that image ϕ_τ is a pure subgroup of $D/D[\tau]$. Replace $G[\tau]$ with $G/G[\tau]$ and apply Lemma 3.1.9(a) to safely assume that $G[\tau] = 0$ and $\phi = \phi_\tau$. It suffices to prove that if p is a prime and $x \in G \backslash pG$, then $\phi(x) \notin pD$. There is $0 \neq f : G \to \mathbb{Q}$ with $f(x) \notin pf(G)$ and $f(G)_p = \mathbb{Z}_p f(x) = f(G)_p$. This holds because G_p is a finite direct sum of copies of $\mathbb{Z}_p$ and $\mathbb{Q}$, being an epimorphic image of a direct sum $C_p = (X_1)_p \oplus \cdots \oplus (X_n)_p$ of rank-1 p-local groups, and if $X = \langle x \rangle_*$, then $X_p = Z_p X$ is a pure submodule, hence a summand of, G_p.

As above, there is $I \in S$ with cosupport $f \subseteq I$ and type $f_I(G) \leq$ type $f(G) \leq \tau$. If $f_I(x) \notin pf_I(G)$, then $\phi(x) \notin pD$ by the definition of ϕ. Now assume that $f_I(x) \in pf_I(G)$. Observe that $f_I(x) \in pf_I(G)_p \subseteq f(G)_p = \mathbb{Z}_p f(x)$, and if $i \notin I$, then $f(\pi X_i)_p \neq 0$. Thus, there is some $i \notin I$ and integers a and b such that $0 \neq h = af_I - bf : G \to Q, h(\pi X_i) = 0$, the p-height of $bf(x)$ is 0 in $bf(G)$, and the p-height of $af_I(x)$ is not equal to 0 in $af_I(G)$. Hence, $h(x) \notin ph(G), h(G)_p = \mathbb{Z}_p h(x)$, and cosupport $h \subset$ cosupport f. By induction on the cardinality of the cosupport of f, $f_{I'}(x) \notin pf_{I'}(G)$ for some $I' \in S$. Thus, $\phi(x) \notin pD$, as desired.

A *finite rank Butler group* is a pure subgroup of a completely decomposable group, equivalently, a torsion-free homomorphic image of a completely decomposable group.

Corollary 3.2.4 [Butler 65] *The class of finite rank Butler groups is closed under finite direct sums, pure subgroups, and torsion-free homomorphic images. Moreover, if G is a torsion-free abelian group of finite rank and H is a finite rank Butler group quasi-isomorphic to G, then G is a Butler group.*

PROOF. Suppose G is quasi-isomorphic to H. Then there is a monomorphism $f : H \to G$ with $G/f(H)$ finite, say $G = H + \mathbb{Z}x_1 + \cdots + \mathbb{Z}x_n$. Since H is a Butler group, there is a completely decomposable group C and an epimorphism $g : C \to H$. There is also an epimorphism $h : \mathbb{Z}^n \to \mathbb{Z}x_1 + \cdots + \mathbb{Z}x_n$ defined by $h(a_1, \ldots, a_n) = a_1 x_1 + \cdots + a_n x_n$. Then $(fg, h) : C \oplus \mathbb{Z}^n \to G$ is an epimorphism and $C \oplus \mathbb{Z}^n$ is a completely decomposable group. The remainder of the corollary follows from Theorem 3.2.2.

The next theorem, a characterization of Butler groups among the class of torsion-free groups of finite rank, is one of the most useful tools in the subject. This is so because the finiteness conditions of this theorem are amenable to induction arguments.

Theorem 3.2.5 *Let G be a torsion-free abelian group of finite rank. The following statements are equivalent:*

(a) *The group G is a Butler group;*

(b) [Butler 65]

    (i) *The typeset of G is finite and*

    (ii) *for each type τ in typeset G, there is a decomposition $G(\tau) = G_\tau \oplus G^{\#}(\tau)$ with G_τ a τ-homogeneous completely decomposable group and $G^{\#}(\tau)/G^*(\tau)$ bounded;*

(c) (i) *The cotypeset of G is finite and*

    (ii) *for each type $\tau \in$ cotypeset G, $G^*[\tau]/G[\tau]$ is a τ-homogeneous completely decomposable summand of $G/G[\tau]$ and if H is the image of $G/G^*[\tau]$ in $\oplus \{G/G[\sigma] : \sigma < \tau\}$, then $\langle H \rangle_* / H$ is bounded.*

PROOF. (a) $\Rightarrow$ (b)

(i) If G is a pure subgroup of a completely decomposable group, then typeset G is finite by Example 3.1.5(b).

(ii) By Lemma 3.2.3, there exist $C = X_1 \oplus \cdots \oplus X_n$, with each X_i of rank 1, and a balanced epimorphism $f : C \to G$. As a consequence of Example 3.1.5(a), $C(\tau) = C_\tau \oplus C^*(\tau)$ with $C_\tau = \oplus \{X_i : \text{type } X_i = \tau\}$ a τ-homogeneous completely decomposable group. Then $D_\tau = f(C_\tau)$ is τ-homogeneous completely decomposable as a consequence of Lemma 3.1.6. Moreover, $D_\tau \cap G^{\#}(\tau)$ is a pure subgroup, hence a summand of D_τ.

Write $D_\tau = G_\tau \oplus (D_\tau \cap G^{\#}(\tau))$ with G_τ τ-homogeneous completely decomposable. Then $G(\tau) = f(C(\tau)) = D_\tau + f(C^*(\tau)) = D_\tau + G^*(\tau) = {}^*D_\tau + G^{\#}(\tau) = G_\tau \oplus G^{\#}(\tau)$. Furthermore, $G^{\#}(\tau)$ is a Butler group, being a pure subgroup of the Butler group G. Hence, $G^{\#}(\tau) = Y_1 + \cdots + Y_m + G^*(\tau)$, with each Y_i a pure rank-1 subgroup of G of type τ. Then $G^{\#}(\tau)/G^*(\tau)$ is torsion, since $G^{\#}(\tau) = \langle G^*(\tau) \rangle_*$. Now, $Y_i \cap G^*(\tau)$, being a pure subgroup of $G^*(\tau)$, is a rank-1 group of type $\tau = \text{type } Y_i$. Consequently, each $Y_i / Y_i \cap G^*(\tau)$ is bounded, from which it follows that $G^{\#}(\tau)/G^*(\tau)$ is bounded.

(b) $\Rightarrow$ (a) The proof is an induction on the length of a chain from an element of typeset G to a maximal element of typeset G. By (i), such a chain must have finite length. Let τ be a maximal type in typeset G. Then $G(\tau) = G_\tau$ is τ-homogeneous completely decomposable, in particular a Butler group. Let $\tau \in$ typeset G and suppose, by way of induction, that $G(\sigma)$ is a Butler group for each σ in typeset G with $\sigma > \tau$. Then $G^*(\tau)$ is a Butler group, being a homomorphic image of the Butler group $\oplus \{G(\sigma) : \sigma > \tau\}$. Moreover, $G(\tau)/(G_\tau \oplus G^*(\tau))$ is bounded by (ii). As a consequence of Corollary 3.2.4, $G(\tau)$ is a Butler group for each τ in typeset G. Let $\tau = \text{IT}(G) \in$ typeset G. Then $G(\tau) = G$ is a Butler group.

(c) $\Rightarrow$ (a) Let τ be a minimal element of cotypeset G. Then $G^*[\tau] = G$, and by (ii), $G/G[\tau] = G^*[\tau]/G[\tau]$ is a τ-homogeneous completely decomposable group, hence a Butler group. Now assume that $\tau \in$ cotypeset G and, by way of induction, that $G/G[\sigma]$ is a Butler group for each $\sigma \in$ cotypeset G with $\sigma < \tau$. By (ii), $G/G^*[\tau]$ is quasi-isomorphic to a pure subgroup of the Butler group $\oplus \{G/G[\sigma] : \tau > \sigma \in$ cotypeset $G\}$. Thus, $G/G^*[\tau]$ is a Butler group as a consequence of Corollary 3.2.4. Moreover, $G^*[\tau]/G[\tau]$ is τ-homogeneous completely decomposable and $G/G[\tau]$ is isomorphic to $G^*[\tau]/G[\tau] \oplus G/G^*[\tau]$ by (ii). This shows that $G/G[\tau]$ is a Butler group for each $\tau \in$ cotypeset G. Let $\tau = \text{OT}(G) \in$ cotypeset G. Then $G[\tau] = 0$, and $G = G/G[\tau]$ is a Butler group.

(a) $\Rightarrow$ (c)

(i) Since G is a Butler group, there is a completely decomposable group C and an epimorphism $f : C \to G$. Then cotypeset $G \subseteq$ cotypeset C, a finite set by Example 3.1.10(a).

(ii) In view of Lemma 3.2.3, there is a completely decomposable group C such that G is a pure cobalanced subgroup of C. It follows that if τ is a type, then $G^*[\tau]/G[\tau]$ is a pure subgroup of $C^*[\tau]/C[\tau]$. But $C^*[\tau]/C[\tau]$ is a τ-homogeneous completely decomposable group by Example 3.1.10(a), so that $G^*[\tau]/G[\tau]$ is a τ-homogeneous completely decomposable group by Lemma 3.1.6(b) and (c). Moreover, $G^*[\tau]/G[\tau]$ is a pure subgroup of $G/G[\tau]$, since $G^*[\tau]$ is a pure subgroup of G. Now $(G^*[\tau]/G[\tau])(\tau) = G^*[\tau]/G[\tau]$ and $(G/G[\tau])[\tau] = 0$ as a consequence of Lemma 3.1.9(a). Thus, $G^*[\tau]/G[\tau]$ is a τ-homogeneous completely decomposable summand of $G/G[\tau]$ by Proposition 3.1.15(b).

As for the final statement, let H be the image of the monomorphism $f_G : G/G^*[\tau] \to \oplus\{G/G[\sigma] : \sigma < \tau\}$ defined by $f_G(x + G^*[\tau]) = \oplus\{x + G[\sigma] : \sigma < \tau\}$, recalling that $G^*[\tau] = \cap\{G[\sigma] : \sigma < \tau\}$. Since G is a cobalanced subgroup of C, $G/G[\sigma]$ is a pure subgroup of $C/C[\sigma]$ for each type σ. Hence, there is a monomorphism $h : \oplus \{G/G[\sigma] : \sigma < \tau\} \to \oplus\{C/C[\sigma] : \sigma < \tau\}$, induced by $h(x + G[\sigma]) = x + C[\sigma]$, with pure image. Define $g : G/G^*[\tau] \to C/C^*[\tau]$ by $g(x + G^*[\tau]) = x + C^*[\tau]$. Then $hf_G = f_C g$, where $f_C : C/C^*[\tau] \to \oplus\{C/C[\sigma] : \sigma < \tau\}$ is a monomorphism. By Example 3.1.10(a), the image of f_C is a pure subgroup of $\oplus\{C/C[\sigma] : \sigma < \tau\}$.

It now suffices to show that if B is the image of g, then $\langle B \rangle_* / B$ is bounded. In this case, since $H = $ image f_G, $\langle H \rangle_* / H$ would also be bounded, as desired. There is a commutative diagram with exact rows and columns:

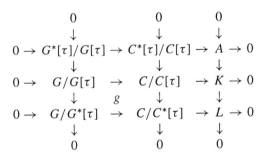

with A a τ-homogeneous completely decomposable group. Now, $K[\tau] = 0$, since $(C/C[\tau])[\tau] = 0$ by Lemma 3.1.9(a), and if A' is the pure subgroup of K generated by A, then $A'(\tau) = A'$, since $A(\tau) = A$. By Proposition 3.1.15(b), A' is τ-homogeneous completely decomposable. It follows that A'/A is bounded, whence the torsion subgroup of L is bounded, and since $B = $ image g, $\langle B \rangle^*/B$ must also be bounded, as desired.

Corollary 3.2.6 *Assume that G is a finite rank Butler group and T is the finite lattice of types generated by the typeset of G.*

(a) *The typeset of G is the meet closure of the critical typeset of G.*

(b) *The cotypeset of G is the join closure of the critical cotypeset of G.*
(c) *The lattice T is the join closure of the typeset of G and the meet closure of the cotypeset of G.*

PROOF. (a) The typeset of G is closed under finite meets by Lemma 3.1.3. Conversely, first assume that τ is a maximal type in typeset G. Then $\tau \in$ critical typeset G. Next assume that $\tau \in$ typeset G, $\tau \notin$ critical typeset G, and for each $t < \sigma \in$ typeset G there is a subset S_σ of the critical typeset of G with $\sigma = \cap\{\delta : \delta \in S_\sigma\}$. Since $\tau \notin$ critical typeset G, $G(\tau) = G^{\#}(\tau)$, whence $G(\tau)/G^{*}(\tau)$ is bounded. Consequently, there is $x \in G$ with type $x = \tau$ and $x = \Sigma\{x_\sigma : \sigma > \tau\}$ for some $x_\sigma \in G$ with $x_\sigma = 0$ or type $x_\sigma = \sigma$. Then $\tau = \text{type } x \geq \cap\{\text{type } x_\sigma : \sigma > \tau\} \geq \cap\{\sigma : \sigma > \tau\} \geq \tau$. By induction on the length of a chain from τ to a maximal type in typeset G, $\tau = \cap\{\delta : \delta \in S\}$ for $S = \cup\{S_\sigma : \sigma > \tau\}$, a subset of the critical typeset of G. Thus, typeset G is contained in the meet closure of the critical typeset of G.

(b) The proof is analogous to that of (a); just use induction on the length of a chain from an element of the cotypeset of G to a minimal element of the cotypeset of G.

(c) The join closure of the typeset of G and the meet closure of the cotypeset of G is contained in T. Conversely, the lattice of all types is a distributive lattice. It follows that elements of T are joins of meets of elements of the typeset of G. But the typeset of G is closed under meets by Lemma 3.1.3. Hence, T is the join closure of the typeset of G. Analogously, T is the meet closure of the cotypeset of G, since the cotypeset of G is closed under joins by Lemma 3.1.9(e).

As a consequence of the following corollary, homogeneous Butler groups are completely decomposable.

Corollary 3.2.7 *Assume that G is a finite rank Butler group.*

(a) *If the critical typeset of G is linearly ordered, then* typeset $G =$ critical typeset G *and G is completely decomposable.*
(b) *If the critical cotypeset of G is linearly ordered, then* cotypeset $G =$ critical cotypeset G *and G is completely decomposable.*
(c) *Homogeneous Butler groups are completely decomposable.*
(d) *If all rank-1 torsion-free quotients of G are isomorphic, then G is a homogeneous completely decomposable group.*

PROOF. (a) If the critical typeset is linearly ordered, then the critical typeset is equal to the typeset by Corollary 3.2.6(a). Hence, the typeset of G is linearly ordered. Write typeset $G = \{\tau_0 < \cdots < \tau_n\}$. Then $G = G(\tau_0) = G_0 \oplus G^{\#}(\tau_0)$ with G_0 τ_0-homogeneous completely decomposable by Theorem 3.2.5(b). Moreover, $G^{\#}(\tau_0) = G(\tau_1)$, and typeset $G(\tau_1) = \{\tau_1 < \cdots < \tau_n\}$. An induction on n shows that $G(\tau_1)$ is completely decomposable, noting that $G(\tau_n)$ is completely decomposable. Then G must be completely decomposable.

(b) The proof is analogous to that of (a) using Corollary 3.2.6(b) and Theorem 3.2.5(c).

Assertions (c) and (d) are consequences of (a) and (b), respectively.

Radicals and socles are intimately related for finite rank Butler groups.

Proposition 3.2.8 [Lady 79] *If G is a finite rank Butler group and τ and δ are types, then*

(a) $G[\tau] = \langle G(\sigma) : \sigma \not\leq \tau \rangle_*$ *and* $G^*[\tau] = \langle G(\sigma) : \sigma \not< \tau \rangle_*$;

(b) $G(\tau) = \cap \{G[\sigma] : \sigma \not\geq \tau\}$ *and* $G^\#(\tau) = \cap \{G[\sigma] : \sigma \not> \tau\}$;

(c) $G^\#(\tau) \subseteq G[\tau]$ *and* $G(\tau) \subseteq G^*[\tau]$; *and*

(d) $G[\tau \cap \delta] = \langle G[\tau] + G[\sigma] \rangle_*$.

PROOF. (a) Define $H = \langle G(\sigma) : \sigma \not\leq \tau \rangle_*$. Then $H \subseteq G[\tau]$, since if $f : G \to \mathbb{Q}$ with type $f(G) \leq \tau$, and $\sigma \not\leq \tau$, then $f(G(\sigma)) = 0$; otherwise, $\sigma = \text{type} \, x \leq \text{type} \, f(x) \leq \tau$. By Lemma 3.2.3(a), there is a completely decomposable group C and a balanced epimorphism $g : C \to G$. Now, $g(C/C[\tau]) \subseteq G/G[\tau]$, $C[\tau] = \Sigma\{C(\sigma) : \sigma \not\leq \tau\}$ by Examples 3.1.5 and 3.1.10, and $g : C/C[\tau] \to G/H$ is onto, noting that $g(C[\tau]) \subseteq H$, since $g(C(\sigma)) \subseteq G(\sigma)$ for each σ. Then $\text{OT}(G/H) \leq \text{OT}(C/C[\tau]) \leq \tau$ by Example 3.1.10(a). Hence, $(G/H)[\tau] = 0$ by Proposition 3.1.13(b). But $G[\tau]/H \subseteq (G/H)[\tau]$, as a consequence of Lemma 3.1.9(b), and so $H = G[\tau]$. A similar argument shows that $G^*[\tau] = \langle G(\sigma) : \sigma \not< \tau \rangle_*$.

(b) Observe that $G(\tau) \subseteq \cap \{G[\sigma] : \sigma \not\geq \tau\}$. This holds because if $x \in G(\tau)$ and $f : G \to \mathbb{Q}$ with type $f(G) = \sigma \not\geq \tau$, then $f(x) = 0$, otherwise, $\sigma = \text{type} \, f(x) \geq \text{type} \, x \geq \tau$. Conversely, G is a pure cobalanced subgroup of a completely decomposable group $C = X_1 \oplus \cdots \oplus X_n$ with rank $X_i = 1$ for each i by Lemma 3.2.3(b). Thus, $\cap \{G[\sigma] : \sigma \not\geq \tau\} \subseteq \cap \{C[\sigma] : \sigma \not\geq \tau\} = C(\tau)$, as a consequence of Example 3.1.10(a). Since $\cap \{G[\sigma] : \sigma \not\geq \tau\} \subseteq G \cap C(\tau) = G(\tau)$, the proof is complete. Analogously, $G_*(\tau) = \cap \{G[\sigma] : \sigma \not> \tau\}$.

Assertion (c) follows from (a) and (b), and (d) follows from (a) and Lemma 3.1.4(b).

Corollary 3.2.9 *Assume that G is a finite rank Butler group. If G has no rank-1 summands, then:*

(a) *the critical typeset of G has neither a least nor a greatest element; and*

(b) *the critical cotypeset of G has neither a least nor a greatest element.*

PROOF. (a) Suppose $\tau \in$ critical typeset G with $\tau \leq \sigma$ for each $\sigma \in$ critical typeset G. By Theorem 3.2.5(b), $G = G(\tau) = G_\tau \oplus G^\#(\tau)$ with G_τ a nonzero τ-homogeneous completely decomposable group. This is a contradiction to the assumption that G has no rank-1 summands.

Next suppose $\tau \in$ critical typeset G with $\tau \geq \sigma$ for each σ in critical typeset G. In view of Corollary 3.2.6(a), Proposition 3.1.13(b), and Proposition 3.2.8(a), $G[\tau] = 0$ and $G^*[\tau] \neq 0$. Then $G^*[\tau]/G[\tau] = G^*[\tau]$ is a nonzero τ-homogeneous completely decomposable summand of $G/G[\tau] = G$, by Theorem 3.2.5(c), again a contradiction.

(b) The proof is analogous to that of (a).

Socles and radicals can be used to identify completely decomposable groups among finite rank Butler groups. If G is a finite rank Butler group and τ is a type, then $G(\tau)/G^\#(\tau)$ and $G^\#[\tau]/G[\tau]$ are both τ-homogeneous completely decomposable. By Proposition 3.2.8(c), inclusion induces a homomorphism $\phi_\tau : G(\tau)/G^\#(\tau) \to G^*[\tau]/G[\tau]$ with image $\phi_\tau = (G(\tau) + G[\tau])/G[\tau]$ and ker $\phi_\tau = (G(\tau) \cap G[\tau])/G^\#(\tau)$. As a consequence of the next proposition, the range of ϕ_τ determines whether G is quasi-isomorphic to a completely decomposable group.

Proposition 3.2.10 *If G is a finite rank Butler group, then*

(a) *G is a completely decomposable group if and only if $G^*[\tau] = G(\tau) + G[\tau]$ for each type τ; and*
(b) *G is quasi-isomorphic to a completely decomposable group if and only if $G^*[\tau]/(G(\tau) + G[\tau])$ is bounded for each type τ.*

PROOF. If C is completely decomposable, then $C^*[\tau] = C(\tau) + C[\tau]$ for each type τ by Examples 3.1.5 and 3.1.10. Moreover, if C is a subgroup of G with G/C bounded, then $G(\tau)/C(\tau)$, $G^*(\tau)/C^*(\tau)$, $G[\tau]/C[\tau]$, and $G^*[\tau]/C^*[\tau]$ are all bounded. It now follows that $G^*[\tau]/(G(\tau) + G[\tau])$ is bounded.

Conversely, assume that $G^*[\tau]/(G(\tau) + G[\tau])$ is bounded for each type τ. In view of the hypotheses and Theorem 3.2.5, if τ is a minimal element of critical cotypeset G, then $G^*[\tau] = G$ and $G/(G(\tau) + G[\tau]) = G/(G_\tau \oplus G^\#(\tau) + G[\tau]) = G/(G_\tau \oplus G[\tau])$ is bounded for some nonzero τ-homogeneous completely decomposable group G_τ. If $H = G[\tau]$, then rank $H <$ rank G and $H^*[\sigma]/(H(\sigma) + H[\sigma])$ is also bounded for each type σ as a consequence of Proposition 3.2.8. By induction on the rank of G, $G[\tau]$ is quasi-isomorphic to a completely decomposable group. Hence, G is also quasi-isomorphic to a completely decomposable group. This proves (b).

As for (a), the same argument shows that if, in addition, $G^*[\tau]/(G(\tau) + G[\tau]) = 0$, then G is completely decomposable.

Corollary 3.2.11

(a) *Given a type τ, there is a functor F_τ from the category of finite rank Butler groups to the category of abelian groups defined by $F_\tau(G) = G^*[\tau]/(G(\tau) + G[\tau])$.*
(b) *A finite rank Butler group G is completely decomposable if and only if $F_\tau(G) = 0$ for each type τ. Moreover, G is quasi-isomorphic to a completely decomposable group if and only if $F_\tau(G)$ is a bounded group for each τ.*
(c) *For each type τ and finite rank Butler group G, image ϕ_τ is isomorphic to a τ-homogeneous completely decomposable quasi-summand of G with maximal rank.*

(d) *A finite rank Butler group G has no rank-1 quasi-summands if and only if $G(\tau)$ is contained in $G[\tau]$ for each type τ.*

PROOF. Exercise 3.2.4.

Let G be a finite rank Butler group. By Theorem 3.2.5(b), $G(\tau) = G_\tau \oplus G^\#(\tau)$ with G_τ a homogeneous completely decomposable group of type τ for each τ in the critical typeset of G. Define a *regulating subgroup* of G to be $B = \Sigma\{G_\tau : \tau \in \text{critical typeset } G\}$. Notice that B need not be unique, since the G_τ's are unique only up to isomorphism.

The following proposition is due to A. Mader and K.J. Krapf; see [Mutzbauer 93].

Proposition 3.2.12 *Let G be a finite rank Butler group.*

(a) *If B is a regulating subgroup of G, then G/B is bounded.*

(b) *There are only finitely many regulating subgroups of G.*

PROOF. (a) Write $B = \Sigma\{G_\tau : \tau \in \text{critical typeset } G\}$. The proof is by induction on the length of a chain from a type to a maximal type in the typeset of G. If τ is maximal in typeset G, then $G(\tau) = G_\tau$. So assume that τ is not maximal and $G(\sigma)/\Sigma\{G_\delta : \delta \geq \sigma\}$ is bounded for each $\tau < \sigma \in$ typeset G. Then $G(\tau) = G_\tau \oplus G^\#(\tau)$, $G^\#(\tau)/G^*(\tau)$ is bounded, and $G^*(\tau)/\Sigma\{G_\delta : \delta > \tau\}$ is bounded by induction. Hence, $G(\tau)/\Sigma\{G_\sigma : \sigma \geq \tau\}$ is finite for each $\tau \in$ typeset G. Let $\tau = \text{IT}(G)$. Then $G = G(\tau)$ and G/B is finite.

(b) For each τ in typeset G, choose $n(\tau)$ with $n(\tau)(G^\#(\tau)/G^*(\tau)) = 0$. It suffices to show that if $e = \text{lcm}\{n(\tau) : \tau \in$ typeset $G\}$, d is the length of a longest chain in typeset G, and $B = \Sigma\{G_\tau : \tau \in$ critical typeset $G\}$ is a regulating subgroup of G, then $e^d G \subseteq B \subseteq G$. In this case, there are only finitely many regulating subgroups, since $G/e^d G$ is bounded, hence finite by Exercise 2.1.1.

For $0 \neq x$ in G, let $d(x)$ denote the length of a longest chain in typeset G from type x to a maximal type. It suffices to prove that $e^{d(x)}x \in B$.

If $d(x) = 0$, then type x is maximal, $G^\#(\text{type } x) = 0$, and $x \in G_\tau \subseteq B$ for $\tau = \text{type } x$. Now assume that $d(x) > 0$ and assume, by way of induction, that $e^{d(y)}y \in B$ for each y in G with type $y > \text{type } x = \tau$. Then $G(\tau) = G_\tau \oplus G^\#(\tau)$ and $x = c + y$ for some c in G_τ and $y \in G^\#(\tau)$. If type $y > \text{type } x$, then $d(y) < d(x)$ and $e^{d(x)}x = e^{d(x)}c + e^{d(x)}y \in B$.

Finally, assume $\tau = \text{type } x = \text{type } y$ and $d(y) = d(x)$. In view of the choice of e, $ey = \Sigma\{y_\sigma : \text{type } y_\sigma = \sigma > \tau\} \in G^*(\tau)$. By induction, $e^{d(y)}y = \Sigma\{e^{d(y)-1}y_\sigma : \sigma > \tau\} \in B$, since $d(y_\sigma) < d(y)$ for each σ. But $d(y) = d(x)$, so that $e^{d(x)}x = e^{d(x)}c + e^{d(y)}y \in B$, as desired.

Given a Butler group G, define the *regulator subgroup $R(G)$* to be the intersection of all regulating subgroups of G. A finite partially ordered set S is an *inverted forest* if for each s in S, $\{\tau \in S : \tau \geq s \text{ in } S\}$ is linearly ordered. Part (c) of the next corollary is an unpublished result of O. Mutzbauer, wherein an inverted forest is called a *V-tree*.

Corollary 3.2.13 *Let G be a finite rank Butler group.*

(a) *The group $G/R(G)$ is finite.*
(b) *If $f : G \to H$ is an isomorphism, then $f(R(G)) = R(H)$ and $G/R(G)$ is isomorphic to $H/R(H)$.*
(c) *If the critical typeset of G is an inverted forest, then $R(G) = \Sigma\{G(\tau) : \tau \in$ critical typeset $G\}$ is a unique regulating subgroup of G, and $G(\tau)$ is completely decomposable with linearly ordered typeset for each τ in the critical typeset of G.*

PROOF. Assertion (a) follows from Proposition 3.2.12, since G has only finite many regulating subgroups B and each G/B is bounded, hence finite.

(b) Notice that $f(B)$ is a regulating subgroup of H if and only if B is a regulating subgroup of G. This follows from the fact that if f is an isomorphism, then $f(G(\tau)) = f(G_\tau) \oplus f(G^\#(\tau)) = f(G_\tau) \oplus H^\#(\tau)$. Hence, $f(R(G)) = R(H)$, since the regulator subgroup is the intersection of all the regulating subgroups.

(c) Let $B = \Sigma\{G_\tau : \tau \in$ critical typeset $G\}$ be a regulating subgroup of G. Then the typeset of $G(\tau)$ is a chain for each τ in the critical typeset by hypothesis. Hence, $G(\tau)$ is completely decomposable by Corollary 3.2.7(a), and $G(\tau) = \oplus\{G_\sigma : \sigma \geq \tau\}$ by Example 3.1.5(a). Thus, $G(\tau) \subseteq B$ for each τ in the critical typeset of G. Since $B \subseteq B' = \Sigma\{G(\tau) : \tau \in$ critical typeset $G\}$, $B = B'$ must be a unique regulating subgroup of G.

The following proposition is a "dual" of the definition of a regulating subgroup and Proposition 3.2.12(a).

Proposition 3.2.14 *Let G be a finite rank Butler group. For each type τ in the critical cotypeset of G, choose an H with $G/G[\tau] = (G^*[\tau]/G[\tau]) \oplus H$ and let $f_\tau : G/G[\tau] \to G^*[\tau]/G[\tau]$ be a projection with $\ker f_\tau = H$. Then $f = \oplus f_\tau : G \to \oplus\{G^*[\tau]/G[\tau] : \tau \in$ critical cotypeset $G\}$ is a monomorphism with $\langle f(G)\rangle_*/f(G)$ bounded, hence finite.*

PROOF. Exercise 3.2.3.

This section concludes with another characterization of finite rank Butler groups that will be used in Theorem 3.4.6. Let G be a torsion-free abelian group of finite rank. Given a nonempty subset S of Π, the set of primes of $\mathbb{Z}$, define $G_S = \cap\{G_p : p \in S\}$. Observe that $G = \cap\{G_p; p \in \Pi\}$ is a subgroup of G_S.

Proposition 3.2.15 *[Bican 70] Let G be a torsion-free abelian group of finite rank. Then G is a finite rank Butler group if and only if there is a partition $\Pi = \cup\{S(i) : 1 \leq i \leq n\}$ such that each $G_{S(i)}$ is a completely decomposable group with linearly ordered typeset.*

PROOF. Assume that G is a finite rank Butler group and write $G = X_1 + \cdots + X_n$ for pure rank-1 subgroups X_i of G. Choose $0 \neq x_i \in X_i$. Given a permutation σ of $n^+ = \{1, 2, \ldots, n\}$, define $S(\sigma)$ to be the set of primes p with p-height $x_{\sigma(1)} \geq \cdots \geq p$-height $x_{\sigma(n)}$. Then $\Pi = \cup\{S(\sigma) : \sigma$ permutation of $n^+\}$, since for each prime p there is a permutation σ of n^+ with p-height $x_{\sigma(1)} \geq \cdots \geq p$-height $x_{\sigma(n)}$. Now, $G_{S(\sigma)} = (X_1)_{S(\sigma)} + \cdots + (X_n)_{S(\sigma)}$ has a linearly ordered typeset, since type $x_{\sigma(1)} \geq \cdots \geq$ type $x_{\sigma(n)}$. But $G_{S(\sigma)}$ is a finite rank Butler group, so that $G_{S(\sigma)}$ is completely decomposable by Corollary 3.2.7(a). To complete the proof, choose subsets $S'(\sigma)$ of $S(\sigma)$ such that the $S'(\sigma)$'s partition Π, and observe that each $G_{S'(\sigma)}$ is completely decomposable with linearly ordered typeset for the same reason that $G_{S(\sigma)}$ is.

Conversely, assume that $\Pi = \cup\{S(i) : 1 \leq i \leq n\}$ is a partition such that each $G_{S(i)}$ is a completely decomposable group with linearly ordered typeset. Define a homomorphism $f : G \to \oplus\{G_{S(i)} : 1 \leq i \leq n\}$ by $f(x) = (x, \ldots, x)$. Since $\Pi = \cup\{S(i) : 1 \leq i \leq n\}$, $G = \cap\{G_{S(i)} : 1 \leq i \leq n\}$ is a pure subgroup of the completely decomposable group $\oplus\{G_{S(i)} : 1 \leq i \leq n\}$.

EXERCISES

1. Suppose G is a finite rank Butler group, $C = A_1 \oplus \cdots \oplus A_n$ with rank $A_i = 1$ for each i, and $f : C \to G$ is an epimorphism with $f(A_i) \neq 0$ for each i. For each nonempty subset S of $\{1, 2, \ldots, n\}$, define $\tau_S = \cap\{$type $A_i : i \in S\}$. Prove that if X is a pure rank-1 subgroup of G, then type $X = \cup\{\tau_S : X \cap (\Sigma\{f(A_i) : i \in S\}) \neq 0\}$.

2. Prove that if G is a finite rank Butler group, then G is completely decomposable if and only if $G[\tau] = \Sigma\{G(\sigma) : \sigma \not\leq \tau\}$ for each $\tau \in$ cotypeset G.

3. Give a proof for Proposition 3.2.14.

4. Give a proof for Corollary 3.2.11.

5. Prove that if G is a τ-homogeneous completely decomposable group and H is a σ-homogeneous completely decomposable group, then $G \otimes_{\mathbb{Z}} H$ is a δ-homogeneous completely decomposable group, where if τ has height sequence (m_p) and σ has height sequence (n_p), then δ has height sequence $(m_p + n_p)$, agreeing that $n + \infty = \infty$ for each $0 \leq n \leq \infty$.

6. A torsion-free abelian group G is *locally completely decomposable* if G_p is completely decomposable for each prime p. Prove that a finite rank Butler group is locally completely decomposable. Give an example of a locally completely decomposable group of finite rank that is not a Butler group.

7. [Koehler 65] Let G be a torsion-free abelian group of finite rank with finite typeset.
   (a) Prove that if T is a nonempty subset of typeset G and $\sigma = \cap\{\tau : \tau \in T\}$, then there is a maximal independent subset B of $G(\sigma)$ such that each $b \in B$ has type σ in G.
   (b) Show that there is a finite rank Butler group H such that H is a subgroup of G, $G(\tau)/H(\tau)$ is torsion for each type τ, and typeset $H =$ typeset G. Conclude that H is unique up to quasi-equality.

8. Compute the cotypeset of the group B given in Example 3.2.1. Is $B \to X_1 \oplus X_2 \oplus X_3$ a cobalanced monomorphism?

3.3 Quasi-isomorphism and $\mathbb{Q}$-Representations of Posets

There are equivalences between categories of finite rank Butler groups and representations of finite posets over the field $\mathbb{Q}$ of rational numbers. These equivalences
can be used to transfer properties of representations developed in Chapter 1 to the
quasi-isomorphism category of finite rank Butler groups.

If G is a finite rank Butler group, then typeset G is finite. Hence, the finite
sublattice of the lattice of all types generated by typeset G is finite. Given a finite
lattice T of types, define $B(T)$ to be the category of finite rank Butler groups G
with typeset G contained in T. Clearly, each finite rank Butler group is in $B(T)$
for some T.

As a consequence of the next lemma, $B(T)$ is the smallest category containing
all rank-1 groups with types in T that is closed under isomorphism, finite direct
sums, pure subgroups, and torsion-free homomorphic images.

Lemma 3.3.1 [Butler 65] *Let T be a finite lattice of all types and G a finite
rank Butler group. The following statements are equivalent:*

(a) G *is in* $B(T)$;
(b) G *is a pure (cobalanced) subgroup of a completely decomposable group in*
    $B(T)$;
(c) G *is a (balanced) homomorphic image of a completely decomposable group
    in* $B(T)$.

PROOF. (a) $\Rightarrow$ (c) By Lemma 3.2.3(a), there is a balanced epimorphism $C \to G$,
where C is a direct sum of rank-1 groups X_I with type $X_I \in$ typeset $G \subseteq T$.
Then typeset $C \subseteq T$ by Example 3.1.5(b) and the fact that T is closed under finite
meets. This shows that $C \in B(T)$.

(c) $\Rightarrow$ (b) In this case, G is a pure cobalanced subgroup of a finite direct sum
C of rank-1 quotients Y_I of G by Lemma 3.2.3(b). Then type $Y_I \in T$, since T is
closed under finite joins. Hence, typeset $C \subseteq T$ and $C \in B(T)$.

(b) $\Rightarrow$ (a) follows from the fact that if G is a pure subgroup of $C \in B(T)$, then
typeset $G \subseteq$ typeset $C \subseteq T$.

Let T be a finite lattice of types with least element τ_0 and $B(T)_{\mathbb{Q}}$ the quasi-
isomorphism category of $B(T)$. Define $JI(T)$ to be the poset of *join irreducible*
elements of T, i.e., those $\tau > \tau_0$ in T such that if $\tau = \sigma \cup \delta$, then either $\sigma = \tau$
or $\delta = \tau$. Notice that if $\tau_0 < \tau \in T$, then $\sigma \leq \tau$ for some $\sigma \in JI(T)$, and if
$\tau \in T \backslash JI(T)$, then $\tau = \cup \{\sigma \in JI(T) : \sigma < \tau\}$. Define S_T to be $JI(T)^{op}$, the poset
with elements those of $JI(T)$ but with reverse ordering. In particular, if $\sigma, \tau \in S_T$,
then $\sigma \leq \tau$ in S_T if and only if $\sigma \geq \tau$ as types.

Theorem 3.3.2 [Butler 68] *Let T be a finite lattice of types. There is a category
equivalence* $H : B(T)_{\mathbb{Q}} \to rep(S_T, \mathbb{Q})$ *given by* $H(G) = (\mathbb{Q}G, \mathbb{Q}G(\tau) : \tau \in
S_T)$.

PROOF. The strategy is to define a fully faithful functor $H : B(T)_{\mathbb{Q}} \to \text{rep}(T^{\text{op}}, \mathbb{Q})$ by $H(G) = (\mathbb{Q}G, \mathbb{Q}G(\tau) : \tau \in T)$ and then prove that the image of H coincides with the image of the functor $F^- : \text{rep}(S_T, \mathbb{Q}) \to \text{rep}(T^{\text{op}}, \mathbb{Q})$ given in Proposition 1.3.1(b).

Observe that $H(G) \in \text{rep}(T^{\text{op}}, \mathbb{Q})$, since if σ and τ are elements of T with $\sigma \leq \tau$ as types, then $\tau \leq \sigma$ as elements of T^{op} and $G(\tau) \subseteq G(\sigma)$ by Lemma 3.1.4(a). For $f \in \text{Hom}(G, K)$ define $H(f) : \mathbb{Q}G \to \mathbb{Q}K$ to be the unique extension of f. Then $H(f)(\mathbb{Q}G(\tau)) \subseteq \mathbb{Q}H(\tau)$ for each $\tau \in S_T$, since $f(G(\tau)) \subseteq H(\tau)$ by Lemma 3.1.4(d). Hence, $H(f)$ is a representation morphism. It follows that H is a faithful functor.

To see that H is a full functor, let $g : \mathbb{Q}G \to \mathbb{Q}K$ be a representation morphism from $H(G)$ to $H(K)$. Since G is a finite rank Butler group, there are finitely many pure rank-1 subgroups $X_1, \ldots, X_n$ of G with $G = X_1 + \cdots + X_n$. Since $g : \mathbb{Q}G(\tau) \to \mathbb{Q}K(\tau)$ for each type τ, there is a nonzero integer m_i with $m_i g : X_i \to K$ for each $1 \leq i \leq n$. Let $m = m_1 \cdots m_n$. Then $mg \in \text{Hom}(G, K)$ with $H(mf) = g$, as desired.

In order to confirm that image H is contained in the image of F^-, let $\tau_0 \neq \tau \in T$. Then $\tau = \cup \{\sigma \in \text{JI}(T) : \sigma \leq \tau \text{ in } T\}$. As a consequence of Lemma 3.1.4(b), $\mathbb{Q}G(\tau) = \cap \{\mathbb{Q}G(\sigma) : \sigma \in \text{JI}(T), \sigma \leq \tau \text{ in } T\}$. Hence, $(\mathbb{Q}G, \mathbb{Q}G(\sigma) : \sigma \in S_T) \in \text{rep}(S_T, \mathbb{Q})$ with $H(G) = F^-(\mathbb{Q}G, \mathbb{Q}G(\sigma) : \sigma \in S_T)$.

It remains to prove that if $U = (U_0, U_\tau : \tau \in T^{\text{op}}) \in \text{rep}(T^{\text{op}}, \mathbb{Q})$ is in the image of F^-, then $H(G) = U$ for some $G \in B(T)$. Observe that U is in the image of F^- if and only if $U_{\tau \cup \sigma} = U_\sigma \cap U_\tau$ for each $\sigma, \tau \in T$.

Choose a free subgroup V_0 of U_0 with $\mathbb{Q}V_0 = U_0$ and let $V_\tau = V_0 \cap U_\tau$ for each $\tau \in T$. Each V_τ is a pure subgroup, hence a summand of V_0, and $\mathbb{Q}V_\tau = U_\tau$. For each $\tau \in S_T$, let X_τ be a subgroup of $\mathbb{Q}$ with type $X_\tau = \tau$ and let X_0 be a subgroup of $\mathbb{Q}$ with type $X_0 = \tau_0$. Define

$$G = X_0 V_0 + \Sigma \{X_\tau V_\tau : \tau \in S_T\},$$

a subgroup of $\mathbb{Q}V_0 = U_0$ with $\mathbb{Q}G = U_0$. There exist a completely decomposable group C, a finite direct sum of groups isomorphic to X_0 or X_τ for $\tau \in S_T \subseteq T$, and an epimorphism $\pi : C \to G$ as a consequence of the definition of G. By Lemma 3.3.1, $G \in B(T)$.

The proof is concluded by showing that $\mathbb{Q}G(\tau) = U_\tau$ for each $\tau \in T$. Now, $U_\tau = \mathbb{Q}X_\tau V_\tau \subseteq \mathbb{Q}G(\tau)$ for each $\tau \in S_T$, since type $X_\tau = \tau$, V_τ is a free group, and $X_\tau V_\tau$ is τ-homogeneous completely decomposable. Because U is in the image of F^- and the elements of S_T are the join irreducible elements of T, U_τ is contained in $\mathbb{Q}G(\tau)$ for each $\tau \in T$.

As for the reverse inclusion, let X be a pure rank-1 subgroup of G with type $X = \tau \in T$. Then $\pi^{-1}(X)$ is a Butler group, being a pure subgroup of C. Write $\pi^{-1}(X) = Y_1 + \cdots + Y_n$ with each Y_i a pure rank-1 subgroup of C and $\pi(Y_i) \neq 0$. Now, π restricts to an epimorphism $Y_1 \oplus \cdots \oplus Y_n \to X$ and $\tau = \text{type } X = \sigma(1) \cup \cdots \cup \sigma(n)$, where $\sigma(i) = \text{type } Y_i \in T$. Then $\mathbb{Q}X = \mathbb{Q}\pi(Y_i) \subseteq \mathbb{Q}G(\sigma_i) \subseteq U_{\sigma(i)}$ for each i. Consequently, $\mathbb{Q}X \subseteq U_\tau = \cap \{U_{\sigma(i)} : 1 \leq i \leq n\}$, since U is in the image of F^-. Therefore, $\mathbb{Q}G(\tau)$ is contained in U_τ for each type $\tau \in T$, as desired.

Theorem 3.3.2 gives an explicit method of constructing groups in $B(T)$ from representations in rep$(S_T, \mathbb{Q})$ as illustrated by the following example:

Example 3.3.3 *If T is a finite lattice, and $S_T = \{\text{type } X_1, \text{type } X_2, \text{type } X_3\}$ is an antichain with each X_i a subgroup of $\mathbb{Q}$ and $X_0 = X_i \cap X_j$ for each $i \neq j$, then a strongly indecomposable group in $B(T)$ is quasi-isomorphic to one of the following rank-1 groups with types in T,*

$$X_0, X_1, X_2, X_3, X_1 + X_2, X_2 + X_3, X_1 + X_3, X_1 + X_2 + X_3,$$

or a rank-2 group

$$X_1 \oplus X_2 + (1+1)X_3.$$

In each case, the endomorphism ring is isomorphic to a subring of $\mathbb{Q}$.

PROOF. By Example 1.1.5, the indecomposable representations in rep$(S_3, \mathbb{Q})$ are $(\mathbb{Q}, 0, 0, 0)$, $(\mathbb{Q}, \mathbb{Q}, 0, 0)$, $(\mathbb{Q}, 0, \mathbb{Q}, 0)$, $(\mathbb{Q}, 0, 0, \mathbb{Q})$, $(\mathbb{Q}, \mathbb{Q}, \mathbb{Q}, 0)$, $(\mathbb{Q}, 0, \mathbb{Q}, \mathbb{Q})$, $(\mathbb{Q}, \mathbb{Q}, 0, \mathbb{Q})$, $(\mathbb{Q}, \mathbb{Q}, \mathbb{Q}, \mathbb{Q})$, and $(\mathbb{Q} \oplus \mathbb{Q}, \mathbb{Q} \oplus 0, 0 \oplus \mathbb{Q}, (1+1)\mathbb{Q})$. In each case, the representation endomorphism ring is isomorphic to $\mathbb{Q}$. Now apply Theorem 3.3.2. For example, if $U = (U_0, U_1, U_2, U_3)$ is an indecomposable representation with $\mathbb{Q}$-dimension $U_0 = 1$, then $G = X_0 + \Sigma\{X_i : U_i \neq 0\}$ is a rank-1 group with type $G = \cup\{\text{type } X_i : U_i \neq 0\}$.

The category $B(T)_\mathbb{Q}$ has *finite, tame, or wild representation type* as, respectively, rep$(S_T, \mathbb{Q})$ has finite, tame, or wild representation type. Since the functor of Theorem 3.3.2 is a category equivalence, the representation type of $B(T)_\mathbb{Q}$ is well-defined and is determined by the structure of S_T.

Corollary 3.3.4 *Suppose T is a finite lattice of types.*

(a) *If $B(T)_\mathbb{Q}$ has infinite representation type, then there are strongly indecomposable groups in $B(T)$ of arbitrarily large rank.*

(b) *If $B(T)_\mathbb{Q}$ has wild representation type, then for each finite-dimensional $\mathbb{Q}$-algebra A, there is a G in $B(T)$ with $\mathbb{Q}$End G isomorphic to A.*

PROOF. Assertion (a) is a consequence of Theorem 3.3.2, while (b) follows from Theorem 3.3.2, Proposition 1.4.5, and Corollary 1.4.3.

Corollary 3.3.5 *Let T be a finite lattice of types. Then $B(T)_\mathbb{Q}$ has:*

(a) *finite representation type if and only if S_T does not contain S_4, $(2, 2, 2)$, $(1, 3, 3)$, $(1, 2, 5)$, or $(N, 4)$ as a subposet;*

(b) *wild representation type if and only if S_T contains S_5, $(1, 1, 1, 2)$, $(2, 2, 3)$, $(1, 3, 4)$, $(N, 5)$, or $(1, 2, 6)$ as a subposet;*

(c) *tame representation type if and only if S_T contains S_4, $(2, 2, 2)$, $(1, 3, 3)$, $(1, 2, 5)$, or $(N, 4)$ but does not contain S_5, $(1, 1, 1, 2)$, $(2, 2, 3)$, $(1, 3, 4)$, $(N, 5)$, or $(1, 2, 6)$ as a subposet.*

PROOF. Apply Theorems 3.3.2, 1.3.6, 1.4.4, and 1.4.6.

The tools of representation theory can be used to provide some striking results for finite rank Butler groups.

Corollary 3.3.6 *Suppose T is a finite lattice of types. Then $B(T)_{\mathbb{Q}}$ has finite representation type if and only if* End G *is a subring of $\mathbb{Q}$ for each strongly indecomposable group G in $B(T)$. In this case, each strongly indecomposable group G in $B(T)$ has rank less than or equal to 6 and G is determined uniquely up to quasi-isomorphism by the vector* (rank G, rank $G(\tau) : \tau \in \mathrm{JI}(T)$).

PROOF. A consequence of Theorem 3.3.2, Corollary 1.3.9, and Theorem 6.2.7.

The next corollary provides analogues of sincere representations and sincere posets.

Corollary 3.3.7 *Let T be a finite lattice of types. If $B(T)_{\mathbb{Q}}$ has finite representation type, then there is a strongly indecomposable group G in $B(T)_{\mathbb{Q}}$ with rank $G(\tau)/G^{\#}(\tau) \neq 0$ for each $\tau \in S_T$ if and only if S_T is one of*

(1), $(1, 1)$, $(1, 1, 1)$, $(1, 1, 2)$, $(1, 2, 2)$, $(1, 2, 3)$, $(1, 2, 4)$, $(N, 2)$, $(N, 3)$,

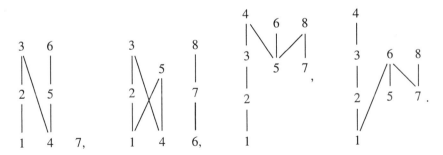

PROOF. Apply Theorems 3.3.2 and 1.3.7.

Given a poset S with finite representation type, all of the indecomposable representations in rep$(S, \mathbb{Q})$ can be found by listing the indecomposable representations for each sincere subposet of S. In view of Theorem 3.3.2, this same process carries over verbatim to $B(T)_{\mathbb{Q}}$.

Example 3.3.8 *Let T be a finite lattice of types with $S_T = (1, 1, 2) = \{$type X_1, type X_2, type $X_3 >$ type $X_4\}$, each X_i a subgroup of $\mathbb{Q}$. If G is a strongly*

indecomposable group in $B(T)$, then G is quasi-isomorphic to a rank-1 group with type in T or one of the rank-2 groups

$$((X_1 + X_4) \oplus X_3) + (1 + 1)X_2,$$

$$(X_1 \oplus X_2) + (1 + 1)X_3,$$

$$(X_1 \oplus X_2) + (1 + 1)X_4.$$

In each case, End G *is a subring of* $\mathbb{Q}$.

PROOF. A consequence of Theorems 3.3.2 and 1.3.7 and Examples 1.3.8 and 1.1.5. The rank-2 groups arise from a sincere 2-dimensional representation of $(1, 1, 2)$ and Example 3.3.3, noticing that S_3 can be embedded as a subposet of $(1, 1, 2)$ in two different ways.

Example 3.3.9 *Let*

$$
T = \quad
\begin{array}{c}
5 \\
/\ \ \backslash \\
3 \quad 4 \\
/\ \backslash\ / \\
1 \quad 2 \\
\backslash\ / \\
0
\end{array}
$$

be a lattice of types. Then $S_T = \{1, 4 < 2\}$. Moreover, the rank-1 groups with type in T are the only strongly indecomposable groups in $B(T)$.

PROOF. Exercise 3.3.3.

Example 3.3.10 *Assume that T is a finite lattice of types and $S_T = S_4 = \{$type X_1, type X_2, type X_3, type $X_4\}$. There are strongly indecomposable groups of arbitrarily large finite rank in $B(T)$. In fact, for each irreducible polynomial $g(x) \in \mathbb{Q}[x]$ and integer $e \geq 1$, there is a strongly indecomposable group G in $B(T)$ with $\mathbb{Q}$End G isomorphic to $\mathbb{Q}[x]/\langle g(x)^e \rangle$.*

PROOF. Apply Theorem 3.3.2 and Example 1.1.6. Specifically, given an $n \times n$ $\mathbb{Z}$-matrix A that is indecomposable as a $\mathbb{Q}$-matrix,

$$G = X_1^n \oplus X_2^n + (1 + 1)X_3^n + (1 + A)X_4^n$$

is a strongly indecomposable group in $B(T)$ with $\mathbb{Q}$End G isomorphic to $\mathbb{Q}[x]/\langle g(x)^e \rangle$, where $g(x)^e$ is the minimal polynomial of A.

EXERCISES

1. Let T be a finite lattice of types. Prove that $w(S_T) \le 2$ if and only if each group in $B(T)$ is quasi-isomorphic to a completely decomposable group.

2. State and prove Theorem 3.3.2 for modules over an arbitrary principal ideal domain.

3. Confirm Examples 3.3.8 and 3.3.9.

4. Prove that if S is a finite partially ordered set, then there is a finite lattice of types T with $S_T = S$. Show that the elements of T may be chosen to be isomorphism classes of subrings of $\mathbb{Q}$.

3.4   Countable Groups

A countable torsion-free abelian group G is a *Butler group* if each pure finite rank subgroup of G is a Butler group. Such groups are also called *finitely Butler* groups [Fuchs, Metelli 92]. An equivalent characterization of countable Butler groups in terms of balanced exact sequences is given in Theorem 3.4.6.

A countable torsion-free abelian group G is *completely decomposable* if G is a direct sum of rank-1 groups.

Proposition 3.4.1 *Assume that G is a countable torsion-free abelian group.*

(a) *If G is a countable pure subgroup of a completely decomposable group, then G is a Butler group.*

(b) *The group G is a Butler group if and only if G is the union of an ascending chain $0 \subseteq B_1 \subseteq \cdots \subseteq B_n \subseteq B_{n+1} \subseteq \cdots$ of pure subgroups such that each B_i is a finite rank Butler group. The B_n's may be chosen with rank $B_{n+1}/B_n = 1$.*

(c) *If G is a homogeneous Butler group, then G is completely decomposable.*

PROOF. (a) Suppose G is a pure subgroup of a completely decomposable group C and H a pure finite rank subgroup of G. Since H has finite rank, H is a pure subgroup of a finite rank completely decomposable summand of C, hence a finite rank Butler group.

(b) Assume that G is a Butler group and $\{x_1, \ldots, x_n, \ldots\}$ is a set of generators of G. Define B_n to be the pure subgroup of G generated by $\{x_1, \ldots, x_n\}$. Then $B_1 \subseteq \cdots \subseteq B_n \subseteq \cdots$ is an ascending chain of pure finite rank subgroups of G with $G = \cup\{B_n : n \ge 1\}$. Since G is a Butler group, each B_n is a Butler group. Moreover, $B_{n+1}/B_n \subseteq \mathbb{Q}(x_{n+1} + B_n)$, so that rank $B_{n+1}/B_n = 1$.

Conversely, suppose $B_1 \subseteq \cdots \subseteq B_n \subseteq \cdots$ is an ascending chain of pure subgroups of G with $G = \cup\{B_n : n \ge 1\}$ and each B_n a finite rank Butler group. If H is a pure finite rank subgroup of G, then H is a pure subgroup of some B_n. To see this, let $\{x_1, \ldots, x_m\}$ be a maximal $\mathbb{Z}$-independent subset of H. Choose n

sufficiently large with each $x_i \in B_n$. Then $H = \langle x_1, \ldots, x_n \rangle_*$ is a pure subgroup of B_n, whence H is a Butler group by Corollary 3.2.4.

(c) By (b), there is an ascending chain $B_1 \subseteq \cdots \subseteq B_n \subseteq \cdots$ of pure subgroups of G with $G = \cup \{B_n : n \geq 1\}$ and B_n a finite rank Butler group. Since G is homogeneous, B_n is homogeneous, hence completely decomposable by Corollary 3.2.7(a). Each B_n is a summand of B_{n+1} by Lemma 3.1.6(a), say $B_{n+1} = B_n \oplus C_{n+1}$ for each $n \geq 1$. Define $C_1 = B_1$. For each i, C_i is a finite rank homogeneous completely decomposable group, by Lemma 3.1.6(c), and so $G = \oplus \{C_i : i \geq 1\}$ is a homogeneous completely decomposable group.

In contrast to finite rank groups, the converse of Proposition 3.4.1(a) does not hold for countable groups.

Example 3.4.2 *There is a countable Butler group that is not isomorphic to a pure subgroup of a completely decomposable group.*

PROOF. Let S be the set of words on the alphabet $\{0,1\}$. The empty word is denoted by $\emptyset$. For each $s \in S$, choose a proper subgroup X_s of $\mathbb{Q}$ such that if $t = s1$ and $u = s0$, then $X_s = X_t \cap X_u$. In other words, the X_s's for $s \in S$ form a tree with root $X_\emptyset$.

In addition, choose the X_s's such that if τ is a type and for each n there is $s \in S$ with length $s \geq n$ and $\tau \geq$ type X_s, then $\tau =$ type $\mathbb{Q}$. To see that such a choice is possible, let $\pi_0 = \{\Pi\}$, $\pi_1 = \{P_0, P_1\}$, $\pi_2 = \{P_{00}, P_{01}, P_{10}, P_{11}\}, \ldots, \pi_n = \{P_s : s \in S, \text{length } s = n\}, \ldots$ be successive partitions of the set Π of all primes such that for each $s \in S$, P_s is infinite, P_s is the disjoint union of P_{s0} and P_{s1}, and if $s(1), \ldots, s(n), \ldots$ is a collection of elements of S with length $s(i) = i$, then $\cap \{P_{s(i)} : 1 \leq i\}$ is empty. For $s \in S$, let $X_s = \cap \{\mathbb{Z}_p : p \in P_s\}$. These X_s's satisfy the above conditions.

Given a nonnegative integer n, define $G_n = \oplus \{X_s : s \in S, \text{length } s = n\}$. There is a homomorphism $f_n : G_n \to G_{n+1}$ with pure image defined by $f(x_s) = (x_s, x_s) \in X_{s0} \oplus X_{s1}$, since $X_s = X_{s0} \cap X_{s1}$. Define G to be the direct limit of $\{G_n, f_n : n \geq 1\}$. Then G is a countable Butler group by Proposition 3.4.1(b) and is reduced, since each X_s is a proper subgroup of $\mathbb{Q}$.

Assume, by way of contradiction, that G is a pure subgroup of a completely decomposable group $C = C_1 \oplus C_2 \oplus \cdots \oplus C_n \oplus \cdots$ with rank $C_n = 1$ for each n. Since G is reduced, G is a pure subgroup of $C/d(C)$, so that it is sufficient to assume that each C_n is reduced. Then $X_\emptyset$ is a pure subgroup of $C_1 \oplus \cdots \oplus C_m$ for some fixed m.

Let $0 \neq x \in X_\emptyset$ and $n \geq 1$. Then $x = \oplus \{x_s : \text{length } s = n\} \in G_n$ with each $x_s \neq 0$ by the choice of the X_s's. Moreover, $x_s = c(1, s) \oplus \cdots \oplus c(k, s)$ with $c(i, s) \in C_i$ for some $k \geq m$. Now,

$$\text{type } x_s = \cap \{\text{type } c(i, s) : 1 \leq i \leq k\} \leq \cap \{\text{type } c(i, s) : 1 \leq i \leq m\}$$

and

$$x = \oplus\{x_s : \text{length } s = n\}$$
$$= \oplus\{\Sigma\{c(i, s) : \text{length } s = n\} : 1 \le i \le k\} \in C_1 \oplus \cdots \oplus C_m.$$

Since $0 \ne x$, there is some $1 \le j \le m$ with $c(j, s) \ne 0$.

In summary, for each $n \ge 1$ there exist $s \in S$ with length $s = n$ and some $1 \le j \le m$ with type $X_s \le$ type C_j. By the choice of the X_s's, type $C_j =$ type $\mathbb{Q}$, a contradiction.

The next theorem gives a method of constructing countable Butler groups of infinite rank with finite typesets. The proof is a modification of the original proof that a torsion-free homomorphic image of a finite rank completely decomposable group is a Butler group given in [Butler 65], wherein $R = \mathbb{Z}$.

Let R be a ring with free additive group such that $\mathbb{Q}R$ is a division ring. Then $\mathbb{Z} \subseteq R$ and a torsion-free R-module is also torsion-free as an abelian group. For a torsion-free R-module M, R-rank(M) is the cardinality of a maximal R-independent subset of M, equivalently the $\mathbb{Q}R$-dimension of $\mathbb{Q}M$. Let $R_p = \mathbb{Z}_p R$, the localization of R at a prime p of $\mathbb{Z}$.

If R/pR has no zero divisors, then each element of R_p is of the form $p^i u$ for u a unit of R_p. To see this, let $0 \ne x \in R_p$ and write $x = p^i r$ for some $r \in R_p \backslash pR_p$. This is possible, since the additive group of R is free, hence p-reduced. Since $\mathbb{Q}R_p = \mathbb{Q}R$ is a division ring, there are $s \in R_p$ and $j \ge 0$ with $r(s/p^j) = 1 = (s/p^j)r$. Then $rs = p^j = sr$. But $R/pR = R_p/pR_p$ has no zero divisors, so that $s = p^j u$ for some u in R_p. Hence, $ru = 1 = ur$, and r is a unit of R_p, as desired.

Theorem 3.4.3 *Suppose R is a countable ring with free additive group, $\mathbb{Q}R$ is a division ring, and R/pR has no zero divisors for each prime p of $\mathbb{Z}$. If G is a torsion-free R-module with $G = RY_1 + \cdots + RY_n$ for rank-1 subgroups Y_i of G, then G is a countable Butler group with finite typeset.*

PROOF. It is sufficient, via Proposition 3.4.1(a), to prove that G is a pure subgroup of a countable completely decomposable group with finite typeset. The proof is by induction on the R-rank of G.

If R-rank $G = 1$, then $G = RY_1$ is a countable direct sum of copies of Y_1, since R is a countable free abelian group. Hence, G is a countable homogeneous completely decomposable group with type $G =$ type Y_1.

Next assume that R-rank $G > 1$ and let $N_i = (\mathbb{Q}RY_i) \cap G$, a pure subgroup of G containing RY_i. Define an R-homomorphism $f : G \to H = G/N_1 \oplus \cdots \oplus G/N_n$ by $f(x) = (x + N_1, \ldots, x + N_n)$. Since N_i is a pure subgroup of G, G/N_i is a torsion-free abelian group that is generated by the rank-1 groups $(Y_j + N_i)/N_i$ for $j \ne i$ as an R-module. Moreover, ker $f = N_1 \cap \cdots \cap N_n = 0$, since R-rank $G > 1$, so that at least two of the RY_i's are R-independent.

The next step in the proof is to prove, by contradiction, that the image of f is a pure subgroup of H. If not, there is a prime p, $y \in G$, and $y_i \in G$ such that $p(y_1 + N_1, \ldots, y_n + N_n) = (y + N_1, \ldots, y + N_n)$ and $y - py_i = x_i \in N_i \backslash pN_i$ for each i. Now, $(N_i)_p = R_p x_i$ since $x_i \notin pN_i$, Y_i is a rank-1 group, and $\mathbb{Q}R$ is a division algebra. Thus, $G_p / pG_p = (R_p / pR_p)(y + pG_p)$ is cyclic, since $G_p = (N_1)_p + \cdots + (N_n)_p = R_p x_1 + \cdots + R_p x_n$, and $y - x_i \in pG$ for each i.

Let $S = \{x_1, \ldots, x_m\}$ be a maximal R_p-independent subset of $\{x_1, \ldots, x_n\}$ with $m = R_p$-rank $G_p = R$-rank $G > 1$ and $y = x_1$. Given $m + 1 \le i \le n$, there is some r in R_p with rx_i in $K = R_p x_1 \oplus \cdots \oplus R_p x_m$, since $\{x_1, \ldots, x_m\}$ is a maximal R_p-independent subset of $\{x_1, \ldots, x_n\}$, a set of R_p-generators for G_p. Write $r = p^j u$ for some unit u of R_p. Then $p^j u x_i \in K \subseteq G_p$, whence $p^j x_i \in K$, since u is a unit. Consequently, there is some e with $p^e x_i \in K$ for each i, and so $p^e G_p = p^e (R_p x_1 + \cdots + R_p x_n) \subseteq K$. Thus, $K = G_p$, since $G_p / p^e G_p$ is generated by $y + p^e G_p$ and $y = x_1 \in K$. In particular, $K/pK = G_p / pG_p$ is cyclic, which contradicts $K/pK = (R_p / pR_p)x_1 \oplus \cdots \oplus (R_p / pR_p)x_m$ with $m > 1$.

By induction on the R-rank of G, each G/N_i is a pure subgroup of a countable completely decomposable group with finite typeset. This shows that H, and consequently G, is a pure subgroup of a countable completely decomposable group with finite typeset.

Let T be a finite lattice of types with $S_T = \{\tau_1, \tau_2\} = S_2$. As a consequence of Theorem 3.3.2 and Example 1.1.3, each strongly indecomposable group in $B(T)$ has rank 1. Hence, each group in $B(T)$ is quasi-isomorphic to a completely decomposable group. The next example demonstrates that countable Butler groups with typeset contained in T are much more complicated.

Example 3.4.4 *Assume that $T_2 = \{$type $X_1 \cap X_2$, type X_1, type $X_2\}$, X_i is a subgroup of $\mathbb{Q}$ containing 1 with type X_1 and type X_2 incomparable, p is a prime with $1/p \notin X_i$, and $R = \text{End} X_i$ for $i = 1, 2$. There is a strongly indecomposable Butler group G with countably infinite rank such that the typeset of G is contained in T_2 and $\text{End} G$ is isomorphic to R.*

PROOF. The first step is to construct $U = (U_0, U_1, U_2) \in \text{Rep}(S_2, R)$ with U_0 a countable free R-module, each U_i a pure submodule of U_0, $U_0 / (U_1 \oplus U_2)$ a p-group, and $\text{End} U = \{f : U_0 \to U_0 : f(U_i) \subseteq U_i\}$ isomorphic to R.

A slight modification of the proof of Theorem 2.4.6 can be used to find a torsion-free R-module M of countably infinite rank containing a free R-submodule L with M/L a p-group, $\text{End} M = R$, and $\text{Hom}(M, R) = 0$. Specifically, let $\{\pi_i : i \ge 1\}$ be a countably infinite subset of the p-adic integers R^* that are algebraically independent over $\mathbb{Q}$, L a free R-submodule of R^* with basis $\{1, \pi_i : i \ge 1\}$, and M the pure subgroup of R^* generated by L. Then M is a torsion-free R-module with countably infinite rank, and M/L is a p-group. Since each $f \in \text{End} M$ extends uniquely to multiplication by an element of R^* and $\{\pi_i : i \ge 1\}$ is an algebraically independent set over $\mathbb{Q}$, it follows that $\text{End} M = R$ and $\text{Hom}(M, R) = 0$.

Let $0 \to U_1 \to U_0 \to M \to 0$ be a countable free R-resolution of M. Then U_0 is a countably free R-module, and U_1 is a pure free R-submodule of U_0, since M is torsion-free, U_0 is a free module, and R is a principal ideal domain. Let $U_1 \oplus L'$ be the preimage of L in U_0 with L' isomorphic to the free R-module L, $\{e_i : i \geq 1\}$ a basis of L', and $\{(n(i), u(i)) : i \geq 1\}$ an indexing of the countable set $N \times U_1$, N the set of natural numbers. Define U_2 to be the pure R-submodule of U_0 generated by $\{p^{n(i)}e_i + u(i) : i \geq 1\}$. Then $U_1 \cap U_2 = 0$, since $L' \cap U_1 = 0$, and $U_0/(U_1 \oplus U_2)$ is a p-group, since M/L is a p-group.

To see that End U is isomorphic to R, let $f \in$ End U. There is $r \in R$ with $g = f - r : U_0 \to U_1$. This is so because $f : U_0 \to U_0$, $f(U_1) \subseteq U_1$, $M = U_0/U_1$, and End $M = R$. Assume, by way of contradiction, that $g \neq 0$. Then $g(U_1) \neq 0$, since U_1 is a free R-module and $\mathrm{Hom}(M, R) = 0$. But U_0 is p-reduced so that $0 \neq g(p^{n(i)}e_i + u(i)) \in g(U_2)$ for some i. Hence, $g(U_2) \subseteq U_2 \cap U_1 = 0$, a contradiction. It follows that each endomorphism of U is multiplication by an element of R, whence End U is isomorphic to R.

Define $G = X_1 U_1 \oplus X_2 U_2 + U_0$, an R-submodule of $\mathbb{Q}U_0 = \mathbb{Q}U_1 \oplus \mathbb{Q}U_2$. A routine computation, using the assumptions that U_i is pure in U_0 and $1/p \notin X_i$, shows that if $\tau_i = \mathrm{type} X_i$, then $G(\tau_i) = X_i U_i$. Moreover, G is a Butler group, since if H is a pure finite rank subgroup of G, then $H = X_1 V_1 \oplus X_2 V_2 + V_0$, with each V_i a finite rank free R-summand of U_i. Hence, H is a Butler group by Corollary 3.2.4. Finally, since $R = $ End $X_1 = $ End X_2, it follows that $R \subseteq$ End $G \subseteq$ End $U = R$.

In the previous example, $G/(G(\tau_1) \oplus G(\tau_2))$ is an unbounded group. However, if G is indecomposable with $G/(G(\tau_1) \oplus G(\tau_2))$ bounded, then rank $G \leq 2$ as a consequence of the next theorem, the proof of which is not included. The finite rank case is proved in Corollary 5.1.12.

Theorem 3.4.5 [Files, Göbel 98] *If G is a countable Butler group with typeset $G \subseteq T_2 = \{\tau_0, \tau_1, \tau_2\}$ and $G/(G(\tau_1) \oplus G(\tau_2))$ is a bounded group, then G is a direct sum of groups of rank less than or equal to 2.*

An exact sequence $0 \to K \to H \overset{f}{\to} G \to 0$ of abelian groups with G torsion-free is *balanced* if rank-1 torsion-free abelian groups are projective with respect to this sequence, i.e., if X is a rank-1 torsion-free abelian group and $g : X \to G$, there is $h : X \to G$ with $fh = g$. If H is also torsion-free, then this definition of balanced coincides with that of Section 3.2 for exact sequences of finite rank torsion-free abelian groups (Exercise 3.4.3).

Theorem 3.4.6 [Bican, Salce 83] *If G is a countable torsion-free abelian group, then G is a Butler group if and only if each balanced exact sequence $0 \to T \to H \to G \to 0$ of abelian groups with T a torsion group is split exact.*

PROOF. [Mines, Vinsonhaler 92] Assume that G is a Butler group and

$$0 \to T \to H \overset{f}{\to} G \to 0$$

is a balanced exact sequence with T a torsion group. By Proposition 3.4.1(b), G is the union of an ascending chain $0 \subseteq G_1 \subseteq \cdots \subseteq G_n \subseteq G_{n+1} \subseteq \cdots$ of pure subgroups such that each G_i is a finite rank Butler group with rank$(G_{i+1}/G_i) = 1$. Assume, by way of induction, that $f^{-1}(G_n) = T \oplus A_n$. Let $K = G_{n+1}$. By Proposition 3.2.15, there is a partition $\Pi = \cup\{S(i) : 1 \leq i \leq m\}$ such that each $K_{S(i)}$ is a completely decomposable group with linearly ordered typeset and K is a pure subgroup of $\oplus\{K_{S(i)} : 1 \leq i \leq m\}$. Since rank $K/G_n = 1$, the partition may be refined to assume that $K_{S(i)} = (G_n)_{S(i)} \oplus D_{S(i)}$ for some rank-1 group D isomorphic to K/G_n.

The sequence $0 \to T_{S(i)} \to (f^{-1}(K))_{S(i)} \to K_{S(i)} = (G_n)_{S(i)} \oplus D_{S(i)} \to 0$ is balanced exact, hence split exact since $K_{S(i)}$ is completely decomposable (see Exercise 3.4.3). Since T is a torsion group and the S(i)'s partition Π, $T = \oplus\{T_{S(i)} : 1 \leq i \leq m\}$. Use the composite of the homomorphisms $f^{-1}(K) \to \oplus\{f^{-1}(K)_{S(i)} : 1 \leq i \leq m\} \to \oplus\{T_{S(i)} : 1 \leq i \leq m\} \to T$ to see that $f^{-1}(K) = f^{-1}(G_{n+1}) = T \oplus A_{n+1}$ for some $A_{n+1} \supseteq A_n$. By transfinite induction, $H = T \oplus A$ for some A isomorphic to G.

Conversely, assume that G is a countable torsion-free abelian group and each balanced exact sequence $0 \to T \to B \to G \to 0$ with T a torsion group is split exact. Let $\{x_1, \ldots, x_i, \ldots\}$ be a maximal $\mathbb{Z}$-independent subset of G. For each n, define $F_n = \mathbb{Z}x_1 \oplus \cdots \oplus \mathbb{Z}x_n$. Enumerate all pure rank-1 subgroups X of G with $X \cap F_n \neq 0$ by $A_{n1}, \ldots, A_{ni}, \ldots$ by listing all linear combinations $a_1 x_1 + \cdots + a_n x_n$ with $a_i \in \mathbb{Z}$ and $\gcd\{a_1, \ldots, a_n\} = 1$ and taking the pure subgroup of G generated by each such linear combination. Define $F = \oplus\{\mathbb{Z}x_i : 1 \leq i\}$, $M_{n,i} = F_n \oplus A_{n,1} \oplus \cdots \oplus A_{n,i}$ and $M = \cup\{M_{n,i} : 1 \leq i, n\}$. Then $M = F \oplus (\oplus\{A_{n,i} : 1 \leq i, n\})$.

There is an exact sequence $0 \to K \to M \to G \to 0$ with $K = \ker f$ and $f : M \to G$ induced by inclusion. Since each pure rank-1 subgroup X of G with $X \cap F_n \neq 0$ is equal to $A_{n,i}$ for some n and i, f is onto and, the sequence is balanced. For each n and i, there is an induced exact sequence $0 \to K \cap M_{n,i} \to M_{n,i} \to G_{n,i} \to 0$. Notice that each $G_{n,i}$ is a finite rank Butler group, since $M_{n,i}$ is a finite rank completely decomposable group. Furthermore, $G_1 \subseteq \cdots \subseteq G_n \subseteq \cdots$ is an ascending chain of pure subgroups of G, where $G_n = \cup\{G_{n,i} : i \geq 1\}$, and $G_{n,i} = G_{n,i-1} + A_{n,i} = A_{n,1} + \cdots + A_{n,i}$.

The next step in the proof is to define a set of elements $p(n, i)$ that are either 1 or a prime and a set of positive exponents $e(n, i)$. If $G_{n,i-1} = G_{n,i}$, define $p(n, i) = 1 = e(n, i)$. If $G_{n,i-1} \neq G_{n,i}$, choose a prime $p = p(n, i)$ with $(G_{n,i-1})_p \neq (G_{n,i})_p$ and a least positive integer $e = e(n, i)$ with $p^e(G_{n,i-1}) \subseteq (F_n)_p + d((G_{n,i-1})_p) \subseteq (G_{n,i})_p$, where $(F_n)_p$ is a free $\mathbb{Z}_p$-module and $d((G_{n,i-1})_p)$ is the divisible subgroup of $(G_{n,i-1})_p$. Such a choice is possible because $(A_{n,j})_p$ is isomorphic to either $\mathbb{Z}_p$ or $\mathbb{Q}$ with $(A_{n,j})_p \cap (F_n)_p \neq 0$ for each $j \leq i$, and $G_{n,i} = A_{n,1} + \cdots + A_{n,i}$.

It is now sufficient to prove that given n, $P_n = \{p(n, i) : p(n, i)^{e(n,i)} \neq 1\}$ is finite. In this case, $G_{n,i-1} \neq G_{n,i}$ for at most finitely many i, and so $G_n = G_{n,i}$ is a finite rank Butler group for some sufficiently large i. An application of Proposition 3.4.1(b) shows that G is a Butler group.

For each $i, n \geq 1$, define $m_{n,i} = \text{lcm}\{p(n, i)^{2e(n,j)} : 1 \leq j \leq i\}$ and $L = \Sigma\{m_{n,i}(K \cap M_{n,i}) : 1 \leq i, n\}$. Then K/L is torsion, since $K = \cup\{K \cap M_{n,i} : 1 \leq i, n\}$, and the induced sequence $0 \to K/L \to M/L \to G \to 0$ is a balanced exact sequence. By hypothesis, this sequence is split exact, so choose $g : G \to M/L$ with $fg = 1_G$.

Because $fg = 1_G$, $g(x_i) - (x_i + L) \in \ker f = K/L$ for each i. Since K/L is torsion and $\{x_1, \ldots, x_n\}$ is a basis of F_n, there is a nonzero integer m with $m(g(y) - (y + L)) = 0 \in K/L$ for each $y \in F_n$. In this case, $h_G(my) = h_{M/L}(my + L)$, where $h_G(my)$ denotes the p-height of my in G. In view of the fact that $L = \Sigma\{m_{n,i}(K \cap M_{n,i}) : 1 \leq i, n\}$ and $m_{n,i} = \mathrm{lcm}\{p(n, i)^{z \in (n, j)} : 1 \leq j \leq i\}$, it suffices to assume that all prime divisors of m are in P_n.

It remains to show that each prime $p \in P_n$ divides m. In this case, P_n must be finite, as desired. Let $1 \neq p \in P_n$. Then $(G_{n,i-1})_p \neq (G_{n,i})_p$ so there exist $x \in A_{n,i}$ with $px \in G_{n,i-1}$ and $p^{h+1}x = y + d \in (F_n)_p + d((G_{n,i-1})_p)$ for some $y \in (F_n)_p$ and $d \in d((G_{n,i-1})_p)$ with $h \leq e(n, i)$, $h = h_{G(n,i-1)}(y)$, and $h_{G(n,i)}(xy) \geq h + 1$. Then $h_{M/L}(my + L) = h_{G(n,i)}(my) \geq h_{\mathbb{Z}}(m) + h + 1$.

Choose $a = a_1 \oplus a_2 \in L$ with $a_1 \in \Sigma\{m_{n,j}(K \cap M_{n,j}) : j \leq i - 1\}$, $a_2 \in \Sigma\{m_{n,j}(K \cap M_{n,j}) : j \geq i\}$, and $h_G(my + a) \geq h_{\mathbb{Z}}(m) + e(n, i) = h_{\mathbb{Z}}(m) + h + 1$. By the definition of the $m_{n,i}$'s, $h_M(a_2) \geq 2e(n, i) > e(n, i) + h$. It follows that $pe(n, i)$ divides m, as desired.

EXERCISES

1. Let G and H be countable completely decomposable groups. Prove that G and H are isomorphic if and only if rank $G(\tau)/G^\#(\tau) =$ rank $H(\tau)/H^\#(\tau)$ for each type τ.

2. Prove that if G is a countable completely decomposable group and H is a summand of G, then H is a completely decomposable group.

3. Let $0 \to K \to H \xrightarrow{f} G \to 0$ be an exact sequence of torsion-free abelian groups. Prove that $f : H(\tau) \to G(\tau)$ is onto for each type τ if and only if for each rank-1 torsion-free abelian group X and $g : X \to G$, there is $h : X \to H$ with $fh = g$.

3.5   Quasi-Generic Groups

There are countable Butler groups of infinite rank, called quasi-generic groups, that determine the representation type of quasi-homomorphism categories of finite rank Butler groups (Theorems 3.5.5 and 3.5.11). In fact, for the tame representation type case, all strongly indecomposable finite rank Butler groups can be constructed from quasi-generic groups. Quasi-generic groups are related to generic representations (Section 1.5). However, the connection between countable Butler groups and countable $\mathbb{Q}$-representations of finite posets is more complicated than the connection between finite rank Butler groups and finite-dimensional $\mathbb{Q}$-representations given in Section 3.3.

The first lemma consists of examples of strongly indecomposable torsion-free abelian groups of countably infinite rank with finite typesets. For each critical poset S_T and associated group G, in Lemma 3.5.1, $(\mathbb{Q}(x)G, \mathbb{Q}(x)G(\tau) : \tau \in S_T)$ is the representation of S_T given in Theorem 1.5.3 with $k(x)$ replaced by $\mathbb{Q}[x]$.

Lemma 3.5.1 *Let T be a finite lattice of types and S_T the opposite of the poset of join irreducible elements of T.*

(a) *For each of the following posets S_T, there is a strongly indecomposable torsion-free abelian group G with countably infinite rank,* typeset $G \subseteq T$, *and* $\mathbb{Q}\mathrm{End}\, G = \mathbb{Q}[x]$. *Write R for $\mathbb{Z}[x]$ and A_i for a subgroup of $\mathbb{Q}$ with* $1 \in A_i$ *and type $A_i = i$.*

(i) $S_T = S_4$

$$G = (A_1 R \oplus 0) + (0 \oplus A_2 R) + ((1+1)A_3 R) + ((1+x)A_4 R) \subseteq \mathbb{Q}[x]^2.$$

$$
\begin{array}{ccc}
2 & 4 & 6 \\
| & | & | \\
\end{array}
$$
(ii) $S_T = (2,2,2) = 1 \quad 3 \quad 5$

$$G = (0 \oplus 0 \oplus A_1 R) + ((1+x)A_2 R \oplus A_2 R) + ((1+1)A_3 R \oplus 0)$$
$$+ (A_4 R \oplus A_4 R \oplus 0) + (0 \oplus (1+1)A_5 R)$$
$$+ (A_6 R \oplus (1+1)A_6 R) \subseteq \mathbb{Q}[x]^3.$$

$$
\begin{array}{cc}
4 & 7 \\
| & | \\
\end{array}
$$
$$
\begin{array}{ccc}
1 & 3 & 6 \\
& | & | \\
\end{array}
$$
(iii) $S_T = (1,3,3) = \quad 2 \quad 5$

$$G = ((1+1)A_1 R \oplus (1+1)A_1 R) + (0 \oplus (1+1)A_2 R \oplus 0)$$
$$+ (0 \oplus A_3 R \oplus A_3 R \oplus 0) + (0 \oplus A_4 R \oplus A_4 R \oplus A_4 R)$$
$$+ (A_5 R \oplus 0 \oplus 0 \oplus 0) + (A_6 R \oplus 0 \oplus 0 \oplus A_6 R)$$
$$+ (A_7 R \oplus (1+x)A_7 R \oplus A_7 R) \subseteq \mathbb{Q}[x]^4.$$

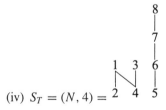

(iv) $S_T = (N,4) =$

$$G = ((1+1)A_2 R \oplus (1+1)A_2 R \oplus 0) + (A_1 R \oplus A_1 R \oplus A_1 R \oplus A_1 R \oplus 0)$$
$$+ (0 \oplus (1+1)A_4 R \oplus 0 \oplus 0) + (0 \oplus A_3 R \oplus A_3 R \oplus (1+1)A_3 R)$$
$$+ (0 \oplus 0 \oplus 0 \oplus 0 \oplus A_5 R) + (A_6 R \oplus 0 \oplus 0 \oplus 0 \oplus A_6 R)$$
$$+ (A_7 R \oplus 0 \oplus 0 \oplus A_7 R \oplus A_7 R)$$
$$+ (A_8 R \oplus (1+x)A_8 R \oplus A_8 R \oplus A_8 R) \subseteq \mathbb{Q}[x]^5.$$

$$
\begin{array}{c}
8 \\
| \\
7 \\
|
\end{array}
$$

$$
\begin{array}{ccc}
1 & 3 & 6 \\
  & | & | \\
  & 2 & 5 \\
  &   & |
\end{array}
$$

(v) $S_T = (1, 2, 5) = \qquad 4,$

$$
\begin{aligned}
G = {} & (0+1+1+0+0+0)A_1 R + (0+0+1+0+0+1)A_1 R \\
& + (0+0+0+1+1+0)A_1 R + ((1+1)A_2 R \oplus (1+1)A_2 R \\
& \oplus 0 \oplus 0) + (A_3 R \oplus A_3 R \oplus A_3 R \oplus A_3 R \oplus 0 \oplus 0) + (0 \oplus 0 \oplus 0 \\
& \oplus 0 \oplus 0 \oplus A_4 R) + (0 \oplus 0 \oplus 0 \oplus 0 \oplus A_5 R \oplus A_5 R) + (A_6 R \oplus 0 \\
& \oplus 0 \oplus 0 \oplus A_6 R \oplus A_6 R) + (A_7 R \oplus 0 \oplus 0 \oplus A_7 R \oplus A_7 R \oplus A_7 R) \\
& + (A_8 R \oplus (1+x)A_8 R \oplus A_8 R \oplus A_8 R \oplus A_8 R) \subseteq Q[x]^6.
\end{aligned}
$$

(b) *If $n \geq 2$ is a positive integer, G is one of the groups constructed in (a) with typeset $G \subseteq T$, and $H = G/Gx^n$, then H is a strongly indecomposable group in $B(T)$ with $\mathbb{Q}\mathrm{End}(H) = \mathbb{Q}[x]/\langle x^n \rangle$.*

PROOF. (a) In each case, G is a torsion-free abelian group with countably infinite rank and $\mathbb{Z}[x] \subseteq \mathrm{End}\, G$. It suffices to prove that $\mathrm{End}\, G$ is contained in $\mathbb{Q}[x]$, in which case G is strongly indecomposable, since $\mathbb{Q}\mathrm{End}\, G = \mathbb{Q}[x]$ has no nontrivial idempotents.

The proof is analogous to that of Theorem 1.5.3. As an illustration, let G be as defined in (i). A brief argument shows that if $\tau_i = \mathrm{type}\, A_i$, then $G(\tau_1) = A_1 \mathbb{Z}[x]$, $G(\tau_2) = A_2 \mathbb{Z}[x]$, $G(\tau_3) = A_3 \mathbb{Z}[x](1, 1)$, and $G(\tau_4) = A_4 \mathbb{Z}[x](1, x)$. Let $f \in \mathrm{End}\, G$. Since $f : G(\tau_i) \to G(\tau_i)$ for $1 \leq i \leq 3$, $f = (g, g)$ with $g \in \mathrm{End}\, \mathbb{Q}[x]$. On the other hand, $f : G(\tau_4) \to G(\tau_4)$, so that g commutes with x. Thus, $g \in \mathrm{End}_{\mathbb{Q}[x]} \mathbb{Q}[x] = \mathbb{Q}[x]$, as desired.

(b) Since H is generated as a group by finitely many copies of the A_i's and type $A_i \in S_T \subseteq T$, $H \in B(T)$. The same argument as that for (a), shows that $\mathbb{Q}\mathrm{End}\, H = \mathbb{Q}[x]/\langle x^n \rangle$, and so H is strongly indecomposable. As an illustration, let $S_T = S_4$ and $R_n = \mathbb{Z}[x]/\langle x^n \rangle$. In this case,

$$
H = A_1 R_n \oplus A_2 R_n + A_3 R_n(1, 1) + A_4 R_n(1, x) \subseteq (\mathbb{Q}[x]/\langle x^n \rangle)^2.
$$

A torsion-free abelian group G is *quasi-generic* if G is strongly indecomposable with countably infinite rank and $\mathbb{Q}G$ has finite length as a $\mathbb{Q}\mathrm{End}\, G$-module. This definition parallels that of a generic representation given in Section 1.5.

The groups in Lemma 1 are not quasi-generic, since in each case, $\mathbb{Q}\mathrm{End}\, G = \mathbb{Q}[x]$ and $\mathbb{Q}G$ has infinite length as a $\mathbb{Q}[x]$-module. This problem can be remedied by replacing $\mathbb{Z}[x]$ with a principal ideal domain Λ such that $\mathbb{Q}\Lambda = \mathbb{Q}(x)$, a field. The ring Λ also satisfies the hypotheses of Theorem 3.4.3 and thereby establishes a method for constructing countable Butler groups with finite typesets that are quasi-generic.

If $f(x) \in \mathbb{Z}[x]$, the ring of polynomials with coefficients in $\mathbb{Z}$, then the *content of* f, denoted by $c(f)$, is the greatest common divisor of the coefficients of f. The polynomial $f(x)$ is called a *primitive polynomial* if $c(f) = 1$. Since $c(fg) = c(f)c(g)$, by Gauss's lemma [Hungerford, 74], the set S of primitive polynomials in $\mathbb{Z}[x]$ is a multiplicatively closed set. Define $\Lambda = \mathbb{Z}[x]_S$, the localization of $\mathbb{Z}[x]$ at S, a subring of the field of quotients $\mathbb{Q}(x)$ of $\mathbb{Z}[x]$. Elements of Λ are of the form $f(x)/g(x)$ with $f(x), g(x) \in \mathbb{Z}[x]$ and $g(x)$ a primitive polynomial.

Lemma 3.5.2 *The ring Λ has a countable free additive group and is a principal ideal domain containing $\mathbb{Z}[x]$ with $\mathbb{Q}\Lambda = \mathbb{Q}(x)$ and $\Lambda/p\Lambda$ a field for each prime p of $\mathbb{Z}$.*

PROOF. (a) Observe that $\mathbb{Z}[x]$ is a subring of Λ and Λ is a subring of $\mathbb{Q}(x)$ with $\mathbb{Q}\Lambda$ contained in $\mathbb{Q}(x)$. On the other hand, if $f(x)/g(x) \in \mathbb{Q}(x)$ with $f(x), g(x) \in \mathbb{Z}[x]$, then $g(x) = ah(x)$ with $c(g) = a$ and $h(x)$ a primitive polynomial. Thus, $f(x)/g(x) = (1/a)(f(x)/h(x)) \in \mathbb{Q}\Lambda$.

The additive group of Λ is countable and torsion-free. To prove that Λ is free as a group, it suffices, by Corollary 2.4.5, to show that each pure finite rank subgroup of Λ is free. Let H be a pure finite rank subgroup of Λ and suppose H is the pure subgroup of Λ generated by a finite number $f_1(x)/g_1(x), \ldots, f_n(x)/g_n(x)$ of elements of Λ with each $f_i(x), g_i(x) \in \mathbb{Z}[x]$ and $g_i(x)$ a primitive polynomial. Then $g(x) = g_1(x) \cdots g_n(x)$ is a primitive polynomial in $\mathbb{Z}[x]$, since each $g_i(x)$ is a primitive polynomial, and $c(g) = c(g_1) \cdots c(g_n)$. Consequently, the subgroup of Λ generated by the $f_i(x)/g_i(x)$'s for $1 \le i \le n$ is a subgroup of $B = (1/g(x))\mathbb{Z}[x]$. Observe that B is a free abelian group, being isomorphic to $\mathbb{Z}[x]$.

In fact, B is a pure subgroup of Λ. To see this, let $a(x)/b(x) \in \Lambda$ and n a nonzero integer with $n(a(x)/b(x)) = f(x)/g(x) \in B$. Then $na(x)g(x) = b(x)f(x) \in \mathbb{Z}[x]$. Taking the content of both sides yields $nc(a) = c(f) \in \mathbb{Z}$, recalling that $c(g) = 1 = c(b)$, since $g(x)$ and $b(x)$ are assumed to be primitive polynomials. Thus, $f(x) = np(x)$ for some $p(x) \in \mathbb{Z}[x]$, and so $a(x)/b(x) = p(x)/g(x) \in B$. Since B is a pure subgroup of Λ and each $f_i(x)/g_i(x) \in B$, it follows that H is a subgroup of B. Because H is a subgroup of the free group B, H is also a free group, as desired.

To see that Λ is a principal ideal domain, suppose that $y = f(x)/g(x)$ is an element of Λ with $f(x), g(x) \in \mathbb{Z}[x]$ and $g(x)$ a primitive polynomial. Write $f(x) = nh(x)$ for some primitive polynomial $h(x)$, where $n = c(f)$. Then $y = n(h(x)/g(x))$ with $h(x)/g(x)$ a unit of Λ. It now follows that if I is a nonzero ideal of Λ, then $I = n\Lambda$, where n is the least positive integer such that $nu \in I$ for some unit u of Λ. Moreover, if p is a prime of $\mathbb{Z}$, then $\Lambda/p\Lambda$ is a field, since each $y \in \Lambda$ is an integer multiple of a unit of Λ.

Lemma 3.5.3 *Let T be a finite lattice of types, G the group associated with S_T as defined in Lemma 3.5.1, and Λ the ring given in Lemma 3.5.2. Then $G' = G\Lambda$ is a quasi-generic group with $\mathrm{End}\, G' = \Lambda$, $\mathbb{Q}\mathrm{End}\, G' = \mathbb{Q}(x)$, and typeset $G' \subseteq T$.*

PROOF. In each case, G' is a countable Butler group with typeset contained in T, via Lemma 3.5.2 and Theorem 3.4.3. As in Lemma 3.5.1, $\Lambda = \mathrm{End}\, G'$. Then $\mathbb{Q}\mathrm{End}\, G' = \mathbb{Q}\Lambda = \mathbb{Q}(x)$, and so G' is strongly indecomposable. Moreover, $\mathbb{Q}G'$ is a finite-dimensional $\mathbb{Q}(x)$-vector space, whence $\mathbb{Q}'$ has finite length as a $\mathbb{Q}\mathrm{End}\, G'$-module and G' is a quasi-generic group.

The quasi-generic groups G' given in Lemma 3.5.3 have some additional properties. Call a countable torsion-free abelian group G *special* if there are finitely many pure rank-1 subgroups $X_1, \ldots, X_n$ of G such that $G/[(\mathrm{End}\, G)X_1 + \cdots + (\mathrm{End}\, G)X_n]$ is a bounded group.

If G is a countable torsion-free abelian group with typeset G contained in a finite lattice T, then $\mathrm{End}\, U_G = \{f : \mathbb{Q}G \to \mathbb{Q}G : f(\mathbb{Q}G(\tau)) \subseteq \mathbb{Q}G(\tau), \tau \in S_T\}$ is the representation endomorphism ring of $U_G = (\mathbb{Q}G, \mathbb{Q}G(\tau) : \tau \in S_T) \in \mathrm{Rep}(S_T, \mathbb{Q})$. Since each $G(\tau)$ is fully invariant in G, $\mathbb{Q}\mathrm{End}\, G \subseteq \mathrm{End}\, U_G$. A special group G is *central-special* if there are finitely many pure rank-1 subgroups X_i of G such that $G/(C_G X_1 + \cdots + C_G X_n)$ is bounded as a group, where $C_G = (\mathrm{CEnd}\, U_G) \cap (\mathrm{End}\, G)$ and $\mathrm{CEnd}\, U_G$ is the center of $\mathrm{End}\, U_G$. Notice that C_G is contained in $\mathrm{CEnd}(G)$, the center of $\mathrm{End}\, G$.

Each finite rank Butler group is central-special, being generated as a group by finitely many pure rank-1 subgroups. For each countable Butler group G' in Lemma 3.5.3, $\mathrm{End}\, U_{G'} = \mathbb{Q}(x)$. Hence, G' is central-special, since $C_{G'} = \mathbb{Q}(x) \cap \mathrm{End}\, G' = \mathrm{End}\, G' = \Lambda$ and G' is generated as a Λ-module by finitely many pure rank-1 subgroups.

Lemma 3.5.4 *If G is a central-special group, then $\mathbb{Q}\mathrm{End}\, G = \mathrm{End}\, U_G$ and $C_G = \mathrm{CEnd}\, G$. In particular, G is strongly indecomposable if and only if U_G is indecomposable in $\mathrm{Rep}(S_T, \mathbb{Q})$.*

PROOF. Since G is central-special, there are pure rank-1 subgroups X_i of G such that $G/(C_G X_1 + \cdots + C_G X_n)$ is bounded as a group, say $tG \subseteq C_G X_1 + \cdots + C_G X_n$ for some nonzero integer t. If $f \in \mathrm{End}\, U_G$, then $f(X_i) \subseteq \mathbb{Q}G(\tau_i)$, where type $X_i = \tau_i$. Hence, there is a nonzero integer s with $sf : X_i \to G(\tau_i)$ for each i.

Now let $x \in G$. Then $tx = c_1 x_1 + \cdots + c_n x_n$ for some c_i in C_G and $x_i \in X_i$. Hence, $tsf(x) = sf(tx) = sf(c_1 x_1) + \cdots + sf(c_n x_n) = c_1(sf(x_1)) + \cdots + c_n(sf(x_n)) \in G$, since $sf \in \mathrm{End}\, U_G$ and each $c_i \in C_G \subseteq \mathrm{CEnd}\, U_G$. This shows that $tsf \in \mathrm{End}\, G$, whence $f \in \mathbb{Q}\mathrm{End}\, G$ and so $\mathrm{End}\, U_G = \mathbb{Q}\mathrm{End}\, G$.

Consequently, $\mathbb{Q}\mathrm{End}\, G$ has no nontrivial idempotents if and only if $\mathrm{End}\, U_G$ does. In particular, G is a strongly indecomposable group if and only if U_G is an indecomposable representation in $\mathrm{Rep}(S_T, \mathbb{Q})$. Finally, $C_G = \mathrm{CEnd}\, U_G \cap \mathrm{End}\, G = C\mathbb{Q}\mathrm{End}\, G \cap \mathrm{End}\, G = \mathrm{CEnd}\, G$.

For a finite lattice T of types, a torsion-free abelian group G is called a T-*group* if G is a countable Butler group with typeset G contained in T. The next theorem is the Butler group analogue of Theorem 1.5.3.

Theorem 3.5.5 *Let T be a finite lattice of types. The following statements are equivalent:*

(a) *The category $B(T)_{\mathbb{Q}}$ has finite representation type;*
(b) *There are no central-special quasi-generic T-groups;*
(c) *$\mathbb{Q}$End $H = \mathbb{Q}$ for each strongly indecomposable H in $B(T)_{\mathbb{Q}}$.*

PROOF. (a) $\Rightarrow$ (b) Suppose G is a central-special quasi-generic T-group. Then $\mathbb{Q}$End $G = $ EndU_G by Lemma 3.5.4. Now, $U_G = (\mathbb{Q}G, \mathbb{Q}(\tau):\tau \in S_T) \in$ Rep$(S_T, \mathbb{Q})$ and rep$(S_T, \mathbb{Q})$ has finite representation type in view of (a) and Theorem 3.3.2. By Theorem 1.5.3, U_G is the direct sum of finite-dimensional $\mathbb{Q}$-representations of S_T. This contradicts the fact that a quasi-generic group is strongly indecomposable with countably infinite rank.

(b) $\Rightarrow$ (a) Assume that $B(T)_{\mathbb{Q}}$ has infinite representation type. By Corollary 3.3.5(a), S_T contains one of the critical posets S_4, $(2, 2, 2)$, $(1, 3, 3)$, $(N, 4)$, or $(1, 2, 5)$ as a subposet. For each of these critical posets S_T, there is a central-special quasi-generic T-group given by Lemma 3.5.3. This contradicts (b).

(a) $\Rightarrow$ (c) This follows from the category equivalence of $B(T)_{\mathbb{Q}}$ with rep$(S_T, \mathbb{Q})$, Theorem 3.3.2, and the corresponding result for rep$(S_T, \mathbb{Q})$, Corollary 1.3.9.

(c) $\Rightarrow$ (a) Suppose that $B(T)_{\mathbb{Q}}$ has infinite representation type. By Corollary 3.3.5(a), it suffices to show that if S_T is one of the critical posets S_4, $(2, 2, 2)$, $(1, 3, 3)$, $(1, 2, 5)$, or $(N, 4)$, then there is a strongly indecomposable H in $B(T)_{\mathbb{Q}}$ with $\mathbb{Q}$End H not isomorphic to $\mathbb{Q}$. An application of Lemma 3.5.1(b) completes the proof.

The next two lemmas establish the framework for existence conditions for quasi-generic T-groups if $B(T)_{\mathbb{Q}}$ has wild representation type. The groups listed in the next lemma are obtained from the list of $\mathbb{Q}$-representations of the poset S_T given in Theorem 1.4.4. In each case, $U_G = (\mathbb{Q}G, \mathbb{Q}G(\tau): \in S_T)$ is the corresponding representation of S_T.

Lemma 3.5.6 *Let T be a finite lattice of types. For each of the following posets S_T, there is a strongly indecomposable torsion-free T-group G with countably infinite rank and $\mathbb{Q}$End $G = \mathbb{Q}\langle x, y \rangle$.*

Let $R = Z\langle x, y \rangle$ and A_i a subgroup of $\mathbb{Q}$ with $1 \in A_i$ and type $A_i = i$.

$$
\begin{array}{c}
5 \\
| \\
\end{array}
$$

(i) $S_T = S_5$ and $(1, 1, 1, 1, 2) = 1 \quad 2 \quad 3 \quad 4$

$$G = (A_1 R \oplus A_1 R \oplus 0 \oplus 0) + (0 \oplus 0 \oplus A_2 R \oplus A_2 R)$$
$$+ (1 + 0 + 1 + 0)A_3 R \oplus (0 + 1 + 0 + 1)A_3 R)$$
$$+ (1 + 0 + x + 1)A_4 R$$
$$+ ((1 + 0 + x + 1)A_5 R \oplus (0 + 1 + y + 0)A_5 R).$$

$$
\begin{array}{ccc}
 & 5 & \\
 & | & \\
2 & 4 & 6 \\
| & | & | \\
\text{(ii)} \quad S_T = (2, 2, 3) = 1 & 3 & 7 \ ,
\end{array}
$$

$$G = (y + x + 0 + 1 + 0 + 1)A_1 R \oplus (0 + 1 + 1 + 0 + 1 + 0)A_1 R$$
$$+ ((1 + 0 + 1 + 0 + 0 + 0)A_2 R \oplus (0 + 1 + 0 + 1 + 0 + 0)A_2 R$$
$$\oplus (y + x + 0 + 1 + 0 + 1)A_2 R \oplus (0 + 1 + 1 + 0 + 1 + 0)A_2 R)$$
$$+ (0 \oplus 0 \oplus 0 \oplus 0 \oplus A_3 R \oplus A_3 R) + (0 \oplus 0 \oplus A_4 R \oplus A_4 R \oplus A_4 R$$
$$\oplus A_4 R) + (A_7 R \oplus 0 \oplus 0 \oplus 0 \oplus 0 \oplus 0) + (A_6 R \oplus A_6 R \oplus 0 \oplus 0 \oplus 0$$
$$\oplus 0) + (A_5 R \oplus A_5 R \oplus A_5 R \oplus A_5 R \oplus 0 \oplus 0).$$

$$
\begin{array}{cc}
 & 8 \\
 & | \\
4 & 7 \\
| & | \\
3 & 6 \\
| & | \\
\text{(iii)} \quad S_T = (1, 3, 4) = 1 \quad 2 & 5 \ ,
\end{array}
$$

$$G = ((0 + 1 + 1 + 0 + 1 + 0 + 0 + 0)A_1 R$$
$$\oplus (y + x + 0 + 1 + 0 + 1 + 0 + 0)A_1 R$$
$$\oplus (1 + 0 + 1 + 0 + 0 + 0 + 1 + 0)A_1 R$$
$$\oplus (0 + 1 + 0 + 1 + 0 + 0 + 0 + 1)A_1 R) + (0 \oplus 0 \oplus 0 \oplus 0 \oplus 0 \oplus 0$$
$$\oplus A_2 R \oplus A_2 R) + (0 \oplus 0 \oplus 0 \oplus 0 \oplus A_3 R \oplus A_3 R \oplus A_3 R \oplus A_3 R)$$
$$+ (0 \oplus 0 \oplus A_4 R \oplus A_4 R \oplus A_4 R \oplus A_4 R \oplus A_4 R \oplus A_4 R)$$
$$+ (A_5 R \oplus 0 \oplus 0 \oplus 0 \oplus 0 \oplus 0 \oplus 0 \oplus 0) + (A_6 R \oplus A_6 R \oplus A_6 R$$
$$\oplus 0 \oplus 0 \oplus 0 \oplus 0 \oplus 0) + (A_7 R \oplus A_7 R \oplus A_7 R \oplus A_7 R \oplus A_7 R$$
$$\oplus 0 \oplus 0 \oplus 0) + (A_8 R \oplus A_8 R \oplus A_8 R \oplus A_8 R \oplus A_8 R$$
$$\oplus A_8 R \oplus A_8 R \oplus 0).$$

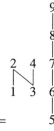

(iv) $T = (N, 5) =$
$$\begin{array}{c} 9 \\ | \\ 8 \\ | \\ 7 \\ | \\ 6 \\ | \\ 5 \end{array}$$

$$G = (0 \oplus 0 \oplus 0 \oplus 0 \oplus 0 \oplus 0 \oplus A_1R \oplus A_1R \oplus A_1R \oplus A_1R)$$
$$+ ((0 \oplus 0 \oplus A_2R \oplus A_2R \oplus 0 \oplus 0 \oplus A_2R \oplus A_2R \oplus A_2R \oplus A_2R)$$
$$\oplus (0+0+0+0+1+0+0+0+1+0)A_2R$$
$$\oplus (0+0+0+0+0+1+0+0+0+1)A_2R)$$
$$+ ((0+0+0+0+1+0+0+0+1+0)A_3R$$
$$\oplus (0+0+0+0+0+1+0+0+0+1)A_3R)$$
$$+ ((0+1+1+0+1+0+0+0+0+0)A_4R$$
$$\oplus (y+x+0+1+0+1+0+0+0+0)A_4R$$
$$\oplus (1+0+1+0+0+0+1+0+0+0)A_4R$$
$$\oplus (0+1+0+1+0+0+0+1+0+0)A_4R$$
$$\oplus (0+0+0+0+1+0+0+0+1+0)A_4R$$
$$\oplus (0+0+0+0+0+1+0+0+0+1)A_4R)$$
$$+ (A_5R \oplus 0 \oplus 0 \oplus 0 \oplus 0 \oplus 0 \oplus 0 \oplus 0 \oplus 0 \oplus 0)$$
$$+ (A_6R \oplus A_6R \oplus A_6R \oplus 0 \oplus 0 \oplus 0 \oplus 0 \oplus 0 \oplus 0 \oplus 0)$$
$$+ (A_7R \oplus A_7R \oplus A_7R \oplus A_7R \oplus A_7R \oplus 0 \oplus 0 \oplus 0 \oplus 0 \oplus 0)$$
$$+ (A_8R \oplus A_8R \oplus A_8R \oplus A_8R \oplus A_8R \oplus A_8R \oplus A_8R \oplus 0 \oplus 0 \oplus 0)$$
$$+ (A_9R \oplus A_9R \oplus A_9R \oplus A_9R \oplus A_9R \oplus A_9R \oplus A_9R \oplus A_9R \oplus 0 \oplus 0).$$

$$\begin{array}{ccc} & & 9 \\ & & | \\ & & 8 \\ & & | \\ 3 & & 7 \\ | & & | \\ 1 \quad 2 & & 6 \\ & & | \\ & & 5 \\ & & | \\ (v) \quad S_T = (1, 2, 6) = & & 4 \end{array}$$

$$G = ((0 + 0 + 0 + 1 + 0 + 0 + 0 + 0 + 0 + 1 + 0 + y)A_1 R$$
$$\oplus (0 + 0 + 1 + 0 + 0 + 0 + 0 + 0 + 1 + 0 + 1 + x)A_1 R$$
$$\oplus (0 + 1 + 0 + 0 + 0 + 1 + 0 + 0 + 0 + 1 + 1 + 0)A_1 R$$
$$\oplus (1 + 0 + 0 + 0 + 1 + 0 + 0 + 0 + 1 + 0 + 0 + 1)A_1 R$$
$$\oplus (0 + 0 + 0 + 0 + 0 + 0 + 0 + 1 + 0 + 0 + 1 + 0)A_1 R$$
$$\oplus (0 + 0 + 0 + 0 + 0 + 0 + 1 + 0 + 0 + 0 + 0 + 1)A_1 R)$$
$$+ (0 \oplus 0 \oplus 0 \oplus 0 \oplus 0 \oplus 0 \oplus 0 \oplus 0 \oplus A_2 R \oplus A_2 R \oplus A_2 R \oplus A_2 R)$$
$$+ (0 \oplus 0 \oplus 0 \oplus 0 \oplus A_3 R \oplus A_3 R \oplus A_3 R \oplus A_3 R \oplus A_3 R \oplus A_3 R$$
$$\oplus A_3 R \oplus A_3 R) + (A_4 R \oplus A_4 R \oplus 0 \oplus 0 \oplus 0 \oplus 0 \oplus 0 \oplus 0 \oplus 0 \oplus 0$$
$$\oplus 0 \oplus 0) + (A_5 R \oplus A_5 R \oplus A_5 R \oplus 0 \oplus 0 \oplus 0 \oplus 0 \oplus 0 \oplus 0 \oplus 0 \oplus 0$$
$$\oplus 0) + (A_6 R \oplus A_6 R \oplus A_6 R \oplus A_6 R \oplus 0 \oplus 0 \oplus 0 \oplus 0 \oplus 0 \oplus 0 \oplus 0$$
$$\oplus 0) + (A_7 R \oplus A_7 R \oplus A_7 R \oplus A_7 R \oplus A_7 R \oplus A_7 R \oplus 0 \oplus 0 \oplus 0 \oplus 0$$
$$\oplus 0 \oplus 0) + (A_8 R \oplus A_8 R \oplus A_8 R \oplus A_8 R \oplus A_8 R \oplus A_8 R$$
$$\oplus A_8 R \oplus A_8 R \oplus 0 \oplus 0 \oplus 0 \oplus 0) + (A_9 R \oplus A_9 R \oplus A_9 R$$
$$\oplus A_9 R \oplus A_9 R \oplus A_9 R \oplus A_9 R \oplus A_9 R \oplus A_9 R \oplus 0 \oplus 0).$$

PROOF. In each case, G is a countable torsion-free group with typeset $G \subseteq T$. It is sufficient to show that End $G \subseteq \mathbb{Q}\langle x, y \rangle$, whence G is strongly indecomposable, since $\mathbb{Q}\langle x, y \rangle = \mathbb{Q}\text{End } G$ has no nontrivial idempotents. The computations to show that End $G \subseteq \mathbb{Q}\langle x, y \rangle$ are as in Theorem 1.4.4. The idea is, using the invariance of the $G(\tau_i)$'s and the fact that $\mathbb{Q}A_i R = \mathbb{Q}\langle x, y \rangle$, to show that if $f \in \text{End}G$, then $f \in \text{End}_{\mathbb{Q}}\mathbb{Q}\langle x, y \rangle$ with $fx = xf$ and $yf = fy$. In this case, $f \in \text{End}_{\mathbb{Q}\langle x, y \rangle}\mathbb{Q}\langle x, y \rangle = \mathbb{Q}\langle x, y \rangle$.

A noncommutative version of Lemma 3.5.2 is needed to construct quasi-generic groups from the groups in Lemma 3.5.6 via Theorem 3.4.3.

Lemma 3.5.7 *There is a ring Δ with countable free additive group containing $\mathbb{Z}\langle x, y \rangle$ such that $\mathbb{Q}\Delta$ is a division algebra and $\Delta / p\Delta$ has no zero divisors for each prime p of $\mathbb{Z}$.*

PROOF. Let $R = \mathbb{Z}[x, t, 2]$ be a skew polynomial ring over $\mathbb{Z}$ in two indeterminates x and t subject to the condition that $tx = xt^2$. Induction arguments show that for positive integers m and n,

$$t^n x = xt^{2n}, \quad t^n x^m = x^m t^u \text{ for } u = 2^m n, \quad \text{and} \quad (xt)^n = x^n t^{\pi(n)},$$

where $\pi(n) = 2^{n-1} + 2^{n-2} + \cdots + 2^2 + 2 + 1$ if $n \geq 2$, $\pi(0) = 0$, and $\pi(1) = 1$.

It follows that a monomial $x^{e(1)}(xt)^{f(1)} \cdots x^{e(k)}(xt)^{f(k)}$ is equal to

$$x^{e(1)+f(1)+\cdots+e(k)+f(k)} t^{\beta(f(1),e(2),\ldots,e(k),f(k))},$$

where

$$\beta(f(1)) = \pi(f(1)) = 2^{f(1)-1} + 2^{f(1)-2} + \cdots + 2 + 1 \quad \text{and}$$

$$\beta(f(1), e(2), \ldots, e(k), f(k))$$
$$= 2^{f(k)+e(k)}\beta(f(1), e(2), \ldots, e(k-1), f(k-1)) + \pi(f(k))$$

is defined recursively for $k \geq 2$. For example, with $k = 2$,

$$\beta(f(1), e(2), f(2)) = 2^{e(2)+f(2)}\pi(f(1)) + \pi(f(2)).$$

Notice that $\beta(f(1), e(2), \ldots, e(k), f(k))$ is a sum of powers of 2.

The next step is to show that $\mathbb{Z}\langle x, y \rangle$ is isomorphic to the subring $\mathbb{Z}\langle x, xt \rangle$ of R. To see this, view $\beta(f(1), e(2), \ldots, e(k), f(k))$ as a 2-adic expansion of a natural number, i.e., sums of powers of 2 with coefficients either 0 or 1. As a consequence of the definition of $\beta(f(1), e(2), \ldots, e(k), f(k))$, this expansion has exactly k blocks of consecutive 1's as coefficients. Since 2-adic expansions of natural numbers are unique, k can be recaptured from $\beta(f(1), e(2), \ldots, e(k), f(k))$. Now suppose that two monomials $x^{e(1)}(xt)^{f(1)} \cdots x^{e(k)}(xt)^{f(k)}$ and $x^{e'(1)}(xt)^{f'(1)} \cdots x^{e'(k')}(xt)^{f'(k')}$ are equal. From the preceding paragraph,

$$\beta(f(1), e(2), \ldots, e(k), f(k)) = \beta(f'(1), e'(2), \ldots, e'(k), f'(k))$$

and

$$e(1) + f(1) + \cdots + e(k) + f(k) = e'(1) + f'(1) + \cdots + e'(k) + f'(k).$$

Hence, $k = k', e(i) = e'(i)$, and $f(i) = f'(i)$ for each i. This shows that $f : \mathbb{Z}\langle x, y \rangle \to \mathbb{Z}\langle x, xt \rangle$ induced by $f(x) = x$ and $f(y) = xt$ is a ring isomorphism, as desired.

Now, R is a *right Ore domain*, i.e., $aR \cap bR \neq \emptyset$ for each pair a, b of nonzero elements of R. This can be easily seen by observing that $\mathbb{Z}[t]$ is a right Ore domain and $g : \mathbb{Z}[t] \to \mathbb{Z}[t]$ induced by $f(t) = t^2$ is a one-to-one ring endomorphism. Hence, R has a right quotient ring D, a division algebra with each element of D of the form $f(x, xt)g(x, xt)^{-1}$ for $f(x, xt), g(x, xt) \in R$ and $g(x, xt) \neq 0$. Note that $\mathbb{Q}\langle x, y \rangle$ is isomorphic to $\mathbb{Q}\langle x, xt \rangle$, a subring of D.

Given $g(x, xt)$ in R, define $c(g(x, xt))$ to be the gcd of the coefficients of $g(x, xt)$. It is not difficult to confirm that $c(g(x, xt)f(x, xt)) = c(g(x, xt))c(f(x, xt))$, just as in Gauss's theorem for polynomials in one variable. Let $T = \{g(x, xt) \in R : c(g(x, xt)) = 1\}$. Then T is a multiplicatively closed set. Hence, $\Delta = RT^{-1} = \{f(x, xt)g(x, xt)^{-1} : f(x, xt) \in R, g(x, xt) \in T\}$ is a subring of D. An argument like that of Lemma 3.5.2 shows that $\mathbb{Q}\Delta = D$, the additive group of Δ, is a countable free abelian group, and $\Delta/p\Delta$ has no zero divisors for each prime p of $\mathbb{Z}$.

The category $B(T)_{\mathbb{Q}}$ is *generically wild* if there is a special quasi-generic T-group G such that $\mathbb{Q}\text{End } G/J\mathbb{Q}\text{End } G$ contains a copy of $\mathbb{Q}\langle x, y \rangle$. Such a G cannot be central-special, since the center of $\mathbb{Q}\langle x, y \rangle$ is $\mathbb{Q}$ and $\mathbb{Q}G$ has infinite $\mathbb{Q}$-dimension. The next theorem is a Butler group version of Theorem 1.5.4 for representations.

Theorem 3.5.8 *For the following statements, $(a) \Rightarrow (b) \Rightarrow (c)$:*

(a) $B(T)_{\mathbb{Q}}$ *has wild representation type;*
(b) $B(T)_{\mathbb{Q}}$ *is generically wild;*
(c) $B(T)_{\mathbb{Q}}$ *is endowild.*

If $S_T = S_n$, then (a), (b), and (c) are all equivalent to the condition that $n \geq 5$.

PROOF. (a) $\Rightarrow$ (b) Assume that $B(T)_{\mathbb{Q}}$ has wild representation type. By Corollary 3.3.5(b), S_T contains one of the posets S_5, $(1, 1, 1, 2)$, $(2, 2, 3)$, $(1, 3, 4)$, $(1, 2, 6)$, or $(N, 5)$ as a subposet. By Lemma 3.5.7 there is a ring Δ with countable free additive group such that $\mathbb{Q}\Delta$ is a division algebra containing $\mathbb{Q}\langle x, y \rangle$ and $\Delta/p\Delta$ has no zero divisors for each prime p of Z.

Given a group G in Lemma 3.5.6, define $H = G\Delta$. Then H is a countable Butler group of infinite rank with typeset contained in T by Theorem 3.4.3. Moreover, H is a special group from the definition of G. Just as in Lemma 3.5.6, $\mathbb{Q}\text{End } H = \mathbb{Q}\Delta$ is a division algebra containing $\mathbb{Q}\langle x, y \rangle$. Thus, H is strongly indecomposable. Since H has finite dimension over $\mathbb{Q}\text{End } H$, H is a quasi-generic T-group.

(b) $\Rightarrow$ (c) Assume that $B(T)_{\mathbb{Q}}$ is generically wild and let G be a special quasi-generic T-group with $\mathbb{Q}\text{End } G/J\mathbb{Q}\text{End } G$ a division $\mathbb{Q}$-algebra containing $\mathbb{Q}\langle x, y \rangle$. Then $\mathbb{Q}\langle x, y \rangle = \mathbb{Q}Z\langle x, y \rangle$ is contained in $\mathbb{Q}\text{End } G$, so that $\mathbb{Z}\langle x, y \rangle \subseteq \text{End } G$. Since G is special, there are pure rank-1 subgroups X_i of G with $G/((\text{End } G)X_1 + \cdots + (\text{End } G)X_n)$ bounded. Define $H = \mathbb{Z}\langle x, y \rangle X_1 + \cdots + \mathbb{Z}\langle x, y \rangle X_n$. Then H is a countable torsion-free abelian group with typeset H contained in T and $\mathbb{Z}\langle x, y \rangle \subseteq \text{End } H$.

As a consequence of Example 1.1.7, for each finite-dimensional $\mathbb{Q}$-algebra X, there are matrices A and B with X isomorphic to $C(A, B)$. Suppose that A and B are two nonzero $m \times m$ $\mathbb{Q}$-matrices with minimal polynomials $f(x)^i$ and $g(y)^j$, respectively, for irreducible polynomials $f(x) \in \mathbb{Q}[x]$ and $g(y) \in \mathbb{Q}[y]$. Let $L = H/KH$, where $K = \langle xy, f(x)^i, g(y)^j \rangle \cap \mathbb{Z}\langle x, y \rangle$ is an ideal of $\mathbb{Z}\langle x, y \rangle$.

Then $L \in B(T)_{\mathbb{Q}}$, since the $\mathbb{Q}$-dimension of $\mathbb{Q}L$ is finite and H is generated as a $\mathbb{Z}\langle x, y \rangle$-module by finitely many pure rank-1 subgroups with types in T. Moreover, $C(A, B)$ is contained in $\mathbb{Q}\text{End}_{\mathbb{Q}\langle x,y \rangle}L$. This is so because x acts on $\mathbb{Q}H/\mathbb{Q}KH$ by A and y acts on $\mathbb{Q}H/\mathbb{Q}KH$ by B. In particular, $C(A, B)$ is a $\mathbb{Q}$-subalgebra of $\mathbb{Q}\text{End } L$. It follows that $B(T)_{\mathbb{Q}}$ is endowild.

If $S_T = S_n$ and $B(T)_\mathbb{Q}$ does not have wild representation type, then $n \leq 4$ by Corollary 3.3.5. Also, if G is a strongly indecomposable group in $B(T)_\mathbb{Q}$, then $\mathbb{Q}\text{End } G = \text{End } U_G$ for U_G an indecomposable in $\text{rep}(S_n, \mathbb{Q})$ by Theorem 3.3.2. However, each such End U_G is a factor algebra of $\mathbb{Q}[x]$ by Example 6.2.7. This contradicts (c). Hence, (c) $\Rightarrow$ (a) if $S_T = S_n$.

There is an interpretation of tame representation type of $B(T)_\mathbb{Q}$ in terms of central-special quasi-generic T-groups. Suppose R is an integral domain with $\mathbb{Q}R = F$, a field. A finite rank torsion-free R-module M is said to be a *finite rank R-Butler module* if there are finitely many pure rank-1 subgroups X_i of M such that $M/(RX_1 + \cdots + RX_n)$ is bounded as a group. A finite rank R-Butler module M is *strongly indecomposable* if $F\text{End}_R M / J F\text{End}_R M$ is a division F-algebra. These definitions directly generalize the definition of a finite rank Butler group.

Quasi-generic Butler groups can be constructed from finite rank R-Butler modules via Theorem 3.4.3. An additional condition on the endomorphism ring yields a central-special quasi-generic group.

Lemma 3.5.9 *Let R be an integral domain with countably free additive group of infinite rank such that pR is a prime ideal of R for each prime p of $\mathbb{Z}$ and $\mathbb{Q}R = F$ is a field. If G is a finite rank strongly indecomposable R-Butler module with $R \subseteq C_G$, then G is a central-special quasi-generic Butler group with $\text{End } G = \text{End}_R G$ and $\mathbb{Q}\text{End } G / J\mathbb{Q}\text{End } G$ a division F-algebra of finite F-dimension.*

PROOF. By Theorem 3.4.3, G is a Butler group with countably infinite rank. Then G is a central-special group, since G is an R-Butler module and $R \subseteq C_G$. Hence, $C_G = \text{CEnd } G$ by Lemma 3.5.4. Since G is a finite rank strongly indecomposable R-Butler module, $\mathbb{Q}\text{End}_R G / J\mathbb{Q}\text{End}_R G = F\text{End}_R G / J F\text{End}_R G$ is a division F-algebra. But $\mathbb{Q}\text{End } G = \mathbb{Q}\text{End}_R G$, since $R \subseteq \text{CEnd } G$, and so $\mathbb{Q}\text{End } G / J\mathbb{Q}\text{End } G$ is a division $\mathbb{Q}$-algebra. Hence, G is strongly indecomposable, and $\mathbb{Q}G$ has finite length as a $\mathbb{Q}\text{End } G$-module. Since G has countably infinite rank, G is quasi-generic.

For $H \in B(T)_\mathbb{Q}$, define cdn $H = (\text{rank } H, \text{rank } H(\tau) : \tau \in S_T)$. Then cdn $H = $ cdn U_H, where $U_H = (\mathbb{Q}H, \mathbb{Q}H(\tau) : \tau \in S_T)$. The next lemma generalizes the construction of central-special quasi-generic groups given in Lemma 3.5.3.

Lemma 3.5.10 *Suppose $N = (N_0, N_i : i \in S_T)$ is an indecomposable in $\text{rep}(S_T, \mathbb{Q}[x])$ with $\text{End}_\mathbb{Q} N = \text{End}_{\mathbb{Q}[x]} N$. There is a central-special quasi-generic T-group G such that $U_G = N\Lambda \in \text{rep}(S_T, \mathbb{Q}(x))$, End $G = \text{End}_\Lambda G$, and $\mathbb{Q}\text{End } G / J\mathbb{Q}\text{End } G$ is a finite-dimensional division $Q(x)$-algebra.*

PROOF. Let Λ denote the principal ideal domain given in Lemma 3.5.2. Then $N\Lambda = (N_0\Lambda, N_i\Lambda : i \in S_T) \in \text{rep}(S_T, \mathbb{Q}(x))$ as $Q\Lambda = Q[x]\Lambda = Q(x)$. In fact, $N\Lambda$ is an indecomposable $\mathbb{Q}(x)$-representation, since N is an indecomposable

$\mathbb{Q}[x]$-representation. This holds because $\text{End}_{\mathbb{Q}(x)}(N\Lambda) = \mathbb{Q}(x)\text{End}_{\mathbb{Q}[x]}N$ since $\mathbb{Q}(x)$ is the quotient field of $\mathbb{Q}[x]$ and N is finitely generated as a $\mathbb{Q}[x]$-module.

Now, $N\Lambda = U_G = (\mathbb{Q}G, \mathbb{Q}G(\tau):\tau \in S_T)$ for some finite rank Λ-Butler module G with $\text{End}_{\mathbb{Q}(x)}N\Lambda$ isomorphic to $\mathbb{Q}\text{End}_\Lambda G$ (Exercise 3.3.2), noting that $\mathbb{Q}G = \mathbb{Q}(x)G$ and $\mathbb{Q}(x)$ is the quotient field of the principal ideal domain Λ. Since $N\Lambda$ is an indecomposable $\mathbb{Q}(x)$-representation, $\mathbb{Q}\text{End}_\Lambda G$ has no nontrivial idempotents. This shows that G is a strongly indecomposable Λ-Butler module.

By hypothesis, $\text{End}_\mathbb{Q} N = \text{End}_{\mathbb{Q}[x]}N$, so that $\mathbb{Q}(x) \subseteq \text{CEnd}_\mathbb{Q} N\Lambda = \text{CEnd}_\mathbb{Q} U_G$ and $\Lambda \subseteq C_G = (\text{CEnd}_\mathbb{Q} U_G) \cap (\text{End } G)$. It follows from Lemma 3.5.9 and Theorem 3.4.3 that G is a central-special quasi-generic T-group with $\text{End } G = \text{End}_\Lambda G$ and $\mathbb{Q}\text{End } G / J\mathbb{Q}\text{End } G$ a finite-dimensional division $Q(x)$-algebra.

Given a finite lattice of types T, a countable torsion-free abelian group H is *pregeneric* if H is a torsion-free $\mathbb{Z}[x]$-module, $H/(\mathbb{Z}[x]X_1 + \cdots + \mathbb{Z}[x]X_n)$ is bounded for pure rank-1 subgroups X_i of H with type $X_i \in T$, and $\text{End } H = \text{End}_{\mathbb{Z}[x]}H$. A pregeneric group H is special, but a strongly indecomposable pregeneric group need not be quasi-generic. For instance, all the groups listed in Lemma 3.5.1 are pregeneric groups.

As a consequence of the next theorem, if $B(T)_\mathbb{Q}$ has tame representation type, then each strongly indecomposable group in $B(T)_\mathbb{Q}$ can be constructed as a factor group of a pregeneric subgroup of a central-special quasi-generic T-group.

Theorem 3.5.11 *The category $B(T)_\mathbb{Q}$ has tame representation type if and only if for each sequence of nonnegative integers $w = (w_0, w_t :t \in S_T)$ there are finitely many strongly indecomposable pregeneric groups $H_1, \ldots, H_n$ such that:*

(a) *For each i, $G_i = H_i\Lambda$ is a central-special quasi-generic T-group with $\text{End}_\Lambda G_i = \text{End } G_i$.*

(b) *If $H \in B(T)_\mathbb{Q}$ is strongly indecomposable with cdn $H = w$, then H is quasi-isomorphic to $H_i A$ for some $A = \mathbb{Z}[x]/\langle f(x)^e \rangle$ and $f(x)$ an irreducible polynomial in $\mathbb{Z}[x]$.*

PROOF. First assume that (a) and (b) hold. Fix w and define $N_i = (\mathbb{Q}H_i, \mathbb{Q}H_i(\tau): \tau \in S_T) \in \text{rep}(S_T, \mathbb{Q}[x])$. If $U \in \text{rep}(S_T, \mathbb{Q})$ is an indecomposable representation with cdn $U = w$, then $U = (\mathbb{Q}H, \mathbb{Q}H(\tau):\tau \in S_T)$ for some strongly indecomposable H in $B(T)_\mathbb{Q}$ and cdn $H = w$ by Theorem 3.3.2. By (b), H is quasi-isomorphic to $K = H_i A$ for some $A = \mathbb{Z}[x]/\langle f(x)^e \rangle$, $f(x)$ an irreducible polynomial in $\mathbb{Z}[x]$. Hence, $\mathbb{Q}A = \mathbb{Q}[x]/\langle f(x)^e \rangle$ with $f(x)$ irreducible in $\mathbb{Q}[x]$. Then $U = (\mathbb{Q}H, \mathbb{Q}H(\tau):\tau \in S_T)$ is isomorphic to $(\mathbb{Q}K, \mathbb{Q}K(\tau):\tau \in S_T) = (\mathbb{Q}H_i A, H_i(\tau)\mathbb{Q}[x]\mathbb{Q}A : \tau \in S_T) = N_i\mathbb{Q}A$, noting that $K(\tau) = H_i(\tau)\mathbb{Z}[x]A$ for each type τ. This proves that $B(T)_\mathbb{Q}$ has tame representation type.

Conversely, assume that $B(T)_\mathbb{Q}$ has tame representation type. Fix w and choose indecomposable $N_1, \ldots, N_m \in \text{rep}(S_T, \mathbb{Q}[x])$ with $\mathbb{Q}[x] \subseteq \text{CEnd } N_i$ such that if $U \in \text{rep}(S_T, k)$ is indecomposable with cdn $U = w$, then U is isomorphic to $N_i B$ for some indecomposable cyclic $\mathbb{Q}[x]$-module B. By Lemma 3.5.10, there is a central-special quasi-generic T-group G_i such that $(\mathbb{Q}G_i, \mathbb{Q}G_i(\tau):\tau \in S_T) =$

$N_i \Lambda \in \text{rep}(S_T, \mathbb{Q}(x))$, End $G_i = \text{End}_\Lambda G_i$, and $\mathbb{Q}E(G_i)/J\mathbb{Q}E(G_i)$ is a finite-dimensional $\mathbb{Q}(x)$-algebra. In fact, G_i is a finite rank strongly indecomposable Λ-Butler module, so there are pure rank-1 subgroups $X_1, \ldots, X_m$ of G_i such that $G_i/(\Lambda X_1 + \cdots + \Lambda X_m)$ is bounded. Hence, G_i is a central-special quasi-generic T-group by Lemma 3.5.9.

Define $H_i = \mathbb{Z}[x]X_1 + \cdots + \mathbb{Z}[x]X_n$, a $\mathbb{Z}[x]$-module with End $H_i = \text{End}_{\mathbb{Z}[x]}H_i$, since End $G_i = \text{End}_\Lambda G_i$. Then H_i is a pregeneric group with $G_i = H_i \otimes_{\mathbb{Z}[x]}\Lambda$. Moreover, H_i is strongly indecomposable, since G_i is strongly indecomposable. This proves (a). As for (b), let H be a strongly indecomposable in $B(T)_\mathbb{Q}$. Then $U_H = (\mathbb{Q}H, \mathbb{Q}H(\tau) : \tau \in S_T)$ is an indecomposable in $\text{rep}(S_T, \mathbb{Q})$, and so $U_H = N_i \otimes_{\mathbb{Q}[x]}B$ for some $B = \mathbb{Q}[x]/\langle g(x)^e \rangle$ by the definition of tame representation type. Moreover, $K = H_i A$ is quasi-isomorphic to H for some $A = \mathbb{Z}[x]/\langle f(x)^e \rangle$ with $\mathbb{Q}A = B$. This is so because $U_K = (\mathbb{Q}K, \mathbb{Q}K(\tau) : \tau \in S_T) = (\mathbb{Q}H_i A, \mathbb{Q}H_i(\tau)A : \tau \in S_T) = N_i B = U_H$, whence K is quasi-isomorphic to H by Theorem 3.3.2. The proof of (b) is now complete.

Open Question: *Are there existence conditions on central-special quasi-generic T-groups, or their endomorphism rings, equivalent to tame representation type for $B(T)_\mathbb{Q}$? This question is the analogue of the open questions in Section 1.5 for k-representations of a finite poset.*

NOTES ON CHAPTER 3

Section 3.1 is an exposition of the traditional tools of the theory of torsion-free abelian groups of finite rank used in this book. Additional properties and examples of such groups are in [Fuchs 73] and [Arnold 82]. The notion of τ-socle dates back to [Baer 37], but the first systematic development of τ-radicals seems to be due to [Lady 79].

Finite rank Butler groups are called quasi-essential groups in [Koehler 65], purely finitely generated groups in [Bican 70, 78], R-groups in [Butler 65], and π-diagrammatic groups in [Butler 87]. One of the attractive features of this category of groups is the number and variety of characterizations of these groups within the category of torsion-free abelian groups of finite rank. The work of [Koehler 65], as acknowledged by [Butler 87], is fundamental to Corollary 3.2.4, Theorem 3.2.5, and Theorem 3.3.2. Some other characterizations are given in [Arnold 81].

Results of [Bican 70, 78], including Proposition 3.2.15, give rise to a characterization of finite rank Butler groups in terms of balanced extensions, Theorem 3.4.6. This characterization provides both a natural definition for Butler groups of arbitrary rank and a host of associated problems for torsion-free abelian groups. The resolution of these problems involves sophisticated set theory and logic; see [Fuchs 94]. Included in [Fuchs, Metelli 92] is a development of properties of countable Butler groups using techniques potentially useful for Butler groups of arbitrary rank.

An error in an argument of [Bican, Salce 83], repeated in [Arnold 86] as observed in [Mines, Vinsonhaler 92], is corrected in [Bican, Rangaswamy 95]. The corrected argument can be used to provide an alternative proof of Theorem 3.4.6. In another direction, the properties listed in Theorem 3.2.5(b) characterizing finite rank Butler groups have been considered for torsion-free abelian groups of infinite rank in [Mader, Mutzbauer, Rangaswamy 94].

In a series of papers, summarized in [Lady 83], the results of Section 3.2 are extended to a class of finite rank torsion-free modules over Dedekind domains called Butler modules. The setting is different from that of this manuscript, since the quasi-homomorphism category of finite rank Butler modules is subdivided into subcategories determined by a generalization of the notion of a splitting field. These subcategories are then related to categories of finitely generated modules over finite-dimensional algebras. Some partial results on representation type of these subcategories are obtained from classifications of representations of species [Ringel 76].

Finite rank Butler modules over 1-dimensional Noetherian domains are considered in [Goeters 99]. Other generalizations include valuated vector spaces, [Richman 84], and groups with Murley groups playing the role of rank-1 groups [Albrecht, Goeters 98].

Quasi-generic groups arose out of the notion of generic representations given in Section 1.5. There are functorial correspondences from categories of infinite-dimensional representations of posets to categories of countable Butler groups, [Dugas, Thomé 91] and [Rangaswamy, Vinsonhaler 94], but the morphisms for groups are "local quasi-homomorphisms," not quasi-homomorphisms as defined herein. This provides an obstacle to the determination of a Butler group analogue of generic representations. The point of Theorem 3.5.11 is that in the tame representation type case, strongly indecomposable Butler groups of finite rank are determined by quasi-generic Butler groups of infinite rank.

4

Representations over a Discrete Valuation Ring

4.1 Finite and Rank-Finite Representation Type

There are several choices for extending the definition of representations of finite posets over fields to representations over discrete valuation rings [Plahotnik 76]. The choice made herein arises naturally from the connection between representations of finite posets over discrete valuation rings and isomorphism at p categories of finite rank Butler groups; Section 4.3. Posets with finite representation type are characterized in Corollary 4.1.7.

A *discrete valuation ring* is a principal ideal domain R with a unique prime ideal pR. If R is a discrete valuation ring, then R is a local ring, the Jacobson radical of R is pR, and R/pR is a field. An important example from abelian group theory is the discrete valuation ring $\mathbb{Z}_p$, p a prime of $\mathbb{Z}$. Specifically, a torsion-free abelian group is a p-local group if and only if it is a torsion-free $\mathbb{Z}_p$-module. Observe that $p\mathbb{Z}_p$ is the unique prime ideal of $\mathbb{Z}_p$ and $\mathbb{Z}_p/p\mathbb{Z}_p = \mathbb{Z}/p\mathbb{Z}$ is a field.

Throughout this chapter, R denotes a discrete valuation ring. If M is a finitely generated free R-module and N is a *pure submodule of* M, i.e., $N \cap pM = pN$, then M/N is a finitely generated torsion-free R-module. Since R is a discrete valuation ring, in particular a principal ideal domain, M/N is a finitely generated free R-module, and so N is a summand of M. Consequently, a submodule of a finitely generated free R-module is a summand if and only if it is a pure submodule.

Let S be a finite poset and define rep(S, R) to be the category with objects $U = (U_0, U_i : i \in S)$ such that U_0 is a finitely generated free R-module, each U_i is a summand of U_0, and U_i is contained in U_j, provided that $i \leq j$ in S. A morphism from $U = (U_0, U_i : i \in S)$ to $V = (V_0, V_i : i \in S)$ is an R-homomorphism

$f : U_0 \to V_0$ with $f(U_i) \subseteq V_i$ for each i in S. Since R is a discrete valuation ring, each U_i is a finitely generated free R-module, being a summand of a free R-module U_0.

The *rank* of a representation $U = (U_0, U_i : i \in S) \in \mathrm{rep}(S, R)$ is the rank of U_0 as a free R-module. Rank-1 representations are easily described. Up to isomorphism, they are of the form $U = (U_0, U_i : i \in S)$ with $U_0 = R$ and $U_i = 0$ or R, subject to the condition that if $i \leq j$ in S and $U_i = R$, then $U_j = R$.

Given an integer $j \geq 0$, let $\mathrm{rep}(S, R, j)$ denote the full subcategory of $\mathrm{rep}(S, R)$ with objects $(U_0, U_i : i \in S)$ such that $p^j U_0 \subseteq \Sigma\{U_i : i \in S\} \subseteq U_0$. In particular, $\mathrm{rep}(S, R, 0)$ consists of representations $U = (U_0, U_i : i \in S)$ with $U_0 = \Sigma\{U_i : i \in S\}$ and each U_i a summand of U_0. The category $\mathrm{rep}(S, R, 0)$ is a direct generalization of the category of elements U in $\mathrm{rep}(S, k)$ with no trivial summands, k a field.

Define $\mathrm{Ind}(S, R, j)$ to be the set of isomorphism classes of indecomposable representations of $\mathrm{rep}(S, R, j)$.

Example 4.1.1 *If S is a finite poset with $w(S) = 1$ and $j \geq 0$ is an integer, then S is a chain and $\mathrm{rep}(S, R, j)$ has finite representation type. Elements of $\mathrm{Ind}(S, R, j)$ are rank-1 representations, and each indecomposable representation has endomorphism ring R.*

PROOF. The argument of Lemma 1.3.3(a) for fields carries over to this setting.

For posets that are not chains, there are more indecomposable representations for discrete valuation rings than there are for fields.

Example 4.1.2 *The category $\mathrm{rep}(S_2, R, j)$ has finite representation type for each $j \geq 0$. The elements of $\mathrm{Ind}(S_2, R, j)$ are:*

(i) *rank-1 representations $(R, 0, 0)$, (R, R, R), $(R, R, 0)$, and $(R, 0, R)$ with endomorphism ring R.*
(b) *rank-2 representations $(R \oplus R + (1 + 1)R(1/p^n), R \oplus 0, 0 \oplus R)$ with endomorphism ring $\{(r, s) \in R \times R : r - s \in p^n R\}$ for each $1 \leq n \leq j$.*

PROOF. Let $U = (U_0, U_1, U_2) \in \mathrm{rep}(S_2, R, j)$ be an indecomposable representation with rank $U_0 > 1$. Notice that each U_i is nonzero and $U_1 \cap U_2 = 0$, since otherwise U is a rank-1 representation as in Example 1.1.3. Hence, $p^n U_0 \subseteq U_1 \oplus U_2 \subseteq U_0$ for some smallest positive integer $n \leq j$, observing that if $n = 0$, then $U_0 = U_1 \oplus U_2$, and U would again be a rank-1 representation.

Now, $V = (U_1 \oplus U_2, U_1, U_2, p^n U_0)$ is an R-representation of S_3 with $p^n U_0$ a free R-module, being isomorphic to U_0, and $p^n U_0 \cap U_i = p^n U_i$, since U_i is a pure submodule of U_0. However, $p^n U_0$ need not a summand of $U_1 \oplus U_2$. There is an isomorphism $\mathrm{End}\, U \to \mathrm{End}\, V$ given by sending f to the restriction of f to $U_1 \oplus U_2$, recalling that $\mathrm{End}\, V$ is the ring of R-endomorphisms f of $U_1 \oplus U_2$ with

$f(U_i) \subseteq U_i$ and $f(p^n U_0) \subseteq U_0$. Then V is an indecomposable representation with rank greater than 1.

The remainder of the proof is a matrix argument generalizing that of Example 1.1.5 for representations of S_3 over a field. The setting is that $U = (U_0, U_1, U_2)$ is indecomposable and $V = (U_1 \oplus U_2, U_1, U_2, p^n U_0)$ is an indecomposable representation with End V isomorphic to End U.

Let $B_1 = \{x_1, \ldots, x_r\}$ be an R-basis of U_1, $B_2 = \{y_1, \ldots, y_s\}$ an R-basis of U_2, and $B_3 = \{z_1, \ldots, z_t\}$ an R-basis of $p^n U_0$. For each i,

$$z_i = \Sigma a_{ij} x_j + \Sigma b_{ij} y_j \text{ for some } a_{ij}, b_{ij} \in R.$$

In particular, $p^n U_0$ can be interpreted as the row space of a $t \times (r + s)$ R-matrix

$$M_V = (A \mid B),$$

where the rows of M_V are labeled by B_3, $A = (a_{ij})$ is a $t \times r$ R-matrix with columns labeled by B_1, and $B = (b_{ij})$ is a $t \times s$ R-matrix with columns labeled by B_2.

The following invertible R-matrix operations on M_V do not change V:

(a) Elementary column operations within A (a basis change for U_1);
(b) Elementary column operations within B (a basis change for U_2);
(c) Elementary row operations on M (a basis change for $p^n U_0$).

Write $M_V \approx N$ if the matrix N can be obtained from M_V by a sequence of operations (a), (b), and (c). Since R is a principal ideal domain, the block matrix A can be reduced by a series of elementary invertible row and column operations to a diagonal matrix [Hungerford 74]. Since n is minimal and R is a discrete valuation ring, this diagonal matrix can be chosen to be

$$\begin{pmatrix} I & 0 \\ 0 & pD \end{pmatrix}$$

with D a diagonal R-matrix and I an identity matrix, agreeing that a 0×0 matrix is empty.

If $x = u_1 \oplus u_2 \in p^n U_0$ with $u_i \in U_i$, and p-height $x \leq n$, then p-height $u_1 = p$-height u_2, since U_1 and U_2 are pure submodules of U_0. It follows that M_V is equivalent to a matrix of the form

$$\begin{pmatrix} I & 0 & \cdots & 0 & B_{11} & B_{12} & \cdots & B_{1n} \\ 0 & pI & \cdots & 0 & pB_{21} & pB_{22} & \cdots & pB_{2n} \\ & & \cdots & & & & \cdots & \\ 0 & 0 & \cdots & p^n I & p^n B_{n1} & p^n B_{n2} & \cdots & p^n B_{nn} \end{pmatrix}$$

with each row of $(p^{i-1} B_{i1}, p^{i-1} B_{i2}, \ldots, p^{i-1} B_{in})$ containing an element of

p-height $i - 1$. If E is an operation as in (c), then

$$(I \quad 0 \quad \cdots \quad 0 \mid B_{11} \quad B_{12} \quad \cdots \quad B_{1n})$$

$$\approx (E \quad 0 \quad \cdots \quad 0 \mid EB_{11} \quad EB_{12} \quad \cdots \quad EB_{1n})$$

$$\approx (I = EE^{-1} \quad 0 \quad \cdots \quad 0 \mid EB_{11} \quad EB_{12} \quad \cdots \quad EB_{1n})$$

as an application of (c) and (a). Applying operations (b) and (c) to diagonalize the matrix $(B_{11} \ B_{12} \ \cdots \ B_{1n})$ yields $M_V \approx N =$

$$\begin{pmatrix} I & 0 & \cdots & 0 & I & 0 & \cdots & 0 \\ 0 & pI & \cdots & 0 & pB_{21} & pB_{22} & \cdots & pB_{2n} \\ & & \cdots & & & & \cdots & \\ 0 & 0 & \cdots & p^n I & p^n B_{n1} & p^n B_{n2} & \cdots & p^n B_{nn} \end{pmatrix}.$$

Now use (c) followed by (a) to see that $N \approx$

$$\begin{pmatrix} I & 0 & \cdots & 0 & I & 0 & \cdots & 0 \\ 0 & pI & \cdots & 0 & 0 & pC_{22} & \cdots & pC_{2n} \\ & & \cdots & & & & \cdots & \\ 0 & 0 & \cdots & p^n I & 0 & p^n C_{n2} & \cdots & p^n C_{nn} \end{pmatrix}.$$

Since n is chosen to be minimal, I must be nonempty.

As in Example 1.1.5, $V = (Rx_1 \oplus Ry_1, Rx_1, Ry_1, R(x_1 + y_1))$, since $V = (U_1 \oplus U_2, U_1, U_2, p^n U_0)$ is indecomposable. Then $U = (Rx_1 \oplus Ry_1 + R(x_1+y_1)/p^n, Rx_1, Ry_1)$ is isomorphic to $W = (R \oplus R + R(1 + 1)/p^n, R \oplus 0, 0 \oplus R)$. A straightforward calculation shows that $f \in \text{End } W$ if and only if $f = (r, s) \in R \times R$ with $r - s \in p^n R$. Hence, $\text{End } U$ is isomorphic to $\{(r, s) \in R \times R : r - s \in p^n R\}$.

For a finite poset S, define S^* to be the disjoint union of S and an element $*$ unrelated to any element of S. For example, If $S = S_n$ is an antichain, then $S^* = S_{n+1}$. Given $n \geq 2$, let C_n denote a chain with n elements. Then C_n^* is the poset $(n, 1)$. For example, $C_1^* = (1, 1) = S_2$ and $C_2^* = (1, 2)$.

Proposition 4.1.3 *Let $n \geq 2$ be an integer.*

(a) *There is a fully faithful functor $F_j : \text{rep}(C_{n-1}^*, R, j) \to \text{rep}(C_n^*, R, 0)$ for each $j \geq 0$.*

(b) *If $V \in \text{rep}(C_n^*, R, 0)$ is indecomposable with rank > 1, then there is some integer $j \geq 0$ and indecomposable $U \in \text{rep}(C_{n-1}^*, R, j)$ with $F_j(U) = V$.*

PROOF. (a) Write $C_{n-1} = \{1 < 2 < \cdots < n - 1\}$ and let $U = (U_0, U_1 \subseteq \cdots \subseteq U_{n-1}, U_*) \in \text{rep}(C_{n-1}^*, R, j)$. Define $F_j(U) = (U_0, U_1 \subseteq \cdots \subseteq U_{n-1} \subseteq U_0, U_*) \in \text{rep}(C_n^*, R, 0)$. It is routine to verify that with $F_j(f) = f$, $F_j : \text{rep}(C_{n-1}^*, R, j) \to \text{rep}(C_n^*, R, 0)$ is a fully faithful functor.

(b) Let $V = (V_0, V_1 \subseteq \cdots \subseteq V_n, V_*)$ be an indecomposable representation in $\text{rep}(C_n^*, R, 0)$ with rank > 1. Then $V_0 = V_n + V_*$, and so $V_* = W_1 \oplus (V_* \cap V_n)$

for some W_1. This holds because $V_* \cap V_n$ is a pure submodule, hence a summand, of V_*, since V_n is a pure submodule of V_0. Therefore,

$$V_0 = V_n + V_* = W_1 \oplus V_n$$

and

$$V = (W_1, 0, \ldots, 0, 0, W_1) \oplus (V_n, V_1, \ldots, V_{n-1}, V_n, V_n \cap V_*)$$

is a representation direct sum. But W_1 is a free R-module. Since V is indecomposable with rank > 1, it follows that $W_1 = 0$, $V_0 = V_n$ contains V_*, and $V = (V_n, V_1, \ldots, V_{n-1}, V_n, V_*)$.

Furthermore, $p^j V_n$ is contained in $V_{n-1} + V_*$ for some j. To see this, let W be the pure submodule of V_n generated by $V_{n-1} + V_*$. Then $V_n = W' \oplus W$ for some W' and

$$V = (W', 0, \ldots, 0, W', 0) \oplus (W, V_1, \ldots, V_{n-1}, W, V_*).$$

Again, $W' = 0$, since V is indecomposable of rank greater than 1, and so $V_n = W$. Thus, there is some j with $p^j V_n$ contained in $V_{n-1} + V_*$, since V_n is a finitely generated free R-module with rank $V_n = \text{rank } V_{n-1} + V_*$. Hence, $U = (V_n, V_1, \ldots, V_{n-1}, V_*) \in \text{rep}(C_{n-1}^*, R, j)$ with $V = F_j(U)$. As a consequence of (a), U must be indecomposable.

Corollary 4.1.4 *If $S = (1, 2)$, then $\text{rep}(S, R, 0)$ has infinite representation type but each indecomposable has rank less than or equal to 2. Rank-2 elements of $\text{Ind}(S, R, 0)$ are $(U_0, 0 \oplus R, R \oplus 0 \subseteq U_0)$ for $j \geq 1$, where $U_0 = R \oplus R + R(1 + 1)/p^j$.*

PROOF. As mentioned above, $C_2^* = (1, 2)$ and $C_1^* = (1, 1) = S_2$. By Theorem 4.1.3, each indecomposable in $\text{rep}((1, 2), R, 0)$ is of the form $F_j(U)$ for some indecomposable $U \in \text{rep}(S_2, R, j)$. Now apply Example 4.1.2.

Lemma 4.1.5 *If S is a finite poset, S' is a subposet of S, and $j' \leq j$, then there is a fully faithful functor $F^- : \text{rep}(S', R, j') \to \text{rep}(S, R, j)$.*

PROOF. Given $U = (U_0, U_i : i \in S') \in \text{rep}(S', R, j')$, define $F^-(U) = (V_0, V_i : i \in S) \in \text{rep}(S, \mathbb{Z}_p, j)$ by $V_0 = U_0$, $V_i = U_i$ if $i \in S'$, $V_i = \cap \{U_j : j \in S', j > i\}$ if $i \in S \backslash S'$ and there is some $j \in S$ with $j > i$, and $V_i = U_0$ otherwise. Each V_i is a pure submodule, hence a summand, of V_0, and $p^j V_0 \subseteq \Sigma_i V_i$, since $j' \leq j$ and $p^{j'} U_0 \subseteq \Sigma_i U_i$. Then F^- is a fully faithful functor, just as in Proposition 1.3.1(b).

The functor F^+, as defined in Proposition 1.3.1(a), is not directly relevant for representations over discrete valuation rings, since if U_i and U_i are pure submodules of U_0, then $U_i + U_j$ need not be a pure submodule of U_0.

Proposition 4.1.6 *Let S be a finite poset of width 2 that does not contain* (1, 2) *as a subposet. For each* $j \geq 0$, rep(S, R, j) *has finite representation type, and each indecomposable has rank* ≤ 2.

PROOF. The proof is by induction on $|S|$, the cardinality of S. If $|S| = 2$, then $S = S_2$, rep(S, R, j) has finite representation type, and each indecomposable has rank ≤ 2 by Example 4.1.2.

Now assume $|S| \geq 3$ and let a and b be incomparable elements of S. Since $w(S) = 2$ and S does not contain (1, 2) as a subposet, it follows that if $s \in S$ with $s \neq a$ and $s \neq b$, then either $s < a$ and $s < b$ or else $s > a$ and $s > b$. Hence, there is a partition $S = A \cup B \cup C$ of S with $C = \{s \in S : s < a, s < b\}$, $A = \{a, b\}$, and $B = \{s \in S : s > a, s > b\}$.

If the pair (a, b) of incomparable elements of S is chosen to be minimal in the sense that $w(C) = 1$, then C is a chain. In this case, the partition $S = A \cup B \cup C$ is a splitting decomposition, as defined in Section 1.3, since $s < t$ for each $s \in A$ and $t \in B$. Use the proof of Proposition 1.3.5 to see that each indecomposable U in rep(S, R, j) is isomorphic to $F^-(V)$ for some indecomposable representation V in either rep$(A \cup B, R, j)$ or rep$(B \cup C, R, j)$, with F^- defined as in Lemma 4.1.5.

Both $A \cup B$ and $B \cup C$ are posets of width 2 that do not contain (1, 2) as a subposet, being subposets of S. By induction on $|S|$, $A \cup B$ and $B \cup C$ have finite representation type, and each indecomposable in rep$(A \cup B, R, j)$ or rep$(B \cup C, R, j)$ has rank less than or equal to 2. Consequently, rep(S, R, j) has finite representation type and each indecomposable has rank less than or equal to 2.

A finite poset S is a *garland* if $w(S) = 2$ and S does not contain (1, 2) as a subposet. Garlands also arise for representations of finite posets over fields [Simson 92]. The terminology is motivated by the fact that if S is a garland, then S is a subposet of the poset G_n, a disjoint union of two chains $C_n = \{1 < 2 < \cdots < n\}$ and $C'_n = \{1' < 2' < \cdots < n'\}$ subject to the additional relations that $i < (i + 1)'$ and $i' < i + 1$ for each $1 \leq i \leq n - 1$.

Corollary 4.1.7 *Let S be a finite poset. Then* rep(S, R, j) *has finite representation type if and only if* $w(S) = 1$ *or S is a garland.*

PROOF. If $w(S) = 1$ or S is a garland, then rep(S, R, j) has finite representation type by Example 4.1.1 and Proposition 4.1.6. The converse follows from Corollary 4.1.4 and Lemma 4.1.5 in the case $w(S) = 2$ and from Example 4.2.4 and Lemma 4.1.5 if $w(S) \geq 3$.

An additive category has *rank-finite* representation type if there is a bound for ranks of indecomposables in the category and *rank-infinite* representation type if there is no such bound. Finite representation type implies rank-finite representation type, but Corollary 4.1.4 illustrates that the converse is not true in general.

The next proposition relates rep$(S_{n+2}, R/pR)$ to rep$(S, R, 2n + 1)$ for $S = (1, 2)$. This provides a systematic construction of indecomposable representations in rep(S, R, j) for sufficiently large j from indecomposable representations of S_n over the field $k = R/pR$. If M is an R-matrix, then $M(\bmod\ p)$ is the R/pR-matrix obtained by reducing each element of M modulo p.

Define $E(n, k)$ to be the subcategory of rep(S_{n+2}, k), consisting of representations $V = (V_0, V_i : 1 \leq i \leq n + 2)$ such that $V_0 = V_1 \oplus \cdots \oplus V_n$, there is a fixed s with $s = k$-dimension V_i for each $i \leq n$, $V_{n+1} \oplus V_{n+2} \subseteq V_0$, and $V_{n+1} = (1 + 1 + \cdots + 1)k^s$ is the image of the diagonal embedding of k^s into V_0.

Proposition 4.1.8 *Let $S = (1, 2)$, $n \geq 0$, and $k = R/pR$. There is a correspondence $\phi : E(n, k) \to \text{rep}(S, R, 2n + 1)$ such that if $V \in E(n, k)$ is an indecomposable representation, then $\phi(V)$ is an indecomposable representation.*

PROOF. Let $k = R/pR$ and $V = (V_0, V_i) \in E(n, k)$. Then $V_0 = (V_1 \oplus \cdots \oplus V_{n+1} \oplus V_{n+2})/K$, where K is the row space of a k-matrix

$$M' = \left(\begin{array}{c|c|c|c|c|c} I & I & \cdots & I & I & 0 \\ \hline M'_1 & M'_2 & \cdots & M'_n & 0 & I \end{array} \right)$$

with each M'_i an $s \times s$ k-matrix, s the k-dimension of V_i for each $i \leq n$. The first row of M' gives the relations for V_{n+1}, and the second row for those of V_{n+2}. Choose $s \times s$ R-matrices M_i with $M_i(\bmod\ p) = M'_i$ for $1 \leq i \leq n$.

Let W and X be free R-modules with rank $2s$, and Y and Z free R-modules with rank $(n + 1)s$. Define $\phi(V) = U = (U_0, U_1, U_2 \subseteq U_3) \in \text{rep}(S, R, 2n + 1)$ by

$$U_0 = W \oplus X \oplus Y \oplus Z + (1/p^{2n+1})(1 + 1 + 0 + 0)R^{2s},$$

$$U_1 = W \oplus Z, U_3 = X \oplus Y \oplus Z,$$

and U_2 the row space of the matrix $M_U = (E_1, E_2, I) =$

$$\left(\begin{array}{cccccc|ccccc} pM_1 & p^2 I & p^3 I & 0 & \cdots & 0 & 0 & I & 0 & \cdots & 0 & 0 \\ p^2 M_2 & p^3 I & 0 & p^5 I & \cdots & 0 & 0 & 0 & I & \cdots & 0 & 0 \\ & & & \cdots & & & & & & \cdots & & \\ p^n M_n & p^{n+1} I & 0 & 0 & \cdots & p^{2n+1} I & 0 & 0 & 0 & \cdots & I & 0 \\ p^{n+1} I & 0 & 0 & 0 & \cdots & 0 & p^{2n+3} I & 0 & 0 & \cdots & 0 & I \end{array} \right),$$

a submodule of $U_3 = X \oplus Y \oplus Z$. The columns of M_U are indexed by X, Y, and Z, respectively, the I's stand for $s \times s$ identity matrices, the 0's stand for $s \times s$ matrices of all zeros, and $(1 + 1 + 0 + 0)R^{2s} = \{x \oplus x \oplus 0 \oplus 0 \in U_0 : x \in R^{2s}\}$. Notice that U_2 is a summand of U_3.

The strategy is to show that if $f \in \text{End}\ U$ with R-matrix M_f relative to an R-basis of U_0, then $M_f(\bmod\ p) = aI + N$ for some nilpotent k-matrix N and some k-matrix a representing an endomorphism of V. In this case, it is verified that if f is idempotent, then $f = 0$ or 1 as a consequence of the assumption that V is indecomposable. Hence, $\phi(V) = U$ must be an indecomposable representation of rank $(6 + 2n)s$.

If $f \in \text{End } U$, then $f : U_1 = W \oplus Z \rightarrow U_1 = W \oplus Z$, $f : U_3 = X \oplus Y \oplus Z \rightarrow U_3 = X \oplus Y \oplus Z$, $f : Z \rightarrow Z$ as $U_1 \cap U_3 = Z$, $f : U_2 \rightarrow U_2$, and

$$f : (1 + 1 + 0 + 0)R^{2s}(\text{mod } p^{2n+1}) \rightarrow (1 + 1 + 0 + 0)R^{2s}(\text{mod } p^{2n+1}).$$

Then f can be represented as an endomorphism of $W \oplus X \oplus Y \oplus Z$ with matrix

$$M_f = \begin{pmatrix} f_{11} & 0 & 0 & f_{14} \\ 0 & f_{22} & f_{23} & f_{24} \\ 0 & f_{32} & f_{33} & f_{34} \\ 0 & 0 & 0 & f_{44} \end{pmatrix},$$

where $f_{11} : W \rightarrow W$, $f_{14} : W \rightarrow Z$, $f_{22} : X \rightarrow X$, $f_{23} : X \rightarrow Y$, $f_{24} : X \rightarrow Z$, $f_{32} : Y \rightarrow X$, $f_{33} : Y \rightarrow Y$, $f_{34} : Y \rightarrow Z$, $f_{44} : Z \rightarrow Z$, and $f : U_2 \rightarrow U_2$.
  Since $f : (1 + 1 + 0 + 0)R^{2s} \rightarrow (1 + 1 + 0 + 0)R^{2s} + p^{2n+1}U_0$, it follows that $f_{11} \equiv f_{22}(\text{mod } p^{2n+1})$, $f_{14} \equiv f_{23} \equiv f_{24} \equiv 0 \ (\text{mod } p^{2n+1})$, and

$$M_f \equiv \begin{pmatrix} f_{11} & 0 & 0 & 0 \\ 0 & f_{11} & 0 & 0 \\ 0 & f_{32} & f_{33} & f_{34} \\ 0 & 0 & 0 & f_{44} \end{pmatrix} (\text{mod } p^{2n+1}).$$

Also, $f : U_2 \rightarrow U_2$, and so $M_U M_f \equiv (O, E_1, E_2, I)M_f(\text{mod } p^{2n+1})$

$$(E_1 f_{11} + E_2 f_{32} \mid E_2 f_{33} \mid E_2 f_{34} + f_{44}) \equiv (g E_1 \mid g E_2 \mid g I)(\text{mod } p^{2n+1})$$

for some g. Equating block matrices yields

$$E_2 f_{34} + f_{44}I \equiv gI(\text{mod } p^{2n+1}),$$

$$E_1 f_{11} + E_2 f_{32} \equiv g E_1 \equiv (E_2 f_{34} + f_{44}I)E_1(\text{mod } p^{2n+1}),$$

$$E_2 f_{33} \equiv g E_2 \equiv (E_2 f_{34} + f_{44}I)E_2(\text{mod } p^{2n+1}).$$

  Write matrices $f_{33} = (a_{ij})_{(n+1) \times (n+1)}$ representing an endomorphism of $Y = (R^s)^{n+1}$ and $f_{44} = (b_{ij})_{(n+1) \times (n+1)}$ representing an endomorphism of $Z = (R^s)^{n+1}$ with each a_{ij} and b_{ij} an $s \times s$ R-matrix representing an endomorphism of R^s. Then

$$E_2 f_{33} \equiv (p^{2i+1} a_{ij})(\text{mod } p^{2n+1}),$$

$$f_{44} E_2 \equiv (p^{2j+1} b_{ij})(\text{mod } p^{2n+1}),$$

$$E_2 f_{33} \equiv E_2 f_{34} E_2 + f_{44} E_2(\text{mod } p^{2n+1}).$$

  Write $f_{34} = (x_{ij})$ with each x_{ij} an $s \times s$ R-matrix. Combine equations to see that $p^{2i+1} a_{ij} \equiv p^{2j+2+2i} x_{ij} + p^{2j+1} b_{ij}(\text{mod } p^{2n+1})$ for $1 \leq i, j \leq n + 1$. Then $p^{2i+1} a_{ij} \equiv p^{2j+1} b_{ij}(\text{mod } p^{\max(2i+1,2j+1)})$.
  If $i < j$, then $a_{ij} \equiv 0(\text{mod } p)$, and if $i > j$, then $b_{ij} \equiv 0(\text{mod } p)$. Consequently, $f_{33}(\text{mod } p)$ is lower triangular, $f_{44}(\text{mod } p)$ is upper triangular, and $a_{ii} \equiv b_{ii}(\text{mod } p)$ for each $1 \leq i \leq n + 1$.

Write the matrix $f_{11} = (c_{ij})_{2\times 2}$ representing an endomorphism of $X = (R^s)^2$ with c_{ij} an $s \times s$ R-matrix representing an endomorphism of R^s. Recall from above that $gE_1 \equiv E_1 f_{11} + E_2 f_{32} \equiv (E_2 f_{34} + f_{44}I)E_1 (\text{mod } p^{2n+1})$. From the last row of the latter equation,

$$c_{11} \equiv b_{n+1,n+1} (\text{mod } p) \text{ and } c_{12} \equiv 0 (\text{mod } p^2).$$

The first row yields

$$c_{11} \equiv b_{11} (\text{mod } p) \equiv c_{22} (\text{mod } p).$$

From the first column, $M_i c_{11} \equiv b_{ii} M_i (\text{mod } p)$ for each $i \leq n$. Finally, the second column of this equation gives

$$c_{22} \equiv b_{ii} (\text{mod } p) \text{ for each } i \leq n.$$

Let $a = c_{11} (\text{mod } p)$, an $s \times s$ k-matrix. By the results of the preceding two paragraphs, $a \equiv c_{22} (\text{mod } p) \equiv b_{ii} (\text{mod } p) \equiv a_{ii} (\text{mod } p)$ for each $i \leq n + 1$, and $aM_i \equiv M_i a (\text{mod } p)$ for each $i \leq n$. In particular, $aM' = M'a$, recalling the definition of M' and the fact that $M'_i = M_i (\text{mod } p)$. Consequently, $aI_{(n+2)s \times (n+2)s}$ is a k-matrix representing an endomorphism of $V \in E(n, k)$. This is so because a represents an endomorphism of $V_1 \oplus \cdots \oplus V_{n+2}$ with $a(K) \subseteq K$, the row space of M'.

In summary, $M_f (\text{mod } p) = aI + N$, where

$$N = \begin{pmatrix} N_1 & 0 & 0 & 0 \\ 0 & N_2 & 0 & 0 \\ 0 & f_{32} & N_3 & f_{34} \\ 0 & 0 & 0 & N_4 \end{pmatrix}$$

and $N_i = f_{ii} (\text{mod } p) - aI$ is of the form

$$N_1 = \begin{pmatrix} 0 & 0 \\ * & 0 \end{pmatrix}, \quad N_2 = \begin{pmatrix} 0 & 0 \\ * & 0 \end{pmatrix}, \quad N_3 = \begin{pmatrix} 0 & 0 \\ * & 0 \end{pmatrix}, \quad N_4 = \begin{pmatrix} 0 & * \\ 0 & 0 \end{pmatrix}.$$

A straightforward calculation shows that N is nilpotent.

Finally, assume that V is indecomposable and $f^2 = f \in \text{End } U$. Then $(aI + N)^2 = a^2 I + aN + Na + N^2 = aI + N$. Since N is a nilpotent matrix with zeros on the diagonal, it follows that $a^2 = a$. Thus, $aI_{(n+2)s \times (n+2)s}$ represents an idempotent endomorphism of V. But V is indecomposable, so $a = 0$ or 1. If $a = 0$, then N is both nilpotent and idempotent. In this case, $f \in p\text{End } U \subseteq \text{End}_R(U_0)$ with $f^2 = f$. Since U_0 is a free R-module, $f = 0$. On the other hand if $a = 1$, then $1 - f = 0$ and $f = 1$. This shows that $\phi(V) = U$ is an indecomposable representation.

Corollary 4.1.9 *If $S = (1, 2)$, then* rep(S, R, j) *has rank-infinite representation type for $j \geq 5$.*

PROOF. Let $n = 2$. Given a positive integer s, there is an indecomposable representation

$$V = (V_0, V_1, V_2, V_3, V_4) = (k^s \oplus k^s, k^s \oplus 0, 0 \oplus k^s, (1+1)k^s, (1+A)k^s)$$

in $\operatorname{rep}(S_4, k)$ with A an indecomposable $s \times s$ k-matrix. In the notation of Proposition 4.1.8,

$$M' = \left(\begin{array}{c|c|c|c} I & I & I & 0 \\ I & A & 0 & I \end{array} \right).$$

In fact, $V \in E(2, k)$, since $V_0 = V_1 \oplus V_2$, k-dimension $V_1 = k$-dimension $V_2 = s$, and $V_3 \oplus V_4 \subseteq V_1 \oplus V_2$. By Proposition 4.1.8, $\phi(V)$ is an indecomposable in $\operatorname{rep}(S, R, j)$. Since there are indecomposable k-matrices A with s arbitrarily large, it follows that $\operatorname{rep}(S, R, j)$ contains indecomposable representations of arbitrarily large finite rank for $j \geq 5 = 2 \cdot 2 + 1$.

Example 4.1.10 *If $S = (1, 2)$, then $\operatorname{rep}(S, R, 4)$ has rank-infinite representation type.*

PROOF. Exercise 4.1.1.

Open Question: *For $S = (1, 2)$ and $1 \leq j \leq 3$, does $\operatorname{rep}(S, R, j)$ have rank-finite or rank-infinite representation type? In this case, can indecomposable representations in $\operatorname{rep}(S, R, j)$ be classified up to isomorphism?*

EXERCISE

1. Find a matrix to show directly that if $S = (1, 2)$, then $\operatorname{rep}(S, R, 4)$ has rank-infinite representation type (Proposition 4.1.8 does not apply in this case).

4.2 Wild Modulo p Representation Type

Included in this section is a computation of the representation type of $\operatorname{rep}(S, R, j)$ in terms of S and j (Corollary 4.2.5). The definition of wild modulo p representation type is motivated by the definition of endowild for $\operatorname{rep}(S, k), k = R/pR$ a field. The category $\operatorname{rep}(S, R, j)$ has *wild modulo p representation type* if for each R-algebra Γ that is finitely generated and free as an R-module, there is $U \in \operatorname{rep}(S, R, j)$ such that $\Gamma/p\Gamma$ is a ring-homomorphic image of End U. It remains an open question as to whether if $\operatorname{rep}(S, R, j)$ has wild modulo p representation type, then for each finite-dimensional k-algebra Λ there is $U \in \operatorname{rep}(S, R, j)$ with Λ a homomorphic image of End U (see Open Question 2).

If $\operatorname{rep}(S, R, j)$ has wild modulo p representation type, then it has infinite representation type. Moreover, in view of the diversity of endomorphism rings in this case (Exercise 4.2.1), it is not reasonable to expect to be able to classify the indecomposable representations.

The first lemma gives a computational tool for determining wild modulo p representation type. The idea is that for a given Γ, there are R-matrices A and B with $C(A, B) = \Gamma$ and $C(A(\bmod p), B(\bmod p)) = \Gamma/p\Gamma$. The matrices A and B are then used to construct representations U with $C(A, B) \subseteq U$ and a homomorphism $\text{End}\, U \rightarrow C(A(\bmod p), B(\bmod p))$. It follows that there is an onto ring homomorphism $\text{End}\, U \rightarrow \Gamma/p\Gamma$. The matrices A and B are somewhat special, since $C(A, B) \rightarrow C(A(\bmod p), B(\bmod p))$ need not be onto for arbitrary matrices A and B (Exercise 4.2.3).

Lemma 4.2.1 *Suppose Γ is an R-algebra that is finitely generated and free as an R-module. There are R-matrices A and B such that*

(a) $C(A, B) = \Gamma$ *and* $C(A(\bmod p), B(\bmod p)) = \Gamma/p\Gamma$, *and*
(b) *the ring homomorphism* $C(A, B) \rightarrow C(A(\bmod p), B(\bmod p))$ *given by* $M \mapsto M(\bmod p)$ *is onto.*

PROOF. (a) If $\{1, y_2, \ldots, y_t\}$ is an R-basis of Γ, then $\{1, x_2, \ldots, x_t\}$ is a k-basis for $\Gamma/p\Gamma$, where $k = R/pR$ is a field and $x_i = y_i + p\Gamma$. Define $(t + 2) \times (t + 2)$ Γ-matrices

$$
A = \begin{pmatrix} 0 & 1 & 0 & \cdots & 0 & 0 \\ 0 & 0 & 1 & \cdots & 0 & 0 \\ & & & \cdots & & \\ 0 & 0 & 0 & \cdots & 0 & 1 \\ 0 & 0 & 0 & \cdots & 0 & 0 \end{pmatrix}, \quad
B = \begin{pmatrix} 0 & 0 & 0 & \cdots & 0 & 0 \\ 1 & 0 & 0 & \cdots & 0 & 0 \\ y_1 & 1 & 0 & \cdots & 0 & 0 \\ 0 & y_2 & 1 & \cdots & 0 & 0 \\ & & & \cdots & & \\ 0 & 0 & \cdots & & y_t & 1 & 0 \end{pmatrix}.
$$

Then A and B are R-matrices with $C(A, B)$ isomorphic to $\text{End}_\Gamma(\Gamma) = \Gamma$. The proof is exactly the same as for Example 1.1.7. Similarly, $\Gamma/p\Gamma$ is isomorphic to $C(A(\bmod p), B(\bmod p))$, as desired. Assertion (b) follows from (a).

Corollary 4.2.2 *If $S = (1, 2)$, then* $\text{rep}(S, R, j)$ *has wild modulo p representation type for $j \geq 7$.*

PROOF. Let Γ be an R-algebra that is finitely generated and free as an R-module and, via Lemma 4.2.1, let A and B be $s \times s$ R-matrices with $\Gamma = C(A, B)$ and $\Gamma/p\Gamma = C(A(\bmod p), B(\bmod p))$. Define $V \in \text{rep}(S_3, k)$ by

$$V = (k^s \oplus k^s \oplus k^s, k^s \oplus 0 \oplus 0, 0 \oplus k^s \oplus 0, 0 \oplus 0 \oplus k^s,$$

$$(1 + 1 + 1)k^s, (1 + A' + B')k^s)$$

for $s \times s$ k-matrices $A' = A(\bmod p)$ and $B' = B(\bmod p)$. Then $\text{End}\, V = C(A', B')$, by Exercise 1.2.2(c), whence $\text{End}\, V$ is isomorphic to $\Gamma/p\Gamma$. It follows from the definitions that $V \in E(3, k)$, as defined in Section 4.1. In the notation of

Proposition 4.1.8,

$$M' = \left(\begin{array}{c|c|c|c|c} I & I & I & I & 0 \\ I & A' & B' & 0 & I \end{array} \right).$$

Moreover, $\phi(V) = U = (U_0, U_1, U_2 \subseteq U_3) \in \text{rep}(S, R, 7)$ is given by $U_0 = W \oplus X \oplus Y \oplus Z + (1/p^7)(1 + 1 + 0 + 0)R^{2s}$, $U_1 = W \oplus Z$, $U_3 = X \oplus Y \oplus Z$, and U_2, a submodule of $X \oplus Y \oplus Z$, the free R-module with basis the rows of $M_U = (E_1, E_2, I) =$

$$\left(\begin{array}{cc|cccccc|cccc} pI & p^2I & p^3I & 0 & \cdots & 0 & 0 & I & 0 & \cdots & 0 & 0 \\ p^2A & p^3I & 0 & p^5I & \cdots & 0 & 0 & 0 & I & \cdots & 0 & 0 \\ p^3B & p^4I & 0 & 0 & \cdots & p^7I & 0 & 0 & 0 & \cdots & I & 0 \\ p^4I & 0 & 0 & 0 & \cdots & 0 & p^9I & 0 & 0 & \cdots & 0 & I \end{array} \right).$$

Notice that $C(A, B) \subseteq \text{End } U$, since if $b \in C(A, B)$, then $bM_U = M_U b$, and so $bI \in \text{End } U$. As in Proposition 4.1.8, if $f \in \text{End } U$, then $M_f(\text{mod } p) = aI + N$ for some unique $aI \in \text{End } V = C(A', B')$ and nilpotent k-matrix N. It follows that there is a ring homomorphism $\pi : \text{End } U \to \text{End } V = \Gamma/p\Gamma$ given by $\pi(f) = aI$. Since $C(A, B) \subseteq \text{End } U$, π is onto by Lemma 4.2.1. This shows that $\text{rep}(S, R, 7)$ has wild modulo representation type. Now apply Lemma 4.1.5 to see that $\text{rep}(S, R, j)$ has wild modulo p representation type for $j \geq 7$.

Submodules of finitely generated free R-modules need not be summands, unless they are pure. Another category of representations of finite posets over discrete valuation rings is defined to reflect this distinction. Let $\text{rep}_f(S, R)$ be the category of representations $U = (U_0, U_i : i \in S)$ such that U_0 is a finitely generated free R-module, each U_i is a R-submodule of U_0, $U_i \subseteq U_j$ if $i \leq j$ in S, and $U_0 = \Sigma\{U_i : i \in S\}$. The morphisms in this category are representation morphisms.

Assertion (b) of the next lemma is a generalization of Theorem 1.2.3, while statement (a) for the special case that $n = 2$ was used in the proof of Example 4.1.2. For an antichain S_n, define $\Delta(S_n, R)$ to be the full subcategory of $\text{rep}(S_{n+1}, R, 0)$ with objects of the form $U = (U_0, U_i, U_* : i \in S)$ such that $U_0 = \oplus\{U_i : i \in S\}$, $U_i \cap U_* = 0$, and $U_i + U_*$ is pure in U_0 for each $i \in S_n$. Recall that the poset $S^* = S \cup \{*\}$ is the disjoint union of S and a point $*$.

Lemma 4.2.3 *Suppose S is a finite poset.*

(a) *There is a fully faithful functor $H' : \text{rep}(S, R, j) \to \text{rep}_f(S^*, R)$. The image of H' consists of all $V = (V_0, V_i, V_* : i \in S)$ with $V_0 = \Sigma\{V_i : i \in S\}$, each V_i a pure submodule of V_0, and $V_* \cap V_i = p^j V_i$ for each i in S.*

(b) *There is a category equivalence $H : \Delta(S_n, R) \to \text{rep}(S_n, R, 0)$, where S_n is an antichain.*

PROOF. (a) Define $H'(U) = V = (\Sigma\{U_i : i \in S\}, U_i, p^j U_0)$, where $U = (U_0, U_i : i \in S)$, and define $H'(f)$ to be the restriction of f to $\Sigma\{U_i : i \in S\}$. Then H' is a functor. For each $i \in S$, $p^j U_0 \cap U_i = p^j U_i$ and U_i is a pure submodule of $\Sigma_i U_i$, since U_i is a pure submodule of U_0. Each U_i is a summand of $\Sigma\{U_i : i \in S\}$, but $p^j U_0$ need not be, which is why the image of H' is in $\text{rep}_f(S, R)$ and not necessarily in $\text{rep}(S, R)$. The functor H' is fully faithful, since if $g : H'(U) \to H'(U')$ is a representation morphism, then g extends uniquely to $f : U_0 \to U'_0$ with $H'(f) = g$.

To see that the image of H' is as described, let $V = (V_0, V_i, V_* : i \in S) \in \text{rep}_f(S^*, R)$ with $V_0 = \Sigma\{V_i : i \in S\}$, each V_i a pure submodule of V_0, and $V_* \cap V_i = p^j V_i$ for each i. Then $p^j V_0 = \Sigma\{p^j V_i : i \in S\} \subseteq V_*$. Define $U = (U_0, V_i : i \in S)$ with $U_0 = V_0 + (1/p^j)V_*$. Now, U_0 is a finitely generated torsion-free, hence free, R-module, and $p^j U_0$ is contained in V_0. To see that V_i is pure in U_0, suppose $x \in U_0$ with $px \in V_i$. Then $p^j x \in V_* \cap pV_i = p^{j+1}V_i$, and so $x \in V_i$, as desired. This shows that $U \in \text{rep}(S, R, j)$ with $H'(U) = V$.

(b) Given $U = (U_0, U_i, U_* : i \in S_n) \in \Delta(S_n, R)$ with $U_0 = \oplus\{U_i : i \in S_n\}$, define $H : \Delta(S_n, k) \to \text{rep}(S_n, R, 0)$ by $H(U) = (V_0, V_i : i \in S_n)$, where $V_0 = U_0/U_*$ and $V_i = (U_i + U_*)/U_*$ for each i in S. Then V_0 is a finitely generated free R-module, since U_* is a pure submodule of the finitely generated free R-module U_0, and $V_0 = \Sigma\{V_i : i \in S_n\}$. Because $U_i + U_*$ is pure in U_0, $U_0/(U_i + U_*)$ is a torsion-free module, and so V_i is pure in V_0. Thus, $H(U) \in \text{rep}(S_n, R, 0)$. The proof that H is a category equivalence is the same as that of Theorem 1.2.3.

Example 4.2.4 *The category* $\text{rep}(S_3, R, 0)$ *has wild modulo p representation type.*

PROOF. Let Γ be an R-algebra that is finitely generated and free as an R-module. By Lemma 4.2.1, there are $n \times n$ R-matrices A and B with $\Gamma = C(A, B)$ and $\Gamma/p\Gamma = C(A(\text{mod } p), B(\text{mod } p))$. Define $U = (U_0, U_1, U_2, U_3, U_*) \in \text{rep}(S_4, R, 0)$ by

$$U_0 = R^{7n} \oplus R^{6n} \oplus R^{5n}, U_1 = R^{7n} \oplus 0 \oplus 0, U_2 = 0 \oplus R^{6n} \oplus 0,$$

$$U_3 = 0 \oplus 0 \oplus R^{5n},$$

and U_* the free R-module with the rows of the following matrix as a basis:

$$\begin{pmatrix}
I & 0 & 0 & 0 & 0 & 0 & 0 & I & 0 & 0 & 0 & 0 & 0 & 0 & 0 & 0 & I & 0 \\
0 & I & 0 & 0 & 0 & 0 & 0 & -pA & p^2 I & 0 & 0 & 0 & 0 & I & 0 & 0 & 0 & p^2 I \\
0 & 0 & I & 0 & 0 & 0 & 0 & -p^4 I & 0 & p^6 I & 0 & 0 & 0 & 0 & I & 0 & 0 & p^5 B \\
0 & 0 & 0 & I & 0 & 0 & 0 & p^5 I & 0 & 0 & 0 & 0 & 0 & 0 & 0 & I & 0 & p^6 I \\
0 & 0 & 0 & 0 & p^2 I & 0 & 0 & 0 & 0 & 0 & I & 0 & 0 & I & 0 & 0 & 0 & 0 \\
0 & 0 & 0 & 0 & 0 & p^6 I & 0 & 0 & 0 & 0 & 0 & I & 0 & 0 & I & 0 & 0 & 0 \\
0 & 0 & 0 & 0 & 0 & 0 & p^7 I & 0 & 0 & 0 & 0 & 0 & I & 0 & 0 & I & 0 & 0
\end{pmatrix}$$

A routine computation shows that $U \in \Delta(S_3, R)$.

If $f \in \text{End } U$, then $M_f(\text{mod } p) = aI + N$ for some nilpotent k-matrix N with zeros on the diagonal and $aI \in C(A(\text{mod } p), B(\text{mod } p))$ (Exercise 4.2.2). Then $\pi : \text{End } U \to C(A(\text{mod } p), B(\text{mod } p)) = \Gamma/p\Gamma$ defined by $\pi(f) = aI$ is a ring homomorphism. Since $C(A, B)$ embeds in End U (Exercise 4.2.2), π is onto by Lemma 4.2.1(b). This shows that rep($S_3, R, 0$) has wild modulo p representation type.

Recall that if S is a finite poset with $w(S) = 2$, then S is not a garland if and only if S contains $(1, 2)$ as a subposet.

Corollary 4.2.5 *The category* rep(S, R, j) *has:*

(a) *wild modulo p representation type if either $w(S) \geq 3$ or $w(S) = 2$, S is not a garland, $j \geq 7$;*
(b) *rank-infinite representation type if $w(S) = 2$, S is not a garland, $j \geq 4$;*
(c) *finite representation type if and only if either $w(S) = 1$ or S is a garland.*

PROOF. Statement (a) is a consequence of Corollary 4.2.2, Example 4.2.4, and Lemma 4.1.5. Assertion (b) follows from Corollary 4.1.9, Example 4.1.10 and Lemma 4.1.5. Statement (c) is Corollary 4.1.7.

For antichains the characterizations are complete; there is only finite or wild modulo p representation type.

Corollary 4.2.6 *The category* rep(S_n, R, j) *has*

(a) *wild modulo p representation type if and only if $n \geq 3$, and*
(b) *finite representation type if and only if $n \leq 2$.*

Open Questions
1. *Suppose S is a finite poset with $w(S) = 2$ and S is not a garland. If $6 \geq j \geq 4$, does* rep(S, R, j) *have wild modulo p representation type? In this case, can indecomposable representations in* rep(S, R, j) *be classified up to isomorphism?*
2. *For which finite-dimensional algebras Γ' over a field $k = R/pR$ does there exist an R-algebra Γ that is finitely generated and free as an R-module with $\Gamma/p\Gamma = \Gamma'$?*

EXERCISES

1. Let G be a finite group, not necessarily abelian, and kG the group algebra of G over the field $k = R/pR$. Prove that the group algebra RG of G over R is an R-algebra that is finitely generated and free as an R-module with $RG/pRG = kG$.

2. Complete the proof of Example 4.2.4.

3. Let R be a discrete valuation ring, $A = \left(\begin{smallmatrix} 1 & p \\ 0 & 1 \end{smallmatrix}\right)$ and $B = \left(\begin{smallmatrix} 1 & 0 \\ 0 & 1 \end{smallmatrix}\right)$. Show that the natural homomorphism $C(A, B) \rightarrow C(A(\bmod p), B(\bmod p))$ is not onto.

4. Let rep$_f(S, R, j)$ be the full subcategory of rep$_f(S, R)$ with objects $U = (U_0, U_i : i \in S)$ such that $p^j U_0 \subseteq \Sigma_i U_i$. Use the results of Sections 4.1 and 4.2 to find bounds for the representation type of rep$_f(S, R, j)$ in terms of S and j.

4.3   Finite Rank Butler Groups and Isomorphism at p

Computations of representation type for categories of representations of posets over the discrete valuation ring $\mathbb{Z}_p$ can be transferred directly to isomorphism at p categories of Butler groups (Corollary 4.3.3).

Recall that for a finite lattice T of types, $B(T)$ is the category of finite rank Butler groups G with typeset $G \subseteq T$, and S_T is the opposite of the poset of join irreducible elements of T. For an integer $j \geq 0$, let $B(T, j)$ denote the full subcategory of $B(T)$ consisting of those groups G with $p^j G$ contained in $\Sigma\{G(\tau) : \tau \in S_T\}$. The isomorphism at p category of $B(T, j)$ is denoted by $B(T, j)_p$. Notice that indecomposable groups in $B(T, j)_p$ are indecomposable in $B(T, j)$, i.e., indecomposable as groups.

Up to rank-1 summands and isomorphism at p, there is no loss of generality in considering $B(T, j)$:

Proposition 4.3.1 *If $G \in B(T)$ with no rank-1 summands and p is a prime, then G is isomorphic at p to a group in $B(T, j)$ for some nonnegative integer j.*

PROOF. Let $\mathrm{IT}(G)$ denote the inner type of G. Since G is a Butler group, $\mathrm{IT}(G) \in$ typeset G by Lemma 3.1.12 and Theorem 3.2.5. Then $\Sigma\{G(\sigma) : \sigma \in S_T\} = G^*(\mathrm{IT}(G))$. This holds because if $x \in G$ with type $x = \tau > \mathrm{IT}(G)$ and $\tau \in T \setminus S_T$ is not join irreducible, then $\tau > \sigma$ for some join irreducible, $\sigma \in S_T$ and $x \in G(\tau) \subseteq G(\sigma)$. Since G is a Butler group, $G = G(\mathrm{IT}(G)) = G_{\mathrm{IT}(G)} \oplus G^\#(\mathrm{IT}(G))$ with $G_{\mathrm{IT}(G)}$ an $\mathrm{IT}(G)$-homogeneous completely decomposable group and $G^\#(\mathrm{IT}(G))/G^*(\mathrm{IT}(G))$ finite (Theorem 3.2.5). But G has no rank-1 summands, so that $G_{\mathrm{IT}(G)} = 0$ and $G/G^*(\mathrm{IT}(G))$ is a finite group. Choose a subgroup H of G with $H/G^*(\mathrm{IT}(G))$ the subgroup of $G/G^*(\mathrm{IT}(G))$ consisting of elements with order a power of p. Then G/H is a finite abelian group with order prime to p. Hence, G is isomorphic to H at p. Moreover, for some j, $p^j(H/G^*(\mathrm{IT}(G))) = 0$, so that $p^j H \subseteq G^*(\mathrm{IT}(G)) \subseteq H^*(\mathrm{IT}(G)) \subseteq H$ and $H \in B(T, j)$.

Given a prime p, a finite lattice T of types is *p-locally free* if X_p is isomorphic to $\mathbb{Z}_p$ for each rank-1 group X with type $X \in T$. Restriction to the case that T is p-locally free amounts to the p-reduced case, in that if type $X \in T$ and X_p is not isomorphic to $\mathbb{Z}_p$, then X_p is isomorphic to $\mathbb{Q}$ (Exercise 2.1.3). A torsion-free abelian group G is *p-locally free* if G_p is a free $\mathbb{Z}_p$-module.

Corollary 4.3.2 [Richman 95] *If T is a p-locally free finite lattice of types and j is a nonnegative integer, then there is a category equivalence $F : B(T, j)_p \to \mathrm{rep}(S_T, \mathbb{Z}_p, j)$ given by $F(G) = (G_p, G(\tau)_p : \tau \in S_T)$.*

PROOF. The proof is analogous to that of Theorem 3.3.2, bearing in mind that integers prime to p are units in $\mathbb{Z}_p$.

Corollary 4.3.3 *If T is a finite p-locally free finite lattice of types, then $B(T, j)_p$ has:*

(a) *wild modulo p representation type if either $w(S_T) \geq 3$ or $w(S_T) = 2$, S_T is not a garland, $j \geq 7$;*
(b) *rank-infinite representation type if $w(S_T) = 2$, S_T is not a garland, $j \geq 4$; and*
(c) *finite representation type if and only if either $w(S_T) = 1$ or S_T is a garland.*

PROOF. A consequence of Corollaries 4.3.2 and 4.2.5.

Following are illustrations of how representations over discrete valuation rings can be used to construct finite rank Butler groups. If G is a finite rank Butler group with linearly ordered typeset, then G is a completely decomposable group by Corollary 3.2.7(a). This is, in the isomorphism at p category, a manifestation of the fact that representations of chains over a discrete valuation ring are direct sums of rank-1 representations. Observe that a finite lattice of types T is a chain if and only if S_T is a chain.

Example 4.3.4 *If T is a finite lattice of types and S_T is a chain, then rank-1 groups are the only indecomposable groups in $B(T, j)_p$.*

PROOF. Corollary 4.3.2 and Example 4.1.1

Example 4.3.5 *Let T be a finite lattice of types with $S_T = \{\text{type } X_1, \text{type } X_2\} = S_2$ for subgroups X_i of $\mathbb{Q}$ with $1 \in X_i$ and $1/p \notin X_i$ for each i. Indecomposable groups in $B(T, j)_p$ are:*

(i) *rank-1 groups $G = X_1 \cap X_2, X_1, X_2, X_1 + X_2$ with $(\text{End } G)_p = \mathbb{Z}_p$, and*
(ii) *rank-2 groups $G = X_1 \oplus X_2 + (1+1)\mathbb{Z}(1/p^i)$ with $(\text{End } G)_p = \{(r, s) \in \mathbb{Z}_p \times \mathbb{Z}_p : r - s \in p^i\mathbb{Z}_p\}$ for each $1 \leq i \leq j$.*

PROOF. Apply Corollary 4.3.2 and Example 4.1.2, noticing that if G is as given in (ii), then $(G_p, G(\tau_1)_p, G(\tau_2)_p) = (\mathbb{Z}_p \oplus \mathbb{Z}_p + (1+1)\mathbb{Z}_p(1/p^i), \mathbb{Z}_p \oplus 0, 0 \oplus \mathbb{Z}_p) \in \text{rep}(S_2, \mathbb{Z}_p, j)$.

Example 4.3.6 *Let p be a prime,*

a p-locally free finite lattice of types, and X_i a subgroup of $\mathbb{Q}$ with $1 \in X_i$, type $X_i = i$, and $1/p \notin X_i$. Then $B(T, 0)_p$ has infinite representation type, but each indecomposable has rank less than or equal to 2. Indecomposable groups in $B(T, 0)_p$ of rank 2 are of the form $G = X_4 \oplus X_3 + X_2(1, 1)/p^j$ for some $j \geq 1$. Moreover, $B(T, j)_p$ has indecomposable groups of arbitrarily large finite rank for $j \geq 4$.

PROOF. Notice that $S_T = C_2^* = 2 \quad 1$

$\qquad\qquad\qquad\qquad\qquad\qquad |$

$\qquad\qquad\qquad\qquad\qquad\qquad 4$

Apply Corollaries 4.3.2 and 4.1.4 to see that indecomposable groups have rank less than or equal to 2. If G is as defined above, then $(G_p, G(4)_p, G(2)_p, G(1)_p) = (U_0, \mathbb{Z}_p \oplus 0 \subseteq U_0, 0 \oplus \mathbb{Z}_p)$ with $U_0 = \mathbb{Z}_p \oplus \mathbb{Z}_p + \mathbb{Z}_p(1, 1)/p^j$, as required by Corollary 4.1.4.

The last statement of the theorem is a consequence of Corollaries 4.3.2 and 4.1.9 and Example 4.1.10. In particular, for $r = 5$,

$$G = X_1^{2s} \oplus X_2^{2s} \oplus X_2^{3s} \oplus X_3^{3s} + X_4 M + (1/p^5)(1 + 1 + 0 + 0)\mathbb{Z}^{2s},$$

where M is the row space of the matrix

$$\begin{pmatrix} pI & p^2 I & p^3 I & 0 & 0 & I & 0 & 0 \\ p^2 A & p^3 I & 0 & p^5 I & 0 & 0 & I & 0 \\ p^3 I & 0 & 0 & 0 & p^7 I & 0 & 0 & I \end{pmatrix}$$

and A is an $s \times s$ $\mathbb{Z}$-matrix such that $A(\bmod p)$ is an indecomposable $\mathbb{Z}/p\mathbb{Z}$-matrix. Notice that $T_{cr}(G) = \{\text{type } X_i; 1 \leq i \leq 4\}$.

In view of Corollary 4.3.2, the following open question is a group-theoretic version of a special case of open questions in Sections 4.1 and 4.2.

Open Question: *What is the representation type of $B(T, j)_p$ if $S_T = (1, 2)$ and $1 \leq j \leq 3$ or $6 \geq j \geq 4$? Can the indecomposable groups be classified in these cases?*

EXERCISES

1. Let T be a finite lattice of types with $S_T = \{\text{type } X_1, \text{type } X_2, \text{type } X_3\}$, an antichain, for subgroups X_i of $\mathbb{Q}$ with $1/p \notin X_i$ for each i. Prove that the following construction demonstrates that $B(T_3, 2)_p$ has wild modulo p representation type. Given two $n \times n$ $\mathbb{Z}$-matrices A and B, define a subgroup G of $\mathbb{Q}^n \oplus \mathbb{Q}^n \oplus \mathbb{Q}^n \oplus \mathbb{Q}^n$ by

$$G = X_1^{2n} \oplus X_2^{2n} + (1 + 0 + 0 + 1)X_3^n + (0 + 1 + 0 + 1)X_4^n$$
$$+ (1/p^2)(1 + 0 + A + 1)\mathbb{Z}^n + (1/p)(0 + 1 + B + 0)\mathbb{Z}^n.$$

2. Provide a proof for Corollary 4.3.2.

NOTES ON CHAPTER 4

There is very little literature on representations of posets over discrete valuation rings. The contents of this chapter are distilled from [Arnold, Dugas 97, 99]. The methods, albeit ad hoc, are sufficient to determine the representation type of most of the isomorphism at p categories of finite rank Butler groups. In addition, indecomposable representations give rise to groups that are indecomposable in the isomorphism at p category, hence indecomposable as groups. The representation types of some more general categories of representations of posets over discrete valuation rings and their factor rings are given in [Plahotnik 76] and [Simson 96].

5

Almost Completely Decomposable Groups

5.1 Characterizations and Properties

An almost completely decomposable group is a torsion-free abelian group G of finite rank quasi-isomorphic to a completely decomposable group. An almost completely decomposable group is a Butler group, by Corollary 3.2.4, but any strongly indecomposable Butler group with finite rank greater than 1 is not almost completely decomposable.

Almost completely decomposable groups are quite complicated, as a consequence of the variety of ways that a completely decomposable group can be embedded as a subgroup of bounded index in an almost completely decomposable group. Chapter 5 is a brief introduction to this technical subject.

Properties of finite rank Butler groups given in Chapter 3 are assumed. In particular, if C is a subgroup of a finite rank Butler group G with G/C bounded, then G/C is finite by Exercise 2.1.1. Moreover, $T_{\mathrm{cr}}(G) = \{\tau : G(\tau)/G^{\#}(\tau) \neq 0\}$ is the critical typeset of G, and a regulating subgroup of G is $B = \Sigma\{G_\tau : \tau \in T_{\mathrm{cr}}(G)\}$ with $G(\tau) = G_\tau \oplus G^{\#}(\tau)$ and G_τ a τ-homogeneous completely decomposable group.

The first characterization of almost completely decomposable groups is in terms of the ranks of the G_τ's.

Proposition 5.1.1 *A finite rank Butler group G is an almost completely decomposable group if and only if rank $G = \Sigma\{\mathrm{rank}\, G(\tau)/G^{\#}(\tau) : \tau \in T_{\mathrm{cr}}(G)\}$. In this case, each regulating subgroup of G is a completely decomposable group isomorphic to $\oplus\{G(\tau)/G^{\#}(\tau) : \tau \in T_{\mathrm{cr}}(G)\}$.*

Proof. Let G be an almost completely decomposable group and C a completely decomposable subgroup of G with G/C finite. Then rank G = rank C and $T_{cr}(G) = T_{cr}(C)$. Moreover, $G(\tau)/C(\tau)$ and $G^{\#}(\tau)/C^{\#}(\tau)$ are finite, whence rank $G(\tau)/G^{\#}(\tau)$ = rank $C(\tau)/C^{\#}(\tau)$ for each type $\tau \in T_{cr}(G)$. But rank C = $\Sigma\{$rank $C(\tau)/C^{\#}(\tau) : \tau \in T_{cr}(C)\}$ by Example 3.1.5, so that rank $G = \Sigma\{$rank$G(\tau)/G^{\#}(\tau) : \tau \in T_{cr}(G)\}$.

Conversely, assume that G is a finite rank Butler group with rank $G = \Sigma\{$rank $G(\tau)/G^{\#}(\tau) : \tau \in T_{cr}(G)\}$, $G(\tau) = G_{\tau} \oplus G^{\#}(\tau)$, and $B = \Sigma\{G_{\tau} : \tau \in T_{cr}(G)\}$ a regulating subgroup of G. Then G/B is finite by Proposition 3.2.12. Hence, $T_{cr}(G) = T_{cr}(B)$, rank G_{τ} = rank $G(\tau)/G^{\#}(\tau)$ = rank $B(\tau)/B^{\#}(\tau)$ for each type $\tau \in T_{cr}(G)$, and $\Sigma\{$rank $B(\tau)/B^{\#}(\tau) : \tau \in T_{cr}(B)\}$ = rank B. It follows that $B = \oplus\{G_{\tau} : \tau \in T_{cr}(G)\}$ is completely decomposable. Since G/B is finite, G is almost completely decomposable.

There is a criterion for deciding whether or not a completely decomposable subgroup B of an almost completely decomposable group G is a regulating subgroup of G. Define a positive integer $i(G)$ to be the least element of $\{|G/C| : C$ completely decomposable, G/C finite$\}$.

Proposition 5.1.2 [Lady 74B] *Let G be an almost completely decomposable group and B a completely decomposable subgroup of G.*

(a) *The group B is a regulating subgroup of G if and only if $|G/B| = i(G)$.*
(b) *If C is a completely decomposable subgroup of G with G/C finite, then $i(G)$ divides $|G/C|$.*
(c) *The group G is completely decomposable if and only if $i(G) = 1$.*

Proof. (a), (b) The first step is to show that if B and C are distinct regulating subgroups of G, then $|G/B| = |G/C|$. Write $B = \oplus\{B_{\tau} : \tau \in T_{cr}(G)\}$ and $C = \oplus\{C_{\tau} : \tau \in T_{cr}(G)\}$ with $G(\tau) = B_{\tau} \oplus G^{\#}(\tau) = C_{\tau} \oplus G^{\#}(\tau)$ and B_{τ} and C_{τ} τ-homogeneous completely decomposable groups for each $\tau \in T_{cr}(G)$.

The proof that $|G/B| = |G/C|$ is an induction on the cardinality of $\{\sigma \in T_{cr}(G) : B_{\sigma} \neq C_{\sigma}\}$. Since B and C are distinct, there is some $\sigma \in T_{cr}(G)$ with $B_{\sigma} \neq C_{\sigma}$. Define $D = C_{\sigma} \oplus (\oplus\{B_{\tau} : \sigma \neq \tau \in T_{cr}(G)\}) = \oplus\{D_{\tau} : \tau \in T_{cr}(G)\}$, another regulating subgroup of G with $|\{\sigma \in T_{cr}(G) : B_{\sigma} \neq D_{\sigma}\}| < |\{\sigma \in T_{cr}(G) : B_{\sigma} \neq C_{\sigma}\}|$.

By induction, it is now sufficient to prove that $|G/B| = |G/D|$. To this end, define $E = B + C_{\sigma}$. Then $E = B + E(\sigma) = D + E(\sigma)$, since B_{σ} and C_{σ} are subgroups of $E(\sigma)$. Consequently, $|E/B| = |E(\sigma)/B \cap E(\sigma)| = |E(\sigma)/B(\sigma)|$ and, similarly, $|E/D| = |E(\sigma)/D(\sigma)|$.

Notice that $B(\sigma) = \oplus\{B_{\tau} : \tau \geq \sigma\}$ and $D(\sigma) = C_{\sigma} \oplus (\oplus\{B_{\tau} : \tau > \sigma\})$ are regulating subgroups of $E(\sigma)$, and $B^{\#}(\sigma) = D^{\#}(\sigma) = \oplus\{B_{\tau} : \tau > \sigma\}$ is a regulating subgroup of $E^{\#}(\sigma)$. Now,

$$|E(\sigma)/B(\sigma)| = |(B_{\sigma} \oplus E^{\#}(\sigma))/(B_{\sigma} \oplus B^{\#}(\sigma))| = |E^{\#}(\sigma)/B^{\#}(\sigma)|$$

and

$$|E(\sigma)/D(\sigma)| = |(C_\sigma \oplus E^\#(\sigma))/(C_\sigma \oplus D^\#(\sigma))| = |E^\#(\sigma)/D^\#(\sigma)|$$

with $|E^\#(\sigma)/B^\#(\sigma)| = |E^\#(\sigma)/D^\#(\sigma)|$. Thus, $|E(\sigma)/B(\sigma)| = |E(\sigma)/D(\sigma)|$ and

$$|G/B| = |G/E||E/B| = |G/E||E(\sigma)/B(\sigma)|$$
$$= |G/E||E(\sigma)/D(\sigma)| = |G/E||E/D| = |G/D|,$$

as desired.

It remains to prove that if B is a completely decomposable subgroup of G and B is not a regulating subgroup of G, then there is a completely decomposable subgroup C of G such that $|G/C|$ is a proper divisor of $|G/B|$. In this case, if B is a completely decomposable subgroup of G with $|G/B| = i(G)$, then B must be a regulating subgroup of G. Conversely, if B is a regulating subgroup, then $B = i(G)$. Otherwise, there is a completely decomposable subgroup C of G with $|G/C| = i(G) < |G/B|$. Thus, C is a regulating subgroup of G with $|G/C| \neq |G/B|$, a contradiction.

Now assume that B is a completely decomposable subgroup of G and B is not a regulating subgroup of G. Choose σ maximal in $T_{cr}(G)$ such that $B(\sigma) = B_\sigma \oplus B^\#(\sigma)$ but $G(\sigma) \neq B_\sigma \oplus G^\#(\sigma)$. Define $C = G_\sigma \oplus (\oplus\{B_\tau : \sigma \neq \tau \in T_{cr}(G)\})$, where $G(\sigma) = G_\sigma \oplus G^\#(\sigma)$ and $B(\tau) = B_\tau \oplus B^\#(\tau)$ for each type τ. Then C is a completely decomposable subgroup of G. By the choice of σ, $B^\#(\sigma) = C^\#(\sigma)$ is a regulating subgroup of $G^\#(\sigma)$. Define $E = C + B_\sigma = B + G_\sigma$. Then $E = C + D(\sigma) = B + D(\sigma)$, and as above, $|E/C| = |E(\sigma)/C(\sigma)|$ and $|E/B| = |E(\sigma)/B(\sigma)|$. Now,

$$|G(\sigma)/C(\sigma)| = |G(\sigma)/E(\sigma)||E(\sigma)/C(\sigma)|$$

and

$$|G(\sigma)/B(\sigma)| = |G(\sigma)/E(\sigma)||E(\sigma)/B(\sigma)|.$$

However, $|G(\sigma)/C(\sigma)| = |(G_\sigma \oplus G^\#(\sigma))/(G_\sigma \oplus C^\#(\sigma))| = |(G_\sigma \oplus G^\#(\sigma))/(G_\sigma \oplus B^\#(\sigma))|$ is a proper divisor of $|G(\sigma)/B(\sigma)| = |(G_\sigma \oplus G^\#(\sigma))/(B_\sigma \oplus B^\#(\sigma))|$ by the choice of σ. Hence, $|E/C| = |E(\sigma)/C(\sigma)|$ is a proper divisor of $|E(\sigma)/B(\sigma)| = |E/B|$, and so $|G/C| = |G/E||E/C|$ is a proper divisor of $|G/B| = |G/E||E/B|$.

Assertion (c) is an immediate consequence of (a).

If B and B' are two regulating subgroups of an almost completely decomposable group G, then $|G/B| = |G/B'| = i(G)$ by Proposition 5.1.2(a). However, the following example shows that G/B need not be isomorphic to G/B'.

Example 5.1.3 [Lady 74B] *Let p be a prime and choose rank-1 subgroups X_1, X_2, X_3, X_4 of $\mathbb{Q}$ such that $1 \in X_i$, $1/p \notin X_i$, type X_1, type X_2, and type X_3*

are pairwise incomparable, and type $X_4 =$ type $X_1 \cap X_2 \not\leq$ type X_3. *Define* $G = X_1 \oplus X_2 \oplus X_3 \oplus X_4 + \mathbb{Z}(1, 1, 0, 0)/p + \mathbb{Z}(0, 0, 1, 1)/p$. *Then*

(a) $B = X_1 \oplus X_2 \oplus X_3 \oplus X_4$ *is a regulating subgroup of G with G/B isomorphic to* $\mathbb{Z}/p\mathbb{Z} \oplus \mathbb{Z}/p\mathbb{Z}$, *and*

(b) *if* $Y = X_4(1, 1, 0, p)/p$, *then* $C = X_1 \oplus X_2 \oplus X_3 \oplus Y$ *is also a regulating subgroup of G with G/C isomorphic to* $\mathbb{Z}/p^2\mathbb{Z}$.

PROOF. (a) Observe that G is an almost completely decomposable group. Let $\tau_i =$ type X_i. Then τ_3 and τ_4 are incomparable, since $\tau_4 \not\leq \tau_3$ by assumption and $\tau_3 \leq \tau_4 \leq \tau_1$ is a contradiction. Hence, $G(\tau_i) = X_i$ with $G^{\#}(\tau_i) = 0$ for $i = 1, 2, 3$. Since $\tau_4 = \tau_1 \cap \tau_2$, $G(\tau_4) = X_4 \oplus G^{\#}(\tau_4)$ with $G^{\#}(\tau_4) = (X_1 \oplus X_2) + \mathbb{Z}(1, 1, 0, 0)/p$. This shows that B is a regulating subgroup of G. Clearly, G/B is isomorphic to $\mathbb{Z}/p\mathbb{Z} \oplus \mathbb{Z}/p\mathbb{Z}$.

(b) On the other hand,

$$G(\tau_4) = (X_4 \oplus X_1 \oplus X_2) + \mathbb{Z}(1, 1, 0, 0)/p$$

$$= Y \oplus ((X_1 \oplus X_2) + \mathbb{Z}(1, 1, 0, 0)/p) = Y \oplus G^{\#}(\tau_4),$$

observing that $(1, 1, 0, p)/p = (1, 1, 0, 0)/p + (0, 0, 0, 1) \in G$. Hence, C is a regulating subgroup of G. Finally, $G/C = \mathbb{Z}/p^2\mathbb{Z}$, since G/C is generated by $x = (0, 0, 1, 1)/p + C$ with order $x = p^2$. This holds because $px = (0, 0, 0, 1) + C = -(1, 1, 0, 0)/p + C$ with $(0, 0, 0, 1) \notin C$, and $p^2 x = p((0, 0, 0, 1) + C) = (0, 0, 0, p) \in C$.

Corollary 5.1.4 [Lady 74B] *Suppose G and H are almost completely decomposable groups.*

(a) *If G is nearly isomorphic to H, then* $i(G) = i(H)$.

(b) *If* $G = H \oplus K$, *then* $i(G) = i(H)i(K)$.

PROOF. (a) Choose a monomorphism $f : G \to H$ with $|H/f(G)|$ relatively prime to $i(H)$. By Proposition 5.1.2(b), $i(H)$ divides $H/f(B)$, for a regulating subgroup B of G, since $f(B)$ is a completely decomposable subgroup of H. But $|H/f(B)| = |H/f(G)| |f(G)/f(B)|$, so $i(H)$ divides $|f(G)/f(B)| = |G/B| = i(G)$. A symmetric argument shows that $i(G)$ divides $i(H)$.

(b) If $C = \oplus\{C_\tau : \tau \in T_{cr}(H)\}$ is a regulating subgroup of H and $D = \oplus\{D_\tau : \tau \in T_{cr}(K)\}$ is a regulating subgroup of K, then $B = C \oplus D$ is a regulating subgroup of G. This holds because $G(\tau) = H(\tau) \oplus K(\tau) = C_\tau \oplus D_\tau \oplus H^{\#}(\tau) \oplus K^{\#}(\tau) = C_\tau \oplus D_\tau \oplus G^{\#}(\tau)$ for each type τ. By (a), $i(G) = |G/B| = |H/C| |K/D| = i(H)i(K)$.

The *exponent* of a bounded abelian group A, written $e = \exp A$, is the least positive integer e with $eA = 0$. If G is a finite rank Butler group, then $G/R(G)$ is finite by Corollary 3.2.13(a), where $R(G)$ is the intersection of all of the regulating subgroups of G.

Define $e_G = \exp G/R(G)$ and $b_\tau(G) = \exp G^\#(\tau)/R(G^\#(\tau))$ for each type τ, adopting the convention that $b_\tau(G) = 1$ if $G^\#(\tau) = 0$. Notice that $G^\#(\tau)/R(G^\#(\tau))$ is finite for each type τ, since $G^\#(\tau)$ is a Butler group, being a pure subgroup of the Butler group G. Moreover, $b_\tau(G)$ is the least common multiple of $\{\exp G^\#(\tau)/D : D$ regulating subgroup of $G^\#(\tau)\}$, since $R(G^\#(\tau)) = \cap\{D : D$ regulating subgroup of $G^\#(\tau)\}$. The $b_\tau(G)$'s are called the *Burkhardt* invariants of G.

The following lemma gives a procedure for constructing $R(G)$ from the Burkhardt invariants and a given regulating subgroup of an almost completely decomposable group G.

Lemma 5.1.5 [Burkhardt 84] *Let G be an almost completely decomposable group and $B = \oplus\{B_\tau : \tau \in T_{cr}(G)\}$ a regulating subgroup of G. Then $R(G) = \oplus\{b_\tau(G)B_\tau : \tau \in T_{cr}(G)\}$.*

PROOF. Write b_τ for $b_\tau(G)$ and let $C = \oplus\{C_\tau : \tau \in T_{cr}(G)\}$ be an arbitrary regulating subgroup of G. Then $G(\tau) = B_\tau \oplus G^\#(\tau) = C_\tau \oplus G^\#(\tau)$ for each $\tau \in T_{cr}(G)$. If $\tau \in T_{cr}(G)$, then $\oplus\{C_\sigma : \tau < \sigma\}$ is a regulating subgroup of $G^\#(\tau)$, since $G(\sigma) = G(\tau)(\sigma)$ if $\tau < \sigma$. Then $b_\tau B_\tau \subseteq b_\tau G(\tau) = b_\tau C_\tau \oplus b_\tau G^\#(\tau) \subseteq b_\tau C_\tau \oplus R(G^\#(\tau)) \subseteq C_\tau \oplus (\oplus\{C_\sigma : \tau < \sigma\}) = C(\tau) \subseteq C$, using the fact that $b_\tau G^\#(\tau) \subseteq R(G^\#(\tau))$. Since C is an arbitrary regulating subgroup of G, $\oplus\{b_\tau B_\tau : \tau \in T_{cr}(G)\} \subseteq R(G)$.

Conversely, let $x \in R(G)\backslash(\oplus\{b_\tau B_\tau : \tau \in T_{cr}(G)\})$. Since $R(G) \subseteq B$, it suffices to assume that $x = \oplus\{x_\tau \in B_\tau\backslash b_\tau B_\tau : \tau \in S\}$ for some subset S of $T_{cr}(G)$. In particular, $b_\tau \neq 1$ for each $\tau \in S$. Let τ be a maximal element of S and write $B_\tau = A_1 \oplus \cdots \oplus A_n$ with each A_i a rank-1 group of type τ. Then $x_\tau = a_1 \oplus \cdots \oplus a_n$ for some $a_i \in A_i$. Because $x_\tau \notin b_\tau B_\tau$, there is some i, say $i = 1$, with $a_i \notin b_\tau B_\tau$. Since $b_\tau = \exp G^\#(\tau)/R(G^\#(\tau))$, there is some $a \in G^\#(\tau)$ with order $(a + R(G^\#(\tau)) = b_\tau \neq 1$ in $G^\#(\tau)/R(G^\#(\tau))$.

Let $D = \oplus\{D_\tau : \tau \in T_{cr}(G^\#(\tau))\}$ be an arbitrary regulating subgroup of $G^\#(\tau)$ and define $C = \oplus\{C_\sigma : \sigma \in T_{cr}(G)\}$, where $C_\tau = A_1' \oplus A_2 \oplus \cdots \oplus A_n$ with A_1' the pure rank-1 subgroup of G generated by $a_1 + a$, $C_\sigma = D_\sigma$ if $\sigma > \tau$, and $C_\sigma = B_\sigma$ if $\sigma \not> \tau$. Then C is a regulating subgroup of G, so that $x \in R(G) \subseteq C$, say $x = \oplus\{c_\sigma \in C_\sigma : \sigma \in T_{cr}(G)\}$.

Now, $0 = x - x = \Sigma\{x_\sigma - c_\sigma : \sigma \in T_{cr}(G)\}$, so that $x_\tau - c_\tau = \Sigma\{c_\sigma - x_\sigma : \sigma \neq \tau\} \in G(\tau)$. Since τ is maximal in S, $x_\sigma = 0$ if $\sigma > \tau$ and $x_\tau - c_\tau = \Sigma\{c_\sigma : \sigma > \tau\} \in G^\#(\tau)$. Also, $x_\tau - c_\tau = (a_1 \oplus \cdots \oplus a_n) - (a_1' \oplus a_2' \oplus \cdots \oplus a_n') \in B_\tau \oplus C_\tau$ with $a_i' \in A_2 \subseteq B_\tau$ for $2 \leq i$ and $a_1' = q_1(a_1 + a)$ for some $q_1 \in Q$. Then $x_\tau - c_\tau + q_1 a = (a_1 - q_1 a_1) \oplus (\oplus\{a_i - a_i' : 2 \leq i \leq n\}) = \Sigma\{c_\sigma : \sigma > \tau\} + q_1 a \in B_\tau \cap G^\#(\tau) = 0$, since $(a_1 - q_1 a_1) \oplus (\oplus\{a_i - a_i' : 2 \leq i \leq n\}) \in B_\tau$ and $a, c_\sigma G^\#(\tau)$. Consequently, $a_1 = q_1 a_1$, $q_1 = 1$, and $a = \Sigma\{-c_\sigma : \sigma > \tau\} \in D$, recalling that $C_\sigma = D_\sigma$ if $\sigma > \tau$. Since D is an arbitrary regulating subgroup of $G^\#(\tau)$, $a \in R(G^\#(\tau))$, a contradiction to the choice of a.

Theorem 5.1.6

(a) [Burkhardt 84] *If $B = \oplus\{G_\tau : \tau \in T_{cr}(G)\}$ is a regulating subgroup of an almost completely decomposable group G, then $R(G) = \Sigma\{b_\tau(G)G(\tau):$*

$\tau \in T_{cr}(G)\}$ is a completely decomposable fully invariant subgroup of G.

(b) [Mader, Vinsonhaler 94] *If G and H are nearly isomorphic almost completely decomposable groups, then $R(G)$ is isomorphic to $R(H)$ and $G/R(G)$ is isomorphic to $H/R(H)$.*

PROOF. (a) By Lemma 5.1.5, $R(G) = \oplus\{b_\tau B_\tau : \tau \in T_{cr}(G)\}$. Hence, $R(G)$ is a completely decomposable subgroup of G contained in $\Sigma\{b_\tau G(\tau) : \tau \in T_{cr}(G)\}$. Conversely, in view of the definition of b_τ and Lemma 5.1.5, if $\tau \in T_{cr}(G)$, then $b_\tau G(\tau) = b_\tau B_\tau \oplus b_\tau G^\#(\tau) \subseteq b_\tau B_\tau \oplus R(G^\#(\tau)) \subseteq R(G)$ as a consequence of Exercise 5.1.12(b). Consequently, $\Sigma\{b_\tau G(\tau) : \tau \in T_{cr}(G)\} \subseteq R(G)$, as desired. If $f \in \text{End } G$, then $f(R(G)) \subseteq R(G)$, since $R(G) = \Sigma\{b_\tau G(\tau) : \tau \in T_{cr}(G)\}$, b_τ is an integer, and $f(G(\tau)) \subseteq G(\tau)$ for each τ. This shows that $R(G)$ is a fully invariant subgroup of G.

(b) Since $G/R(G)$ is finite, G is quasi-isomorphic to $R(G)$. But G is nearly isomorphic to H, so that $R(G)$ is quasi-isomorphic to $R(H)$ and $T_{cr}(G) = T_{cr}(H)$. By (a), $R(G)$ and $R(H)$ are completely decomposable groups, whence $R(G)$ is isomorphic to $R(H)$ by Theorem 3.1.7(b).

In view of Theorem 2.2.2(a), it is sufficient to assume that there is an integer m relatively prime to $e_G = \exp G/R(G)$ with $mH \subseteq G \subseteq H$. Then $mH(\tau) \subseteq G(\tau) \subseteq H(\tau)$ for each type τ and $b_\tau(G) = b_\tau(H)$ for each $\tau \in T_{cr}(G)$. Hence, $mR(H) \subseteq R(G) = \Sigma\{b_\tau(G)G(\tau) : \tau \in T_{cr}(G)\} \subseteq R(H) = \Sigma\{b_\tau(H)H(\tau) : \tau \in T_{cr}(G)\}$. Since $e_G = \exp G/R(G)$ is relatively prime to m, $G/R(G)$ is isomorphic to $H/R(H)$.

If H and K are almost completely decomposable groups, then $H \oplus K$ is an almost completely decomposable group with $R(H \oplus K) \subseteq R(H) \oplus R(K)$ (Exercise 5.1.12(a)). However, the group constructed in Example 5.1.3 is an example of an almost completely decomposable group $H \oplus K$ with $R(H \oplus K) \neq R(H) \oplus R(K)$.

Example 5.1.3 (continued) *Let G be as defined in Example 5.1.3. Then*

(c) $G = H \oplus K$, where $H = (X_1 \oplus X_2) + \mathbb{Z}(1, 1, 0, 0)/p$ and $K = (X_3 \oplus X_4) + \mathbb{Z}(0, 0, 1, 1)/p$;

(d) $R(H) = X_1 \oplus X_2$ and $R(K) = X_3 \oplus X_4$; and

(e) $R(G) = X_1 \oplus X_2 \oplus X_3 \oplus pX_4$.

PROOF. Statement (c) is immediate from the definition of G. As for (d), the critical typeset of H is $\{\tau_1, \tau_2\}$ with τ_1 and τ_2 incomparable and $H(\tau_i) = X_i$. Then $R(H) = X_1 \oplus X_2$. Alternatively, $R^\#(H(\tau_i)) = 0$, so that the Burkhardt invariants for H are $b_\tau = 1$ for $\tau = \tau_i$. Hence, $R(H) = X_1 \oplus X_2$ by Theorem 5.1.6(a). Similarly, $R(K) = X_3 \oplus X_4$.

(e) In this case, there is not a unique regulating subgroup of G, by Example 5.1.3(a) and (b). However, the Burkhardt invariants b_τ of G can still be computed. Now, $T_{cr}(G) = \{\tau_1, \tau_2, \tau_3, \tau_4\}$ with $G^\#(\tau_i) = 0$ for $i = 1, 2, 3$ and

$G^{\#}(\tau_4) = (X_1 \oplus X_2) + \mathbb{Z}(1, 1, 0, 0)/p$. Hence, $b_\tau = 1$ for $\tau = \tau_i$, $1 \le i \le 3$, and $b_\tau = p$ for $\tau = \tau_4$. By Theorem 5.1.6(a), $R(G) = \Sigma\{b_\tau G(\tau) : \tau \in T_{cr}(G)\} = X_1 \oplus X_2 \oplus X_3 \oplus pX_4$.

The next example completes the computation of the integer $n(G)$ for an almost completely decomposable group G mentioned just after the proof of Lemma 2.2.7.

Example 5.1.7 *If G is an almost completely decomposable group and e_G is the exponent of $G/R(G)$, then $e_G = n(G)$.*

PROOF. Since $C = R(G)$ is a fully invariant completely decomposable subgroup of G, restriction induces a ring embedding $e_G \operatorname{End} C \subseteq \operatorname{End} G \subseteq \operatorname{End} C$ by Theorem 5.1.6(a). Then $e_G(\operatorname{End} C/N\operatorname{End} C) \subseteq \operatorname{End} G/N\operatorname{End} G \subseteq \operatorname{End} C/N\operatorname{End} C$. By Proposition 3.1.8, $\operatorname{End} C/N\operatorname{End} C$ is isomorphic to $\operatorname{Mat}_{n(1)}(R_1) \times \cdots \times \operatorname{Mat}_{n(m)}(R_m)$ with each R_i a subring of $\mathbb{Q}$. Since a subring of $\mathbb{Q}$ is a principal ideal domain, each $\operatorname{Mat}_{n(i)}(R_i)$ is a maximal order over R_i, as noted in Section 2.2. Apply the definition of $n(G)$ to complete the proof.

Parts (b), (c), and (d) of the next corollary are consequences of Corollary 2.2.11, the proof of which uses the fact that maximal orders are hereditary. A group-theoretic proof for the special case of almost completely decomposable groups is a consequence of Exercise 5.1.4 and the proof of Corollary 2.2.11.

Corollary 5.1.8 *Suppose G and H are almost completely decomposable groups.*

(a) *The groups G and H are nearly isomorphic if and only if there is a completely decomposable group C with $G \oplus C$ isomorphic to $H \oplus C$. In this case, C may be chosen to be isomorphic to both $R(G)$ and $R(H)$.*

(b) *The group H is a near summand of G if and only if $G = X \oplus X'$ for some X and X' with X nearly isomorphic to H. In particular, G is indecomposable if and only if 0 and G are the only near summands of G.*

(c) *If $G \oplus K$ is nearly isomorphic to $H \oplus K$ for some torsion-free abelian group K of finite rank, then G is nearly isomorphic to H.*

(d) *If G is nearly isomorphic to $G_1 \oplus \cdots \oplus G_n$, then $G = H_1 \oplus \cdots \oplus H_n$ with each G_i nearly isomorphic to H_i.*

PROOF. (a) Suppose G and H are nearly isomorphic. Then $C = R(G)$ is completely decomposable, C is isomorphic to $R(H)$ by Theorem 5.1.6(b), and $eG \subseteq C$ with $e = \exp G/R(G)$. By Theorem 2.2.2(a), there is $f \in \operatorname{Hom}(G, H)$ and $g \in \operatorname{Hom}(H, G)$ and a nonzero integer m prime to e such that $gf = m1_G$ and $fg = m1_H$. Write $1 = rm + se$ for integers r and s and define $\phi : G \to H \oplus C$ by $\phi(x) = (fx, ex)$ and $\alpha : H \oplus C \to G$ by $\alpha(x, y) = rgx + sy$. Then $\alpha\phi = 1_G$, so that $H \oplus C$ is isomorphic to $G \oplus D$ for some D.

Since G and H are nearly isomorphic, C is nearly isomorphic to D by Corollary 2.2.4(a). Thus, D is isomorphic to a summand of the completely decomposable group $C \oplus C$, by Corollary 2.2.6, hence completely decomposable by Theorem 3.1.7(a). Therefore, C is isomorphic to D, since quasi-isomorphic completely decomposable groups are isomorphic by Theorem 3.1.7(b).

Conversely, if $G \oplus C$ is isomorphic to $H \oplus C$, then G is nearly isomorphic to H by Corollary 2.2.4(a).

Statement (b) is a consequence of Corollary 2.2.11, (c) is a special case of Corollary 2.2.4, and (d) follows from (b) and (c). □

There are examples of almost completely decomposable groups G and H with no rank-1 summands such that $G \oplus H$ has a rank-1 summand (Exercise 5.1.7). However, the next lemma demonstrates that an almost completely decomposable group contains a maximal completely decomposable summand that is unique up to isomorphism.

Proposition 5.1.9 [Lady 74B] *Assume that G is an almost completely decomposable group.*

(a) *Then $G = C \oplus H$, where C is completely decomposable and H has no rank-1 summands.*

(b) *If $G = C' \oplus H'$ is another such decomposition of G, then C is isomorphic to C' and H is nearly isomorphic to H'.*

PROOF. (a) Since G has finite rank, it follows immediately that G has such a decomposition.

(b) First assume that C and C' are τ-homogeneous completely decomposable. Then $H \cap C'$ is a pure subgroup, hence a τ-homogeneous completely decomposable summand, of C' by Lemma 3.1.6. Write $C' = (H \cap C') \oplus D$. Then $G = C' \oplus H' = (H \cap C') \oplus D \oplus H'$, from which it follows that $H \cap C'$ is a summand of H. But H has no rank-1 summands, whence $H \cap C' = 0$. Let $f : G \to C$ be a projection with ker $f = H$. Since $H \cap C' = 0$, $f : C' \to C$ must be a monomorphism. Similarly, there is a monomorphism $g : C \to C'$. This shows that rank $C = $ rank C'. Since C and C' are τ-homogeneous completely decomposable groups of the same rank, they must be isomorphic. Moreover, H is nearly isomorphic to H' by Corollary 5.1.8(c).

For the general case, let τ be a maximal type in the union of the typeset of C and the typeset of C'. Then $C(\tau)$ and $C'(\tau)$ are τ-homogeneous completely decomposable groups by the choice of τ. Write $C = C(\tau) \oplus D$ and $C' = C'(\tau) \oplus D'$. Then $G = C(\tau) \oplus (D \oplus H) = C'(\tau) \oplus (D' \oplus H')$, and the groups $D \oplus H$ and $D' \oplus H'$ have no rank-1 summands of type τ. The argument of the preceding paragraph shows that $C(\tau)$ is isomorphic to $C'(\tau)$ and $D \oplus H$ is nearly isomorphic to $D' \oplus H'$ with D and D' completely decomposable. In view of Corollary 5.1.8(d), it suffices to assume that $D \oplus H = D' \oplus H'$. By induction on the rank of G, D is isomorphic to D' and H is nearly isomorphic to H'. Thus, $C = C(\tau) \oplus D$ is isomorphic to $C' = C'(\tau) \oplus D'$. □

There is a characterization, in terms of the join irreducible elements S_T, of those finite lattices T such that each finite rank Butler group with typeset contained in T is almost completely decomposable.

Corollary 5.1.10 *Let T be a finite lattice of types. Each group in $B(T)$ is an almost completely decomposable group if and only if $w(S_T) \leq 2$.*

PROOF. If $w(S_T) \geq 3$, then there is a strongly indecomposable group G in $B(T)$ with rank $G = 2$ by Example 3.3.3. In this case, G cannot be quasi-isomorphic to a completely decomposable group. Conversely, if $w(S_T) \leq 2$, then each strongly indecomposable group in $B(T)$ has rank 1, as a consequence of Theorem 3.3.2 and Example 1.1.3. Hence, each G in $B(T)$ is quasi-isomorphic to a completely decomposable group.

The remainder of this section is devoted to the case that T is a finite lattice of types with $S_T = S_2 = \{\tau, \sigma\}$. If $G \in B(T)$, then G is almost completely decomposable by Corollary 5.1.10. If, in addition, G has no rank-1 summands and $G(\tau) \cap G(\sigma) = 0$, then $T_{\mathrm{cr}}(G) = S_2$ and $R(G) = G(\tau) \oplus G(\sigma)$ is a unique regulating subgroup of G. It follows from Theorem 3.4.5 that $G = G_1 \oplus \cdots \oplus G_n$ is a finite direct sum of indecomposable groups of rank 2. Moreover, $R(G) = G(\tau) \oplus G(\sigma) = (G_1(\tau) \oplus G_1(\sigma)) \oplus \cdots \oplus (G_n(\tau) \oplus G_n(\sigma)) = R(G_1) \oplus \cdots \oplus R(G_n)$ with $G/R(G) = G_1/R(G_1) \oplus \cdots \oplus G_n/R(G_n)$ and each $G_i/R(G_i)$ a finite cyclic group.
The following proposition is the converse in that a decomposition of the finite group $G/R(G)$ into a direct sum of cyclic groups gives rise to a corresponding direct sum decomposition of G into indecomposable groups of rank less than or equal to 2.

Proposition 5.1.11 [Mader, Vinsonhaler 95] *Suppose T is a finite lattice of types with $S_T = \{\tau, \sigma\}$. Let $G \in B(T)$ with no rank-1 summands and $G(\tau) \cap G(\sigma) = 0$ and write $G/R(G) = C_1 \oplus \cdots \oplus C_n$ with each C_i a cyclic group. Then $G = G_1 \oplus \cdots \oplus G_n$ for some indecomposable almost completely decomposable group G_i of rank 2 with $G_i/R(G_i) = C_i$ for each $1 \leq i \leq n$.*

PROOF. Assume that C_i is a cyclic group of order a_i and let X and Y be subgroups of $\mathbb{Q}$ containing 1 with type $X = \tau$ and type $Y = \sigma$. Choose $x_i \in G(\tau)$ and $y_i \in G(\sigma)$ with $z_i = (x_i + y_i)/a_i \in G$, $z_i + R(G)$ a generator of C_i, $\{x_1, x_2, \ldots, x_n, y_1, \ldots, y_n\}$ a $\mathbb{Z}$-independent subset of G, and $C = (Xx_1 \oplus \cdots \oplus Xx_n) \oplus (Yy_1 \oplus \cdots \oplus Yy_n)$ a pure subgroup, hence a summand, of $R(G) = G(\tau) \oplus G(\sigma)$. This can be done inductively, observing that $G(\tau)$ and $G(\sigma)$ are non-zero homogeneous completely decomposable groups with $G(\tau) \cap G(\sigma) = 0$ and by recalling that G has no rank-1 summands.
Define $G_i = (Xx_i \oplus Yy_i) + \mathbb{Z}z_i$, a subgroup of G. Then $G_i/R(G_i) = C_i$ and $(G_1 \oplus \cdots \oplus G_n) + R(G) = G$ as a consequence of the fact that each z_i is a generator of C_i. Now, $C = (Xx_1 \oplus \cdots \oplus Xx_n) \oplus (Yy_1 \oplus \cdots \oplus Yy_n) = C(\tau) \oplus C(\sigma)$ is a summand of $R(G) = G(\tau) \oplus G(\sigma)$, say $R(G) = C \oplus D$. Since $G/C = G/R(G)$, it follows that D is a summand of G. But G has no rank-1 summands, whence

$R(G) = C \subseteq G_1 \oplus \cdots \oplus G_n$. Thus, $G = G_1 \oplus \cdots \oplus G_n$ with rank $G_i = 2$ and $G_i/R(G_i) = C_i$, as desired.

Corollary 5.1.12 *Suppose G is a finite rank Butler group with typeset $\subseteq T_2 = \{\tau \cap \sigma, \tau, \sigma\}$ and τ and σ incomparable. Then G is a finite direct sum of indecomposable groups with rank less than or equal to 2.*

PROOF. By Proposition 5.1.9, it is sufficient to assume that G has no rank-1 summands. In view of the hypotheses, $G(\tau \cup \sigma) = G(\tau) \cap G(\sigma) = 0$. Since the finite group $G/R(G)$ can be written as a direct sum of cyclic groups, Proposition 5.1.11 applies.

Proposition 5.1.11 can be used to demonstrate that direct sum decompositions of almost completely decomposable groups are not unique up to near-isomorphism even if the critical typeset has 2 elements. As an illustration, suppose G satisfies the hypotheses of Proposition 5.1.11 with $G/R(G) = \mathbb{Z}/30\mathbb{Z}$. Then $G/R(G) = \mathbb{Z}/2\mathbb{Z} \oplus \mathbb{Z}/3\mathbb{Z} \oplus \mathbb{Z}/6\mathbb{Z} = \mathbb{Z}/6\mathbb{Z} \oplus \mathbb{Z}/5\mathbb{Z} = \mathbb{Z}/10\mathbb{Z} \oplus \mathbb{Z}/3\mathbb{Z} = \mathbb{Z}/15\mathbb{Z} \oplus \mathbb{Z}/2\mathbb{Z}$. Each of these decompositions of $G/R(G)$ into cyclic groups induces a direct sum decomposition of G into indecomposable groups of rank 2. These decompositions are not unique up to near-isomorphism by Theorem 5.1.6(b), since if X and X' are nearly isomorphic summands of G, then $X/R(X)$ is isomorphic to $X'/R(X')$.

On the other hand, if $G/R(G)$ is a p-group for some prime p, necessarily finite, then any two decompositions of $G/R(G)$ into cyclic groups are unique up to isomorphism and order. This uniqueness carries over to G, since it is shown in Corollary 5.4.3 that if $G/R(G)$ is a p-group, then direct sum decompositions of an almost completely decomposable group G into indecomposable groups are unique up to isomorphism at p.

EXERCISES

1. Prove that if G is an indecomposable finite rank Butler group with critical typeset $G \subseteq T_2 = \{\tau_1, \tau_2\}$ and rank $G = 2$, then G is isomorphic to $X_1 \oplus X_2 + \mathbb{Z}(1, m)/n$, where m and n are relatively prime integers with $n > 1$ and X_i is a subgroup of $\mathbb{Q}$ containing 1 with type $X_i = \tau_i$.

2. Find a complete set of near isomorphism invariants for finite rank Butler groups G with typeset $G \subseteq T_2$.

3. [Cruddis 70], [Arnold, Dugas 93B] Let G be a finite rank Butler group with $T_{cr}(G) \subseteq T_3 = \{\tau_1, \tau_2, \tau_3\}$ and $\tau_1 + \tau_2 + \tau_3 = $ type $\mathbb{Q}$. Prove that G is the direct sum of groups of rank less than or equal to 2. Find a complete set of near-isomorphism invariants for G.

4. Let G be an almost completely decomposable group with $e_G G \subseteq R(G) \subseteq G$ and K a near summand of G. Prove that if $g : G \to K$, then there is $h : g(G) \to K$ with $ghg = e_G g$. Conclude that $X = g(G)$ is a summand of G nearly isomorphic to K.

5. Let G be an almost completely decomposable group with $e_G G \subseteq R(G) \subseteq G$ and D the unique maximal subgroup of G with $e_G D = D$. Write $R(G) = \oplus\{A_\tau : \tau \in$

$T_{\mathrm{cr}}(G)\}$, A_τ a τ-homogeneous completely decomposable group. Prove that:
(a) $D = \oplus\{A_\tau : \tau \in T_{\mathrm{cr}}(G), e_G A_\tau = A_\tau\}$; and
(b) $G = D \oplus E$, where E is the pure subgroup of G generated by $\oplus\{A_\tau : \tau \in T_{\mathrm{cr}}(G), e_G A_\tau \neq A_\tau\}$.

6. Assume that G is an almost completely decomposable group with $T_{\mathrm{cr}}(G)$ an inverted forest. Compute the Burkhardt invariants of G.

7. (a) [Jónsson 57] Give an example of an almost completely decomposable group $G = C \oplus H = C' \oplus H'$ such that C and C' are completely decomposable groups, H and H' have no rank-1 summands, but H and H' are not isomorphic.
   (b) [Jónsson 57] Give an example of two almost completely decomposable groups G and H with no rank-1 summands such that $G \oplus H$ has a rank-1 summand.

8. Find almost completely decomposable groups G and H and $f \in \mathrm{Hom}(G, H)$ such that $f(R(G))$ is not contained in $R(H)$.

9. [Mader 95] Let G be an almost completely decomposable group and define $G^{(0)} = G$, $G^{(1)} = \Sigma\{G^\#(\tau) : \tau \in T_{\mathrm{cr}}(G)\}$, and, inductively, $G^{(n+1)} = (G^{(n)})^{(1)}$.
   (a) Prove that $G^{(n)} = 0$ for some n and that $G^{(0)} \supseteq G^{(1)} \supseteq \cdots \supseteq G^{(n)} = 0$ is a descending chain of purely invariant subgroups of G.
   (b) Prove that $G(\mathrm{IT}(G))/G^{(1)}$ is isomorphic to $\oplus\{G(\tau)/G^\#(\tau) : \tau$ is minimal in $T_{\mathrm{cr}}(G)\}$.
   (c) Prove that $i(G) = |G/G(\mathrm{IT}(G))| \, |G^{(1)}/G^{(1)}(\mathrm{IT}(G^{(1)}))| \cdots |G^{(n)}/G^{(n)}(\mathrm{IT}(G^{(n)}))|$.

10. [Burkhardt 84] Let G be an almost completely decomposable group and C a completely decomposable subgroup of G with G/C finite. Prove that C is a regulating subgroup of G if and only if there is a decomposition $C = \oplus\{C_\tau : \tau \in T_{\mathrm{cr}}(G)\}$ such that each for each τ, $C_\tau \subseteq (\exp G^\#(\tau)/C^\#(\tau))G(\tau) \subseteq C(\tau)$.

11. [Burkhardt 84] Let G be an almost completely decomposable group. Prove that if G has regulating subgroups B and C such that $G/B = \mathbb{Z}/p^n\mathbb{Z}$ and $G/C = (\mathbb{Z}/p\mathbb{Z})^n$, then for each finite abelian group A with $|A| = p^n$ there is a regulating subgroup D of G with $G/D = A$.

12. Let G and H be almost completely decomposable groups.
    (a) Prove that $R(G \oplus H) \subseteq R(G) \oplus R(H)$.
    (b) Prove that if $\tau \in T_{\mathrm{cr}}(G)$, then $R(G^\#(\tau)) \subseteq R(G)$.

5.2   Isomorphism at p and Representation Type

The representation type of the category of almost completely decomposable groups in $B(T, j)$ is computed in Corollaries 5.2.4 and 5.2.9 by determining the representation type of a subcategory of $\mathrm{rep}(S_T, \mathbb{Z}_p, j)$. Indecomposable representations in $\mathrm{rep}(S_T, \mathbb{Z}_p, j)$ can be used to construct indecomposable almost completely decomposable groups in $B(T, j)$.

Let T be a finite lattice of types, j a nonnegative integer, and $C(T, j)$ the full subcategory of $B(T, j)$ consisting of almost completely decomposable groups. Specifically, the groups in $C(T, j)$ are almost completely decomposable groups G with typeset $G \subseteq T$ and $p^j G \subseteq \Sigma\{G(\tau) : \tau \in S_T\} \subseteq G$. Since S_T consists of the join-irreducible elements of T, $\Sigma\{G(\tau):\tau \in S_T\} = G^*(\mathrm{IT}(G))$. Henceforth, it is assumed that $j \geq 1$, since $C(T, 0)$ consists only of almost completely decomposable groups (Exercise 5.2.2).

For almost completely decomposable groups, there is little loss of generality, up to isomorphism at p, in assuming that T is a p-locally free lattice of types. By Proposition 5.1.9, an almost completely decomposable group G is the direct sum of a completely decomposable group C and an almost completely decomposable group H with no rank-1 summands. Moreover, H is unique up to near isomorphism.

As a consequence of Proposition 4.3.1, if G is an almost completely decomposable group with no rank-1 summands, the typeset of G is contained in T, and p is a prime, then G is isomorphic at p to some $H \in C(T, j)$ with $e_H = \exp H/R(H)$ a power of p. If D is the maximal p-divisible subgroup of H, then D is completely decomposable and $H = D \oplus E$ for some E with typeset E a p-locally free set of types (Exercise 5.1.5). Hence, if H is indecomposable at p with rank $H > 1$, then $H \in C(T', j)$ for some finite p-locally free lattice T'.

Recall from Corollary 4.3.2 that if T is a p-locally free finite lattice of types and $j \geq 1$ an integer, then there is a category equivalence $F_T : B(T, j)_p \to$ rep$(S_T, \mathbb{Z}_p, j)$ given by $F_T(G) = (G_p : G(\tau)_p : \tau \in S_T)$. Define Crep$(S_T, \mathbb{Z}_p, j)$ to be the full subcategory of rep$(S, \mathbb{Z}_p, j)$ with objects of the form $F_T(G)$, $G \in C(T, j)$. Clearly, the categories $C(T, j)_p$ and Crep$(S_T, \mathbb{Z}_p, j)$ are equivalent categories.

Lemma 5.2.1 *If T is a finite lattice, T' is a sublattice of T, and $j' \leq j$, then there is a fully faithful functor $F : C$rep$(S_{T'}, R, j') \to C$rep(S_T, R, j).*

PROOF. A consequence of the definition of Crep$(S_T, \mathbb{Z}_p, j)$ and the observation that $C(T', j)$ is contained in $C(T, j)$. Specifically, $F = F_T(F_{T'})^{-1}$.

It is not all that easy to identify those representations in rep $(S_T, \mathbb{Z}_p, j)$ that are in Crep$(S_T, \mathbb{Z}_p, j)$; see Exercise 5.2.1. By Proposition 5.1.1, $G \in B(T)$ is an almost completely decomposable group if and only if rank $G = \Sigma\{\text{rank } G(\tau)/G^{\#}(\tau) : \tau \in T_{cr}(G)\}$. However, $T_{cr}(G)$ need not be contained in S_T, the join irreducible elements of T. For example, let T be a Boolean algebra with exactly 3 atoms. Then S_T consists of these three atoms. If G is an almost completely decomposable group with critical typeset consisting of three pairwise incomparable elements of T that are not atoms, then the critical typeset of G is clearly not contained in S_T.

Define $C_{crit}(T, j)$ to be the category consisting of all almost completely decomposable groups in $C(T, j)$ with critical typeset contained in S_T, the set of join irreducible elements of T. The image of the functor F_T on groups in $C_{crit}(T, j)$, denoted by Crep$_{crit}(S_T, \mathbb{Z}_p, j)$, consists of those representations $U = (U_0, U_i : i \in S_T) \in Crep(S_T, \mathbb{Z}_p, j)$ such that rank $U_0 = \Sigma\{\text{rank } U_i/U_i^{\#} : i \in S_T\}$, where $U_i^{\#}$ is the purification of $\Sigma\{U_s : s < i \text{ in } S_T\}$ in U_i, as a consequence of Proposition 5.1.1 and the definition of F_T. For each such representation, U_0 is a finitely generated free $\mathbb{Z}_p$-module and each U_i is a summand of U_0.

Notice that if $\text{Crep}_{\text{crit}}(S_T, \mathbb{Z}_p, j)$ has infinite or wild modulo p representation type, then $\text{Crep}(S_T, \mathbb{Z}_p, j)$ has, respectively, infinite or wild modulo p representation type by Lemma 5.2.1. This is because $\text{Crep}_{\text{crit}}(S_T, \mathbb{Z}_p, j)$ embeds as a full subcategory of $\text{Crep}(S_T, \mathbb{Z}_p, j)$. The representations constructed in the following examples are in $\text{Crep}_{\text{crit}}(S_T, \mathbb{Z}_p, j)$, hence automatically in $\text{Crep}(S_T, \mathbb{Z}_p, j)$.

Example 5.2.2 *Let T be a p-locally free finite lattice of types. If either $w(S_T) = 5$, $j \geq 1$; $w(S_T) \geq 4$ and $j \geq 2$; or $w(S_T) = 3$ and $j \geq 3$, then $\text{Crep}(S_T, \mathbb{Z}_p, j)$ has wild modulo p representation type.*

PROOF. Let Γ be a $\mathbb{Z}_p$-algebra that is finitely generated and free as an $\mathbb{Z}_p$-module and A and B two $n \times n$ $\mathbb{Z}_p$-matrices with $C(A, B) = \Gamma$, guaranteed by Lemma 4.2.1(a).

(i) $w(S_T) = 5$ and $j \geq 1$. By Lemma 5.2.1, it is sufficient to prove that $\text{Crep}(S_5, \mathbb{Z}_p, 1)$ has wild modulo p representation type. Define $U = (U_0, U_1, U_2, U_3, U_4, U_5) \in \text{Crep}(S_5, \mathbb{Z}_p, 1)$ by

$$U_0 = \mathbb{Z}_p^n \oplus \mathbb{Z}_p^n \oplus \mathbb{Z}_p^n \oplus \mathbb{Z}_p^n \oplus \mathbb{Z}_p^n + (1/p)(1 + 0 + 1 + A + B)\mathbb{Z}_p^n,$$
$$+ (1/p)(0 + 1 + 1 + 1 + 1)\mathbb{Z}_p^n,$$
$$U_1 = \mathbb{Z}_p^n \oplus 0 \oplus 0 \oplus 0 \oplus 0, \; U_2 = 0 \oplus \mathbb{Z}_p^n \oplus 0 \oplus 0 \oplus 0,$$
$$U_3 = 0 \oplus 0 \oplus \mathbb{Z}_p^n \oplus 0 \oplus 0, \; U_4 = 0 \oplus 0 \oplus 0 \oplus \mathbb{Z}_p^n \oplus 0,$$
$$U_5 = 0 \oplus 0 \oplus 0 \oplus 0 \oplus \mathbb{Z}_p^n.$$

As a consequence of Lemma 4.2.3(a), $\text{End } U$ is isomorphic to $\text{End } V$, with $V = (V_0, U_1, U_2, U_3, U_4, U_5, pU_0) \in \text{rep}_f(S_6, \mathbb{Z}_p)$ and $V_0 = U_1 \oplus U_2 \oplus U_3 \oplus U_4 \oplus U_5 = \mathbb{Z}_p^{5n}$. Moreover, $pU_0 = pV_0 + W$, where W is the free $\mathbb{Z}_p$-module with the rows of the $\mathbb{Z}_p$-matrix $M = \left(\begin{smallmatrix} I & 0 & I & A & B \\ 0 & I & I & I & I \end{smallmatrix} \right)$ as a basis and I an $n \times n$ identity matrix.

To see that there is an onto ring homomorphism $\text{End } V \to C(A(\text{mod } p), B(\text{mod } p))$, let $f \in \text{End } V$. Then f can be written as $(f_1, f_2, f_3, f_4, f_5)$ with each f_i a $\mathbb{Z}_p$-endomorphism of U_i, and $f(pU_0) \subseteq pU_0$. In matrix terms, $fM \equiv Mg(\text{mod } p)$ for some g. This equation implies that $f_2 \equiv f_3 \equiv f_4 \equiv f_5(\text{mod } p)$ (from the second row) and $f_1 \equiv f_3(\text{mod } p)$ with $f_1 A \equiv A f_4(\text{mod } p)$ and $f_1 B \equiv B f_5(\text{mod } p)$ from the first row. Hence, $f(\text{mod } p) = (a, a, a, a, a)$, with $a \in C(A(\text{mod } p), B(\text{mod } p))$.

Then $\phi : \text{End } V \to C(A(\text{mod } p), B(\text{mod } p))$, given by $\phi(f) = a$, is a ring homomorphism. Now, $C(A, B)$ embeds in $\text{End } V$, via $a \mapsto (a, a, a, a, a)$, whence ϕ is onto by Lemma 4.2.1. Since $\text{End } U$ is isomorphic to $\text{End } V$, $\text{Crep}(S_5, \mathbb{Z}_p, 1)$ has wild modulo p representation type.

(ii) $w(S_T) \geq 4$ and $j \geq 2$. Once again, it is sufficient to prove that $\text{Crep}(S_4, \mathbb{Z}_p, 2)$ has wild modulo p representation type. Define $U = (U_0, U_1, U_2, U_3, U_4) \in$

$Crep(S_4, \mathbb{Z}_p, 2)$ by

$$U_0 = \mathbb{Z}_p^n \oplus \mathbb{Z}_p^n \oplus \mathbb{Z}_p^n \oplus \mathbb{Z}_p^n + (1/p^2)(1 + 1 + A + B)\mathbb{Z}_p^n$$
$$+ (1/p)(0 + 1 + 1 + 1)\mathbb{Z}_p^n,$$
$$U_1 = \mathbb{Z}_p^n \oplus 0 \oplus 0 \oplus 0, \; U_2 = 0 \oplus \mathbb{Z}_p^n \oplus 0 \oplus 0,$$
$$U_3 = 0 \oplus 0 \oplus \mathbb{Z}_p^n \oplus 0, \text{ and } U_4 = 0 \oplus 0 \oplus 0 \oplus \mathbb{Z}_p^n.$$

By Lemma 4.2.3(a), End U is isomorphic to End V, with $V = (V_0, U_1, U_2, U_3, U_4, p^2 U_0) \in \text{rep}_f(S_5, \mathbb{Z}_p)$ and $V_0 = U_1 \oplus U_2 \oplus U_3 \oplus U_4 = \mathbb{Z}_p^{4n}$. Then $p^2 U_0 = p^2 V_0 + W$, where W is the free $\mathbb{Z}_p$-module with the rows of the $\mathbb{Z}_p$-matrix

$$M = \begin{pmatrix} I & I & A & B \\ 0 & pI & pI & pI \end{pmatrix}$$

as a basis.

If $f \in \text{End } V$, then $f = (f_1, f_2, f_3, f_4)$ with each f_i a $\mathbb{Z}_p$-endomorphism of U_i and $fM \equiv Mg (\text{mod } p^2)$ for some g. This equation implies, from the first row of M, that $f_1 \equiv f_2 (\text{mod } p)$, $f_3 A \equiv A f_1 (\text{mod } p)$, and $f_4 B \equiv B f_1 (\text{mod } p)$. From the second row, $f_2 \equiv f_3 \equiv f_4 (\text{mod } p)$. Hence, $f(\text{mod } p) = (a, a, a, a)$ with $a \in C(A(\text{mod } p), B(\text{mod } p))$. Then $\phi : \text{End } V \to C(A(\text{mod } p), B(\text{mod } p))$ given by $\phi(f) = a$ is a ring homomorphism and $C(A, B)$ embeds in End V, via $a \mapsto (a, a, a, a)$. Thus, ϕ is onto by Lemma 4.2.1. Since End U is isomorphic to End V, $\text{rep}(S_4, \mathbb{Z}_p, 2)$ has wild modulo p representation type, as desired.

(iii) $w(S_T) = 3$, $j \geq 3$. Define $U \in Crep(S_3, \mathbb{Z}_p, 3)$ by

$$U_0 = \mathbb{Z}_p^{3n} \oplus \mathbb{Z}_p^{3n} \oplus \mathbb{Z}_p^{3n} + (1/p^3)U_*,$$
$$U_1 = \mathbb{Z}_p^{3n} \oplus 0 \oplus 0, \; U_2 = 0 \oplus \mathbb{Z}_p^{3n} \oplus 0, \; U_3 = 0 \oplus 0 \oplus \mathbb{Z}_p^{3n},$$

and U_* the free $\mathbb{Z}_p$-module with the rows of the $\mathbb{Z}_p$-matrix

$$M = \begin{pmatrix} I & 0 & 0 & pI & 0 & 0 & 0 & I & 0 \\ 0 & pI & 0 & -p^2 I & 0 & 0 & pI & 0 & p^2 I \\ 0 & 0 & pI & 0 & I & 0 & A & I & 0 \\ 0 & 0 & 0 & 0 & 0 & pI & pI & 0 & p^2 B \end{pmatrix}.$$

Then End U is isomorphic to End V, where $V = (V_0, U_1, U_2, U_3, p^3 U_0)$ and $V_0 = U_1 \oplus U_2 \oplus U_3$. If $f \in \text{End } V$, then $f = (f_1, f_2, f_3)$ with $f_i : U_i \to U_i$ and $fM \equiv Mg (\text{mod } p^3)$ for some g. Write $f_1 = (a_{ij})$, $f_2 = (b_{ij})$, and $f_3 = (c_{ij})$ as 3×3 matrices with each a_{ij}, b_{ij}, and c_{ij} an $n \times n$ $\mathbb{Z}_p$-matrix. The equation, $fM \equiv Mg (\text{mod } p^3)$, yields, from the first row of M, $a_{11} \equiv b_{11} \equiv c_{22} (\text{mod } p)$ and $a_{12}, a_{13}, c_{21}, c_{23}$ all $\equiv 0$ modulo p. Continuing with each row of M leads to the conclusion that $a_{ii} \equiv b_{ii} \equiv c_{ii} (\text{mod } p) = a$ for each $1 \leq i \leq 3$ with

$a \in C(A(\bmod p), B(\bmod p))$ and $f(\bmod p) = aI + N$ with

$$N = \begin{pmatrix} 0 & 0 & 0 & 0 & b'_{12} & b'_{13} & 0 & 0 & 0 \\ 0 & 0 & 0 & 0 & 0 & 0 & 0 & 0 & 0 \\ a'_{31} & a'_{32} & 0 & 0 & 0 & 0 & 0 & 0 & 0 \end{pmatrix}$$

a nilpotent $\mathbb{Z}/p\mathbb{Z}$-matrix. Complete details of this computation are given in [Arnold, Dugas 97, Appendix].

The ring homomorphism $\phi : \text{End } V \rightarrow C(A(\bmod p), B(\bmod p))$, defined by $\phi(f) = a$, is onto by Lemma 4.2.1, since $C(A, B)$ embeds in End V. Since End U is isomorphic to End V, $\text{Crep}(S_3, \mathbb{Z}_p, 3)$ has wild modulo p representation type. Finally, apply Lemma 5.2.1 to see that $\text{Crep}(S_T, \mathbb{Z}_p, j)$ has wild modulo p representation type for $w(S_T) = 3$ and $j \geq 3$.

Example 5.2.3 *Let T be a p-locally free finite lattice of types. If $w(S_T) = 4$ and $j \geq 1$ or $w(S_T) \geq 3$ and $j \geq 2$, then $\text{Crep}(S_T, \mathbb{Z}_p, j)$ has rank-infinite representation type.*

PROOF. Let A be an $n \times n$ $\mathbb{Z}_p$-matrix with $A(\bmod p)$ an indecomposable $\mathbb{Z}/p\mathbb{Z}$-matrix.

(i) $w(S_T) = 4$ and $j \geq 1$. Define $U = (U_0, U_1, U_2, U_3, U_4) \in \text{Crep}(S_4, \mathbb{Z}_p, 1)$ by

$$U_0 = \mathbb{Z}_p^n \oplus \mathbb{Z}_p^n \oplus \mathbb{Z}_p^n \oplus \mathbb{Z}_p^n + (1/p)(1 + 0 + 1 + 1)\mathbb{Z}_p^n + (1/p)(0 + 1 + 1 + A)\mathbb{Z}_p^n,$$

$$U_1 = \mathbb{Z}_p^n \oplus 0 \oplus 0 \oplus 0, \; U_2 = 0 \oplus \mathbb{Z}_p^n \oplus 0 \oplus 0, \; U_3 = 0 \oplus 0 \oplus \mathbb{Z}_p^n \oplus 0,$$

$$U_4 = 0 \oplus 0 \oplus 0 \oplus \mathbb{Z}_p^n.$$

As a consequence of Lemma 4.2.3(a), End U is isomorphic to End V with $V = (V_0, U_1, U_2, U_3, U_4, pU_0)$ and $V_0 = U_1 \oplus U_2 \oplus U_3 \oplus U_4$. Then $pU_0 = pV_0 + W$, where W is the free R-module with the rows of the R-matrix $M = \left(\begin{smallmatrix} I & 0 & I & I \\ 0 & I & I & A \end{smallmatrix}\right)$ as a basis.

An argument like that of Example 5.2.2 shows that if $f \in \text{End } V$, then $f(\bmod p) = aI$, with $a \in C(A(\bmod p))$. If f is an idempotent, then a is an idempotent. Since $A(\bmod p)$ is an indecomposable matrix, $a = 0$ or 1. If $a = 0$, then $f^2 = f \in p\text{End } V$. Since V_0 is a finitely generated free $\mathbb{Z}_p$-module, $f = 0$. On the other hand, if $a = 1$, then the same arguments show that $1 - f = 0$ and $f = 1$. Hence, V, and consequently U is an indecomposable representation. Since there are indecomposable $\mathbb{Z}/p\mathbb{Z}$-matrices for each n, $\text{Crep}(S_4, \mathbb{Z}_p, 1)$ has rank-infinite representation type. Now apply Lemma 5.2.1.

(ii) $w(S_T) \geq 3$ and $j \geq 2$. Define $U = (U_0, U_1, U_2, U_3) \in \text{Crep}(S_3, \mathbb{Z}_p, 2)$ by

$$U_0 = \mathbb{Z}_p^n \oplus \mathbb{Z}_p^n \oplus \mathbb{Z}_p^n + (1/p^2)(1 + 1 + A)\mathbb{Z}_p^n + (1/p)(0 + 1 + 1)\mathbb{Z}_p^n,$$

$$U_1 = \mathbb{Z}_p^n \oplus 0 \oplus 0, \; U_2 = 0 \oplus \mathbb{Z}_p^n \oplus 0, \; U_3 = 0 \oplus 0 \oplus \mathbb{Z}_p^n.$$

Then End U is isomorphic to End V, with $V = (V_0, U_1, U_2, U_3, p^2U_0)$ and $V_0 =$

$U_1 \oplus U_2 \oplus U_3$. In particular, $p^2 U_0 = p^2 V_0 + W$, where W is the free $\mathbb{Z}_p$-module with the rows of the $\mathbb{Z}_p$-matrix

$$M = \left(\begin{array}{c|c|c} I & I & A \\ 0 & pI & pI \end{array} \right)$$

as a basis. A by now familiar argument shows that if $f \in \mathrm{End}\, V$, then $f \,(\mathrm{mod}\ p) = aI$, with $a \in C(A(\mathrm{mod}\ p))$. As in the proof of (i), $C\mathrm{rep}(S_3, \mathbb{Z}_p, 2)$ has rank-infinite representation type. Again, apply Lemma 5.2.1.

As a consequence of the preceding constructions, the representation type for $C(T, j)_p$ can be computed. The case that $w(S_T) = 3$ and $j = 1$ is addressed in Corollary 5.2.9.

Corollary 5.2.4 *Let T be a p-locally free finite lattice of types. Then $C(T, j)_p$ has:*

(a) *wild modulo p representation type if either*
 (i) $w(S_T) \geq 5$, $j \geq 1$;
 (ii) $w(S_T) = 4$, $j \geq 2$;
 (iii) $w(S_T) = 3$, $j \geq 3$; *or*
 (iv) $w(S_T) = 2$, S_T *is not a garland, and* $j \geq 7$.
(b) *rank-infinite representation type if either*
 (i) $w(S_T) \geq 4$, $j \geq 1$;
 (ii) $w(S_T) = 3$, $j \geq 2$; *or*
 (iii) $w(S_T) = 2$, S_T *is not a garland, and* $j \geq 4$.
(c) *finite representation type if either*
 (i) $w(S_T) = 1$, $j \geq 1$ *or*
 (ii) $w(S_T) = 2$, $j \geq 1$, *and* S_T *is a garland.*

PROOF. As a consequence of Corollary 5.1.10, $C(T, j) = B(T, j)$ if $w(S_T) \leq 2$. Thus, (a)(iv), (b)(iii), and (c) follow from Corollary 4.3.3. The remainder of the corollary is a consequence of the definition of $C\mathrm{rep}(S_T, \mathbb{Z}_p, j)$ and Examples 5.2.2 and 5.2.3, wherein the representations constructed are in $C\mathrm{rep}_{\mathrm{crit}}(S_T, \mathbb{Z}_p, j)$.

An important consequence of Corollary 5.2.4 is the following corollary, demonstrating circumstances under which there are indecomposable almost completely decomposable groups of arbitrarily large finite rank. Representations constructed in this section can be used to give explicit descriptions of such groups; see Exercise 5.2.3.

Corollary 5.2.5 *Assume that T is a p-locally free finite lattice of types for some prime p. There are indecomposable almost completely decomposable groups of arbitrarily large finite rank with typeset contained in T if either $w(S_T) \geq 4$, $j \geq 1$; $w(S_T) = 3$, $j \geq 2$; or $w(S_T) = 2$, S_T is not a garland, and $j \geq 4$.*

PROOF. Use Corollary 5.2.4 and the observation that if a group is indecomposable at p, then it is indecomposable.

Suppose $X_n = \{\tau_1, \ldots, \tau_n\}$ is a p-locally free set of types with $\tau_i \cap \tau_j = \tau_0$ for each $1 \leq i \neq j \leq n$. Define $C(n)$ to be the category of almost completely decomposable groups with typeset $G \subseteq X_n \cup \{\tau_0\}$. Given $j \geq 0$, $C(n, j)$ is the subcategory of groups G in $C(n)$ with $p^j G \subseteq G^*(\tau_0) \subseteq G$. In this case, $G^*(\tau_0) = G(\tau_1) \oplus \cdots \oplus G(\tau_n) = R(G)$. The category $C(n, j)_p$ is the isomorphism at p category of $C(n, j)$.

For example, X_n could be the set of atoms for a p-locally free Boolean algebra T_n. Then $C(n, j) = C_{\text{crit}}(T_n, j)$, those almost completely decomposable groups in $B(T_n, j)$ with critical typeset contained in X_n, the join irreducible elements of T_n.

Corollary 5.2.6 *The category $C(n, j)_p$ has:*

(a) *wild modulo p representation type if either $n \geq 5$, $j \geq 1$; $n = 4$, $j \geq 2$; or $n = 3$, $j \geq 3$.*

(b) *rank-infinite representation type if either $n = 4$, $j \geq 1$ or $n = 3$, $j = 2$.*

(c) *finite representation type if and only if either $n \leq 2$, $j \geq 1$ or $n = 3$, $j = 1$.*

PROOF. A special case of Corollary 5.2.4, where T is the sublattice of the lattice of all types generated by X_n, except for the case $n = 3$, $j = 1$. It is a consequence of Theorem 5.2.8(c) and Example 1.1.5 that $C(3, 1)_p$ has finite representation type.

Remark *Let $(S(n), j-1)$ be the poset defined in Exercise 1.4.1. Observe that by comparison, the representation type of $C(n, j)_p$ is exactly that of $\text{rep}((S(n), j - 1), k)$, k a field, with the possible exception of $n = 3$, $j = 2$. A direct proof of this observation is not known.*

It remains to compute the representation type of $C(T, j)_p$ with $w(S_T) \leq 3$ and $j = 1$. Oddly enough, this apparently simpler case seems to require an additional hypothesis on S_T. A finite poset S is a *forest* if it is a disjoint union of trees, equivalently, $\{s : s \leq i \text{ in } S\}$ is a chain for each i in S. For example, S_n is a forest.

If S_T is a forest, then $C\text{rep}_{\text{crit}}(S_T, \mathbb{Z}_p, j)$ has some special properties as given in (c) of the next lemma, the representation analogue of Theorem 3.2.5(b) for almost completely decomposable groups. The module $\oplus\{U_i^* : i \in S_T\}$ is analogous to a regulating subgroup for almost completely decomposable groups. If G is an almost completely decomposable group with critical typeset that is an inverted forest, then there is a unique regulating subgroup $R(G)$ (Corollary 3.2.13). This property is reflected in $C\text{rep}_{\text{crit}}(S_T, \mathbb{Z}_p, j)$ in the case S_T is a forest, remembering that S_T is the poset of join irreducible elements of T with opposite order.

Lemma 5.2.7 *Suppose T is a p-locally free finite lattice of types and $U = (U_0, U_i : i \in S_T) \in C\text{rep}_{\text{crit}}(S_T, \mathbb{Z}_p, j)$.*

(a) *For each i in S_T, $U_i = U_i^* \oplus U_i^{\#}$, where $U_i^{\#}$ is the pure submodule of U_i generated by $\{U_j : j < i \text{ in } S_T\}$.*
(b) *The $\mathbb{Z}_p$-module $\oplus\{U_i^* : i \in S_T\}$ is a submodule of $\Sigma\{U_i : i \in S_T\}$ with $\Sigma\{U_i : i \in S_T\}/(\oplus\{U_i^* : i \in S_T\})$ and $U_i/ \oplus \{U_s^* : s \leq i\}$ bounded $\mathbb{Z}_p$-modules for each $i \in S_T$.*
(c) *If S_T is a forest, then $\Sigma\{U_i : i \in S_T\} = \oplus\{U_i^* : i \in S_T\}$ and $U_i = \oplus\{U_s^* : s \leq i\}$ for each $i \in S_T$.*

PROOF. (a) The pure submodule $U_i^{\#}$ is a summand of the finitely generated free $\mathbb{Z}_p$-module U_i.

(b) If i is minimal in S_T, then $U_i^{\#} = 0$ and $U_i = U_i^*$. Next assume that i is not minimal in S_T and $U_j/\Sigma\{U_s^* : s \leq j\}$ is bounded for each $j < i$ in S_T. Then $U_i^{\#}/\Sigma\{U_j : j < i\}$ is a bounded module, since $U_i^{\#}$ is a finitely generated free $\mathbb{Z}_p$-module and rank $U_i^{\#} = \text{rank } \Sigma\{U_j : j < i\}$. By induction, $U_i/ \oplus \{U_s^* : s \leq i\}$ is a bounded module for each i in S_T. Hence, $\Sigma\{U_i : i \in S_T\}/\Sigma\{U_i^* : i \in S_T\}$ must be a bounded module.

Finally, $\Sigma\{U_i^* : i \in S_T\} = \oplus\{U_i^* : i \in S_T\}$, since rank $\Sigma_i U_i^* = \text{rank } \Sigma_i U_i = \text{rank } U_0 = \Sigma_i \text{ rank } U_i/U_i^{\#}$ and $U_i/U_i^{\#}$ is isomorphic to U_i^* by (a).

(c) In view of (b), it is sufficient to show that $U_i = \Sigma\{U_s^* : s \leq i\}$ for each i in S_T. If i is minimal in S_T, then $U_i = U_i^*$. Now suppose i is not minimal in S_T. Then $C_i = \{s : s \leq i\}$ is a chain, since S_T is a forest. Thus, $U_i^{\#} = U_t$ for t the maximal element of $C_i\backslash\{i\}$. By induction on the length of C_i, $U_t = \Sigma\{U_s^* : s \leq t\}$. This shows that $U_i = \Sigma\{U_s^* : s \leq i\}$.

There is a connection between $\text{Crep}_{\text{crit}}(S_T, \mathbb{Z}_p, j)$ and representations of S_T^* over the finite ring $\mathbb{Z}/p^j\mathbb{Z}$, recalling that S_T^* is the disjoint union of S_T and a point $*$ unrelated to any element of S_T. For a finite poset S_T, let $\Delta\text{rep}(S_T, \mathbb{Z}/p^j\mathbb{Z})$ be the category of representations $U = (U_0, U_i : i \in S_T^*)$ such that $U_0 = \Sigma\{U_i : i \in S_T\}$ with each U_i a free $\mathbb{Z}/p^j\mathbb{Z}$-module, $U_i \subseteq U_j$ if $i \leq j$ in S_T, and $U_* \cap U_i = 0$ for each $i \in S$. Morphism sets are $\mathbb{Z}/p^j\mathbb{Z}$-homomorphisms $f : U_0 \to U_0'$ with $f(U_i) \subseteq U_i'$ for each $i \in S_T$. Elements U of $\Delta\text{rep}(S_T, \mathbb{Z}/p^j\mathbb{Z})$ with $U_0 = \oplus\{U_i : i \in S_T\}$ can be analyzed by solving matrix problems, as illustrated in Section 4.1 for $\text{rep}(S, \mathbb{Z}_p, j)$.

An additive functor $F : \mathbf{X} \to \mathbf{X}'$ between additive categories is a *representation embedding* if F preserves indecomposables and reflects isomorphisms. The functor F is a *representation equivalence* if F is a representation embedding that is *dense*, that is, for each Y in $\mathbf{X}'$ there is X in $\mathbf{X}$ with $F(X)$ isomorphic to Y. A representation embedding, respectively equivalence, induces an injection, respectively a bijection, from Ind $\mathbf{X}$ to Ind $\mathbf{X}'$.

In the next theorem, (a) relates categories of $\mathbb{Z}_p$-representations of S_T to categories of $\mathbb{Z}/p^j\mathbb{Z}$ representations and corresponding matrix problems, (b) generalizes the category equivalence of Lemma 4.2.3(b) for $S_T = S_n$, and (c) and (d) are applications of (a) and (b) that relate the representation type of categories of almost completely decomposable groups to categories of representations of finite posets over the field $k = \mathbb{Z}/p\mathbb{Z}$. A nonfunctorial partial generalization of (a), without the hypothesis that S_T is a forest, is given in Theorem 5.4.2.

Theorem 5.2.8 *Assume that T is a p-locally free finite lattice of types, S_T is a forest, and $j \geq 1$.*

(a) *There is a full functor $F : \mathrm{Crep}_{\mathrm{crit}}(S_T, \mathbb{Z}_p, j) \to \Delta\,\mathrm{rep}(S_T, \mathbb{Z}/p^j\mathbb{Z})$ that is a representation equivalence.*

(b) *There is a category equivalence $H : \Delta\mathrm{rep}(S_T, \mathbb{Z}/p\mathbb{Z}) \to \mathrm{rep}(S_T, \mathbb{Z}/p\mathbb{Z})$.*

(c) *The category of almost completely decomposable groups $C_{\mathrm{crit}}(T, 1)_p$ has finite representation type if and only if $\mathrm{rep}(S_T, \mathbb{Z}/p\mathbb{Z})$ has finite representation type.*

(d) *If $\mathrm{rep}(S_T, \mathbb{Z}/p\mathbb{Z})$ has wild representation type, then $C_{\mathrm{crit}}(T, 1)$ has wild modulo p representation type.*

PROOF. (a)   For $U = (U_0, U_i : i \in S_T) \in \mathrm{Crep}_{\mathrm{crit}}(S_T, \mathbb{Z}_p, j)$, let $\Sigma_U = \Sigma\{U_i : i \in S\} \subseteq U_0$. Define $F(U) = (V_0 = \Sigma_U/p^j\Sigma_U, V_i = (U_i + p^j\Sigma_U)/p^j\Sigma_U, V_* = p^j U_0/p^j\Sigma_U : i \in S_T)$ and $F(f) : F(U) \to F(V)$, the morphism induced by a morphism $f : U \to V$. Then F is a functor with $F(U) \in \Delta\mathrm{rep}(S_T, \mathbb{Z}/p^j\mathbb{Z})$, noticing that $f(\Sigma_U) \subseteq \Sigma_V$. In particular, for each i, U_i is a pure submodule of U_0. Hence, $V_i = (U_i + p^j U_0)/p^j U_0$ is isomorphic to the free $\mathbb{Z}/p^j\mathbb{Z}$-module $U_i/p^j U_i$ and $V_i \cap V_* = 0$.

To see that F is a full functor, let $g : F(U) \to F(V)$ be a representation morphism. By Lemma 5.2.7, $U_i = U_i^* \oplus U_i^\#$, $\Sigma_U = \oplus\{U_i^* : i \in S_T\}$, and $U_i = \oplus\{U_s^* : s \leq i\}$ for each $i \in S_T$. It follows by induction that g lifts to an homomorphism $f : \Sigma_U \to \Sigma_V$ with $f(U_i) \subseteq V_i$ for each $i \in S_T$ and $f(p^j U_0) \subseteq p^j V_0$. Then f extends to $h : U_0 \to V_0$ with $F(h) = g$, as desired.

To confirm that F is a dense functor let $V = (V_0, V_i, V_* : i \in S_T) \in \Delta\mathrm{rep}(S_T, \mathbb{Z}/p^j\mathbb{Z})$. Each $V_i = V_i^* \oplus V_i^\#$, where $V_i^\#$ is the injective envelope of $\Sigma\{V_s : s < i\}$ in V_i. This is so because V_i is a free $\mathbb{Z}/p^j\mathbb{Z}$-module and $\mathbb{Z}/p^j\mathbb{Z}$ is self-injective. Then $V_i^\#$ is a free $\mathbb{Z}/p^j\mathbb{Z}$-module, and just as in Lemma 5.2.7, $\Sigma\{V_i : i \in S_T\} = \oplus\{V_i^* i \in S_T\}$ and $V_i = \oplus\{V_s^* : s \leq i\}$ for each $i \in S_T$.

Choose a finitely generated free $\mathbb{Z}_p$-module $W_0 = \oplus\{U_i^* : i \in S_T\}$ with $W_0/p^j W_0 = \oplus\{V_i^* : i \in S_T\}$, $(U_i^* + p^j W_0)/p^j W_0 = V_i^*$ for each $i \in S_T$, and a submodule K of W_0 with $K/p^j W_0 = V_*$. Define $U_i = \oplus\{U_s^* : s \leq i\}$ for each i in S_T and $U_0 = W_0 + (1/p^j)K$. Then each U_i is pure in U_0, since $K \cap U_i \subseteq p^j W_0$. Furthermore, $(U_i + p^j W_0)/p^j W_0 = V_i$, $p^j U_0/p^j W_0 = K/p^j W_0 = V_*$, $W_0 = \Sigma_U$, and rank $U_0 = $ rank $W_0 = \Sigma\{\text{rank } U_i^* : i \in S_T\}$. Then $U = (U_0, U_i : i \in S_T) \in \mathrm{Crep}_{\mathrm{crit}}(S_T, \mathbb{Z}_p, j)$ with $F(U) = (V_0, V_i, V_* : i \in S_T) = V$, as desired.

If $F(U)$ is isomorphic to $F(V)$, then, since F is a full functor, there are representation morphisms $f : U \to V$ and $g : V \to U$ with $1_V - fg \in p^j\mathrm{End}\,V$ and $1_U - gf \in p^j\mathrm{End}\,U$. Since U_0 is a free R-module and $p\mathbb{Z}_p = J\mathbb{Z}_p$, the Jacobson radical of $\mathbb{Z}_p$, it follows that $p\mathrm{End}\,U \subseteq J\mathrm{End}\,U$. Then $gf = 1 - (1 - gf)$ is a unit of End U, and similarly, $fg = 1 - (1 - fg)$ is a unit of End V. Consequently, U is isomorphic to V.

Finally, assume that $F(U) = V' \oplus V''$ is a representation direct sum. Since F is a dense functor, there are $U', U'' \in \mathrm{Crep}_{\mathrm{crit}}(S_T, \mathbb{Z}_p, j)$ with $F(U') = V'$ and $F(U'') = V''$. Then $F(U' \oplus U'') = F(U') \oplus F(U'') = F(U)$. But F reflects isomorphisms, so that U is isomorphic to $U' \oplus U''$. If U is indecomposable, then U' or $U'' = 0$. Hence, V' or V'' is 0, and so $F(U)$ is indecomposable. This shows that F preserves indecomposables.

(b) Let $k = \mathbb{Z}/p\mathbb{Z}$, $U = (U_0, U_i, U_* : i \in S_T) \in \Delta\mathrm{rep}(S_T, k)$, and define $H : \Delta\mathrm{rep}(S_T, k) \to \mathrm{rep}(S_T, k)$ by

$$H(U) = ((\oplus\{U_i : i \in S_T\})/U_*, (U_i + U_*)/U_* : i \in S_T).$$

Then H is a fully faithful functor. To see that H is dense, let $V = (V_0, V_i : i \in S_T) \in \mathrm{rep}(S_T, k)$ with no trivial summands. Since S_T is a forest, $V_i = V_i^* \oplus V_i^\#$ and $V_i = \oplus\{V_j^* : j \leq i \text{ in } S_T\}$ for each $i \in S_T$ as a consequence of Lemma 5.2.7. Write $V_0 = \Sigma\{V_i : i \in S_T\} = (\oplus\{V_i : i \in S_T\})/U_*$ for some U_*. Define $U_0 = \oplus\{V_i : i \in S_T\})$ and $U_i = \oplus\{U_j^* : j \leq i\}$, where $(U_i^* \oplus U_*)/U_* = V_i^*$, for each $i \in S_T$. Then $U = (U_0, U_i, U_* : i \in S_T) \in \Delta\mathrm{rep}(S_T, k)$ with $H(U) = V$, as desired.

(c) Observe that $C_{\mathrm{crit}}(T, 1)_p$ has finite representation type if and only if $\mathrm{rep}(S_T, k)$ has finite representation type, by (a) and (b) and the fact that $C_{\mathrm{crit}}(T, 1)_p$ is category equivalent to $\mathrm{Crep}_{\mathrm{crit}}(S_T, \mathbb{Z}_p, 1)$.

(d) Let Γ be an $\mathbb{Z}_p$-algebra that is finitely generated and free as a $\mathbb{Z}_p$-module. Since $\mathrm{rep}(S_T, k)$ has wild representation type, $\mathrm{rep}(S_T, k)$ is endowild by Corollary 1.4.3. In view of the category equivalence (b), there is some V in $\Delta\mathrm{rep}(S_T, \mathbb{Z}/p\mathbb{Z})$ with $\mathrm{End}\, V = \Gamma/p\Gamma$. Since F is dense and full, there is $U \in \mathrm{Crep}_{\mathrm{crit}}(S_T, \mathbb{Z}_p, 1)$ with $F(U) = V$ and $\mathrm{End}\, U \to \mathrm{End}\, V = \Gamma/p\Gamma$ onto. This proves that $\mathrm{Crep}_{\mathrm{crit}}(S_T, \mathbb{Z}_p, 1)$ has wild modulo p representation type.

The next corollary completes the computation of the representation type of $C(T, 1)_p$ with $w(S_T) = 3$ for the case that S_T is a forest.

Corollary 5.2.9 *Suppose T is a p-locally free finite lattice of types and S_T is a forest with $w(S_T) = 3$. Then $C(T, 1)_p$ has:*

(a) *rank-infinite representation type if S_T contains $(2, 2, 2)$, $(1, 3, 3)$, or $(1, 2, 5)$ as a subposet; and*

(b) *wild modulo p representation type if S_T contains $(2, 2, 3)$, $(1, 3, 4)$, or $(1, 2, 6)$ as a subposet.*

PROOF. (a) By Theorem 5.2.8, $C_{\mathrm{crit}}(T, 1)_p$ has rank-infinite representation type if and only if $\mathrm{rep}(S_T, k)$ has infinite representation type. Now apply Theorem 1.3.6, bearing in mind that S_T is a forest with $w(S_T) = 3$, and that $C_{\mathrm{crit}}(T, j)_p$ is a full subcategory of $C(T, j)_p$.

Assertion (b) follows from Theorems 5.2.8 and 1.4.4.

Open Questions

1. *Suppose T is a finite p-locally free lattice of types with $w(S_T) = 2$ and S_T is not a garland. If $6 \geq j \geq 4$, does $C(T, j)_p$ have wild modulo p representation type? If $1 \leq j \leq 3$, does $C(T, j)_p$ have rank-finite or rank-infinite representation type? In either case, can the indecomposable groups in $C(T, j)_p$ be classified?*

2. *What is the representation type of $C(T, 1)_p$ if $w(S_T) = 3$ and S_T is not a forest? In particular, what if $S_T = (N, 4)$?*

3. *Do the categories $C(T, j)_p$ and $C_{\mathrm{crit}}(T, j)_p$ have the same representation type? If $w(S_T) = 3$ and $j = 2$, does $C(T, j)_p$ have wild modulo p representation type? Does $C(3, 2)_p$ have wild modulo p representation type? Can the indecomposable groups be classified?*

4. *Is there a definition of tame representation type such that*
   (i) *if $C(T, j)_p$ has tame representation type, then indecomposables can be classified, and*
   (ii) *each $C(T, j)_p$ has exactly one of finite, tame, or wild modulo p representation type?*

1. Give a description of representations in $Crep(S_T, \mathbb{Z}_p, j)$.

2. Prove that each group in $C(T, 0)$ is completely decomposable.

3. Give explicit constructions of indecomposable almost completely decomposable groups corresponding to the indecomposable representations in $Crep(S_T, \mathbb{Z}_p, j)$ given in Examples 5.2.2 and 5.2.3.

5.3 Uniform Groups

An almost completely decomposable group G is called *uniform* if there is a prime p such that $G/G^*(\mathrm{IT}(G))$ is a free $\mathbb{Z}/p^j\mathbb{Z}$-module, where $\mathrm{IT}(G)$ is the inner type of G. Define $U(T, j)$ to be the subcategory of $C(T, j)$ consisting of the uniform almost completely decomposable groups in $C(T, j)$ and define $U_{\mathrm{crit}}(T, j)$ to be the full subcategory of all uniform almost completely decomposable groups in $C_{\mathrm{crit}}(T, j)$.

Uniform almost completely decomposable groups arise naturally and are intimately related to more general finite rank Butler groups.

Example 5.3.1 *Let T be a finite lattice of types.*

(a) *Then $U(T, 1) = C(T, 1)$ and $U_{\mathrm{crit}}(T, 1) = C_{\mathrm{crit}}(T, 1)$.*
(b) *Let G be a finite rank Butler group in $B(T)$ and write $G = C/K$ for some completely decomposable group $C \in B(T)$. Then $H = C + (1/p^j)K$ is a uniform almost completely decomposable group in $U(T, j)$.*

(c) *Conversely, suppose $H = C + (1/p^j)K$ is a uniform almost completely decomposable group in $U(T, j)$ with C a completely decomposable group and K a pure subgroup of C. Then C/K is a finite rank Butler group in $B(T)$.*

PROOF. (a) Each $\mathbb{Z}/p\mathbb{Z}$-module is free.

(b) The group H is almost completely decomposable, $p^j H \subseteq C$, and H/C is isomorphic to $p^j H / p^j C = (p^j C + K)/p^j C$. But $(p^j C + K)/p^j C$ is isomorphic to the free $\mathbb{Z}/p^j\mathbb{Z}$-module $K/p^j K$, since K is a pure subgroup of C. Furthermore, typeset H = typeset $C \subseteq T$.

(c) Since K is a pure subgroup of C, $G = C/K$ is a finite rank Butler group. Moreover, $G \in B(T)$ by Lemma 3.3.1, since typeset C = typeset $H \subseteq T$ and T is a lattice.

The following three examples demonstrate that representation type changes from $C(T, j)_p$ to $U(T, j)_p$; compare Corollary 5.2.4.

Example 5.3.2 *The category $U_{\mathrm{crit}}(T, j)_p$ has finite representation type if $S_T = S_3$ and $j \leq 2$. In this case, each indecomposable in $U_{\mathrm{crit}}(T, j)_p$ has rank less than or equal to 9.*

PROOF. [Arnold, Dugas 93A]. A complete and explicit list of indecomposables is given in [Mouser 93].

Example 5.3.3 *The category $U_{\mathrm{crit}}(T, j)_p$ has finite representation type if $S_T = (1, 2)$. Each indecomposable has rank less than or equal to 3.*

PROOF. If $G \in U_{\mathrm{crit}}(T, j)_p$, then $F_T(G) = (G_p, G(\tau)_p : \tau \in S_T) = (U_0, U_1, U_2 \subseteq U_3) = U$ is an indecomposable in $\mathrm{Crep}(S_T, \mathbb{Z}_p, 2)$ with $S = (1, 2)$ and $(\mathrm{End}\, G)_p$ isomorphic to End U by Corollary 4.3.2. Then End U is isomorphic to End V for $V = (U_1 \oplus U_3, U_1, U_2 \subseteq U_3, p^2 U_0)$ by Lemma 4.2.3(a). In particular, V is indecomposable.

Assume that rank $V > 1$ and let $A = (M_1 \mid M_2 \mid M_3)$ be a $\mathbb{Z}_p$-matrix with rows a basis of the free $\mathbb{Z}_p$-module $p^2 U_0$ and the columns labeled by bases of U_1, U_2, and U_3^*, respectively. As demonstrated in Section 4.1, allowable operations that do not change the isomorphism class of V are invertible elementary $\mathbb{Z}_p$-column operations in M_i for $1 \leq i \leq 3$, invertible $\mathbb{Z}_p$-column operations from M_2 to M_3 (but not from M_3 to M_2), and invertible $\mathbb{Z}_p$-row operations on M.

Diagonalizing M_1 by invertible row and column operations yields $A \approx$

$$(L \mid M \mid N) = \begin{pmatrix} I & 0 & M_{11} & M_{12} & N_{11} & N_{12} \\ 0 & pD & M_{21} & M_{22} & N_{21} & N_{22} \end{pmatrix}$$

with pD a diagonal matrix having diagonal entries $p^i, i \geq 1$. Since U is uniform, each linear combination of the rows of the matrix $(pD \mid M_{21}\ M_{22} \mid N_{21}\ N_{22})$ must contain a unit in the $(M_{21}\ M_{22})$ block and a unit in the $(N_{21}\ N_{22})$ block.

Furthermore, $(L \mid M \mid N) \approx (I\ 0 \mid M_2 \mid M_3)$. This can be seen by first diagonalizing $(M_{21}\ M_{22})$ with elementary row and column operations. The result is a matrix $(0\ EpD \mid I\ 0 \mid N_{21}\ N_{22})$ with E a product of elementary row operations. Using column operations, this matrix may be assumed to be of the form $(0\ pD_1 \mid I\ 0 \mid N_{21}\ N_{22})$. Next, apply column operations from M_2 to M_3 to clear the M_3 block. This yields the second row block of A equivalent to $(0\ pD \mid I\ 0 \mid 0)$, contradicting the assumption that U is uniform. In particular, the second-row block must be empty.

A similar argument shows that $A = (I\ 0 \mid M_2 \mid M_3)$ is equivalent to

$$\begin{pmatrix} I & 0 & I & 0 & 0 & 0 \\ 0 & I & 0 & pD & 0 & I \end{pmatrix}$$

with D a diagonal matrix. Since V is indecomposable with rank $V > 1$, either rank $V = 3$ and, from the second-row block, $V = (\mathbb{Z}_p x \oplus \mathbb{Z}_p y \oplus \mathbb{Z}_p z, \mathbb{Z}_p x, \mathbb{Z}_p y, \mathbb{Z}_p y \oplus \mathbb{Z}_p z, (1 + p^i + 1)\mathbb{Z}_p), i < j$, or else, from the first row block, rank $V = 2$ and $V = (Rx \oplus Ry, Rx, Ry, Ry, (1 + 1)R)$.

Example 5.3.4 *The category $U(T, j)_p$ has wild modulo p representation type if*

(a) $w(S_T) = 4$, $j \geq 2$, *or*
(b) $w(S_T) = 3$, $j \geq 3$.

PROOF. (a) Suppose $w(S_T) = 4$, $j \geq 2$. Let Γ be a $\mathbb{Z}_p$-algebra that is finitely generated and free as a $\mathbb{Z}_p$-module and let A, B be $n \times n$ $\mathbb{Z}_p$-matrices with $\Gamma = C(A, B)$. Define $U = (U_0, U_1, U_2, U_3, U_4) \in \text{rep}(S_4, \mathbb{Z}_p, 2)$ by

$$U_0 = U_1 \oplus U_2 \oplus U_3 \oplus U_4 + (1/p^2)U_5,$$

$$U_1 = \mathbb{Z}_p^{2n} \oplus 0 \oplus 0 \oplus 0, \quad U_2 = 0 \oplus \mathbb{Z}_p^{2n} \oplus 0 \oplus 0,$$

$$U_3 = 0 \oplus 0 \oplus \mathbb{Z}_p^{2n} \oplus 0, \quad \text{and} \quad U_4 = 0 \oplus 0 \oplus 0 \oplus \mathbb{Z}_p^{2n},$$

where U_5 is the free R-module with the rows of the following matrix as a basis:

$$\begin{pmatrix} I & 0 & 0 & 0 & I & 0 & A & 0 \\ 0 & I & 0 & 0 & 0 & I & 0 & pB \\ 0 & 0 & I & 0 & I & 0 & pI & I \\ 0 & 0 & 0 & I & 0 & I & 0 & pI \end{pmatrix},$$

for I an $n \times n$ identity matrix and 0 an $n \times n$ matrix of 0's. Notice that $U = F_{T'}(G)$ for some $G \in U(T', 2) \subseteq U(T, j)$ with $S_{T'} = S_4$, since $U \in C\text{rep}_{\text{crit}}(S_4, \mathbb{Z}_p, j)$ with $U_0/(U_1 + U_2 + U_3 + U_4)$ a free $\mathbb{Z}/p^2\mathbb{Z}$-module. Thus, $(\text{End } G)_p = \text{End } U$. Now, End U is isomorphic to End V, where $V = (U_1 \oplus U_2 \oplus U_3 \oplus U_4, U_1, U_2, U_3, U_4, p^2U_0)$. A by now familiar computation yields an epimorphism End $V \to C(A(\text{mod } p), B(\text{mod } p))$ and $C(A, B)$ embeds in End V. It follows from

Lemma 4.2.1 that $U(T, j)_p$ has wild modulo p representation type in the case $w(S_T) \geq 4, j \geq 2$.

(b) The case that $w(S) = 3, j \geq 3$ follows from Corollary 4.3.2 and Example 5.2.2, since in that example $U_0/(U_1 + U_2 + U_3)$, being isomorphic to $U_*/p^3 U_*$, is a free $\mathbb{Z}/p^3\mathbb{Z}$-module.

Example 5.3.5 *The category $U(T, j)_p$ has:*

(a) *rank-infinite representation type if $w(S_T) = 2, j \geq 6$, and S_T contains either $(1, 3)$ or $(2, 2)$ as a subposet, and*
(b) *wild modulo p representation type if $w(S_T) = 2, j \geq 11$, and S_T contains either $(1, 3)$ or $(2, 2)$ as a subposet.*

PROOF. The associated representations are Exercises 5.3.3 and 5.3.4.

Corollary 5.3.6 *Let T be a finite p-locally free lattice of types and j a positive integer. Then $U(T, j)_p$ has:*

(a) *wild modulo p representation type if either*
   (i) $w(S_T) = 5, j \geq 1$;
   (ii) $w(S_T) = 4, j \geq 2$;
   (iii) $w(S_T) = 3, j \geq 3$; or
   (iv) $w(S_T) = 2, j \geq 11$, and S_T contains $(1, 3)$ or $(2, 2)$ as a subposet.
(b) *rank-infinite representation type if either*
   (i) $w(S_T) = 4, j \geq 1$; or
   (ii) $w(S_T) = 2, j \geq 6$, and S_T contains $(1, 3)$ or $(2, 2)$ as a subposet.
(c) *finite representation type if either $w(S_T) = 1$; or $w(S_T) = 2, j \geq 1$, and S_T is a garland.*

Moreover, $U_{\text{crit}}(T, j)$ has finite representation type if $w(S_T) = 2$ and S_T is a forest that does not contain $(1, 3)$ or $(2, 2)$ as a subposet.

PROOF. Statements (a) and (b) follow from Examples 5.3.4 and 5.3.5 and Corollary 5.2.4, noticing that $U(T, 1) = C(T, 1)$ by Example 5.3.1(a). As for (c), apply Corollary 4.3.3(c). For the last statement, if S_T is a forest of width 2 that does not contain $(2, 2)$ or $(1, 3)$, then $S_T = (1, 2)$. Now apply Example 5.3.3.

Descriptions of representation type for $U(T, R, j)_p$ remain incomplete for those T with $w(S_T) = 2$ or $w(S_T) = 3$ and $j \leq 2$. This deficiency already arose in Section 5.2 for $C(T, 1)_p$ with $w(S_T) = 3$ and S_T not a garland.

Define $U(n, j)$ to be the full subcategory of uniform almost completely decomposable groups in $C(n, j)$ and $U(n, j)_p$ the isomorphism at p category of $U(n, j)$.

Corollary 5.3.7 *The category $U(n, j)_p$ for $j \geq 1$ has:*

(a) *wild modulo p representation type if either $n \geq 5$; $n = 4, j \geq 2$; or $n = 3$, $j \geq 3$.*

(b) *rank-infinite representation type if* $n = 4$, $j = 1$; *and*
(c) *finite representation type if* $n \leq 2$ *or* $n = 3$, $j \leq 2$;

PROOF. Corollary 5.3.6 and Example 5.3.2.

Open Questions

1. *What is the representation type for the missing cases of Corollary 5.3.6?*
2. *Are there representation embeddings or equivalences from representations over fields that can explain the representation type of* $U(T, j)_p$?
3. *Give a theoretical explanation of why* $C_{\text{crit}}(S_3, 2)_p$ *has rank-infinite representation type but* $U_{\text{crit}}(S_3, 2)_p$ *has finite representation type.*

EXERCISES

1. Prove that there are indecomposable groups G in $B(T_3)$ of arbitrarily large finite rank with $G = G^*(\text{IT}(G))$.

2. Prove that if $S_T = (1, 2)$, then there is a bound on the ranks of indecomposable groups in $B(T)$ with $G = G^*(\text{IT}(G))$.

3. Assume that $S_T = (1, 3)$.
   (a) Use the following representation to show that $U(T, j)_p$, $j \geq 11$, has wild modulo p representation type: $U = (U_0, U_1, U_2 \subseteq U_3 \subseteq U_4)$ with $U_0 = \mathbb{Z}_p^{5n} \oplus \mathbb{Z}_p^{2n} \oplus \mathbb{Z}_p^{5n} \oplus \mathbb{Z}_p^{4n} + (1/p^{11})U_*$, $U_1 = \mathbb{Z}_p^{5n} \oplus 0 \oplus 0 \oplus 0$, $U_2 = 0 \oplus \mathbb{Z}_p^{2n} \oplus 0 \oplus 0 \oplus$, $U_3 = 0 \oplus \mathbb{Z}_p^{2n} \oplus \mathbb{Z}_p^{5n} \oplus 0$, $U_4 = 0 \oplus \mathbb{Z}_p^{2n} \oplus \mathbb{Z}_p^{5n} \oplus \mathbb{Z}_p^{4n}$, and U_* the row space of the matrix contained in $U_1 \oplus U_2 \oplus U_3^* \oplus U_4^*$:

$$\begin{pmatrix} I & 0 & 0 & 0 & 0 & p^5I & p^6I & I & 0 & 0 & 0 & 0 & 0 & 0 & 0 & 0 \\ 0 & I & 0 & 0 & 0 & p^6I & p^7A & 0 & p^2I & 0 & 0 & 0 & I & 0 & 0 & 0 \\ 0 & 0 & I & 0 & 0 & p^7I & p^8B & 0 & 0 & p^4I & 0 & 0 & 0 & I & 0 & 0 \\ 0 & 0 & 0 & I & 0 & p^8I & 0 & 0 & 0 & 0 & p^6I & 0 & 0 & 0 & I & 0 \\ 0 & 0 & 0 & 0 & I & 0 & p^{10}I & 0 & 0 & 0 & 0 & p^8I & 0 & 0 & 0 & I \end{pmatrix}.$$

   (b) Use the following representation to show that $U(T, j)_p$, $j \geq 6$, has infinite representation type: $U = (U_0, U_1, U_2 \subseteq U_3 \subseteq U_4)$ with $U_0 = \mathbb{Z}_p^{3n} \oplus \mathbb{Z}_p^{2n} \oplus \mathbb{Z}_p^{3n} \oplus \mathbb{Z}_p^{2n} + (1/p^6)U_*$, $U_1 = \mathbb{Z}_p^{3n} \oplus 0 \oplus 0 \oplus 0$, $U_2 = 0 \oplus \mathbb{Z}_p^{2n} \oplus 0 \oplus 0 \oplus$, $U_3 = 0 \oplus \mathbb{Z}_p^{2n} \oplus \mathbb{Z}_p^{3n} \oplus 0 \oplus$, $U_4 = 0 \oplus \mathbb{Z}_p^{2n} \oplus \mathbb{Z}_p^{3n} \oplus \mathbb{Z}_p^{2n}$, and U_* the row space of the matrix contained in $U_1 \oplus U_2 \oplus U_3^* \oplus U_4^*$:

$$\begin{pmatrix} I & 0 & 0 & p^3I & p^4A & I & 0 & 0 & 0 & 0 \\ 0 & I & 0 & p^4I & p^5I & 0 & p^2I & 0 & I & 0 \\ 0 & 0 & I & p^5I & 0 & 0 & 0 & p^4I & 0 & I \end{pmatrix}.$$

4. Assume that $S_T = (2, 2)$.
   (a) Use the following representation to show that $U(T, j)_p$, $j \geq 10$, has wild modulo p representation type: $U = (U_0, U_1 \subseteq U_2, U_3 \subseteq U_4) \subseteq$ with $U_0 = \mathbb{Z}_p^{5n} \oplus \mathbb{Z}_p^{5n} \oplus \mathbb{Z}_p^{2n} \oplus \mathbb{Z}_p^{5n} + (1/p^{10})U_*$, $U_1 = \mathbb{Z}_p^{5n} \oplus 0 \oplus 0 \oplus 0$, $U_2 = \mathbb{Z}_p^{5n} \oplus \mathbb{Z}_p^{5n} \oplus 0 \oplus 0$,

$U_3 = 0 \oplus 0 \oplus \mathbb{Z}_p^{2n} \oplus 0$, $U_4 = 0 \oplus 0 \oplus \mathbb{Z}_p^{2n} \oplus \mathbb{Z}_p^{5n}$, and U_* the row space of the matrix contained in $U_1 \oplus U_2^* \oplus U_3 \oplus U_4^*$:

$$
\begin{pmatrix}
pI & 0 & 0 & 0 & 0 & I & 0 & 0 & 0 & 0 & p^4I & p^5I & I & 0 & 0 & 0 & 0 \\
0 & p^2I & 0 & 0 & 0 & 0 & I & 0 & 0 & 0 & p^5I & p^6A & 0 & I & 0 & 0 & 0 \\
0 & 0 & p^3I & 0 & 0 & 0 & 0 & I & 0 & 0 & p^6I & p^7B & 0 & 0 & I & 0 & 0 \\
0 & 0 & 0 & p^4I & 0 & 0 & 0 & 0 & I & 0 & p^7I & 0 & 0 & 0 & 0 & I & 0 \\
0 & 0 & 0 & 0 & p^5I & 0 & 0 & 0 & 0 & I & 0 & p^9I & 0 & 0 & 0 & 0 & I
\end{pmatrix}.
$$

(b) Use the following representation to show that $U(T, j)_p$, $j \geq 6$, has infinite representation type: $U = (U_0, U_1 \subseteq U_2, U_3 \subseteq U_4)$ with $U_0 = \mathbb{Z}_p^{3n} \oplus \mathbb{Z}_p^{3n} \oplus \mathbb{Z}_p^{2n} \oplus \mathbb{Z}_p^{3n} + (1/p^6)U_*$, $U_1 = \mathbb{Z}_p^{3n} \oplus 0 \oplus 0 \oplus 0$, $U_2 = \mathbb{Z}_p^{3n} \oplus \mathbb{Z}_p^{3n} \oplus 0 \oplus 0 \oplus 0$, $U_3 = 0 \oplus 0 \oplus \mathbb{Z}_p^{2n} \oplus 0$, $U_4 = 0 \oplus 0 \oplus \mathbb{Z}_p^{2n} \oplus \mathbb{Z}_p^{3n}$, and U_* the row space of the matrix contained in $U_1 \oplus U_2^* \oplus U_3 \oplus U_4^*$:

$$
\begin{pmatrix}
pI & 0 & 0 & I & 0 & 0 & p^2I & p^3A & I & 0 & 0 \\
0 & p^3I & 0 & 0 & I & 0 & p^3I & p^4I & 0 & I & 0 \\
0 & 0 & p^5I & 0 & 0 & I & p^4I & 0 & 0 & 0 & I
\end{pmatrix}.
$$

5.4   Primary Regulating Quotient Groups

An almost completely decomposable group G is a *p-primary regulating quotient group* if p is a prime and G has a regulating subgroup B such that G/B is a p-group, equivalently, $i(G)$ is a power of p by Proposition 5.1.2. It follows that an almost completely decomposable group is a p-primary regulating quotient group if and only if $e_G = \exp G/R(G)$ is a power of p. Notice that the class of p-primary regulating quotient groups is closed under finite direct sums, since if B is a regulating subgroup of H and C is a regulating subgroup of K with $|H/B|$ and $|K/C|$ both powers of p, then $D = B \oplus C$ is a regulating subgroup of $G = H \oplus K$ with $|G/D|$ a power of p.

The p-primary regulating quotient groups are generalizations of groups in $C(T, j)_p$ in the following sense. If $T_{\mathrm{cr}}(G)$ is an inverted forest and G is an almost completely decomposable group with no rank-1 summands, then $R(G) = G^*(IT(G))$ is a unique regulating subgroup of G by Corollary 3.2.13(c). If, in addition, typeset G is contained in a finite lattice T of types, then $G \in C(T, j)_p$ if and only if G is a p-primary regulating quotient group.

For p-primary regulating quotient groups, near-isomorphism is the same as isomorphism at p:

Lemma 5.4.1 *Let G and H be p-primary regulating quotient groups. Then G and H are nearly isomorphic if and only if G and H are isomorphic at p. Moreover, G is indecomposable if and only if G is indecomposable at p.*

PROOF. Since near-isomorphism is isomorphism at p for each prime p, it suffices to prove that if G and H are isomorphic at p, then they are nearly isomorphic. If G

and H are isomorphic at p, then there is a monomorphism $f : G \to H$ such that $H/f(G)$ is finite and relatively prime to p. Consequently, $R(G)$ is isomorphic to $R(H)$ as in Theorem 5.1.6(b). Since e_G is a power of p, G is isomorphic to $R(G)$ at q for each prime $q \neq p$. Similarly, H is isomorphic to $R(H)$ at q for each prime $q \neq p$. Thus, G is isomorphic to H at q for each prime q, and so G is nearly isomorphic to H. The last statement follows from Corollary 5.1.8(b).

The next theorem is a generalization of Theorem 5.2.8(a), wherein S_T is assumed to be a forest. However, in contrast to that theorem, this construction need not be functorial, since there are almost completely decomposable groups G and H and a homomorphism $f : G \to H$ such that $f(R(G))$ is not contained in $R(H)$; Exercise 5.1.8.

Theorem 5.4.2 [Mader, Vinsonhaler 97] *Suppose G is a p-primary regulating quotient group with $e_G = p^j$. Define a $\mathbb{Z}/p^j\mathbb{Z}$-representation $F(G) = (U_0, U_\tau, U_* : \tau \in T_{cr}(G))$ by $U_0 = R(G)/p^j R(G)$, $U_\tau = (R(G)(\tau)) + p^j R(G))/p^j R(G)$, and $U_* = p^j G/p^j R(G)$.*

(a) *There is a ring isomorphism $\phi : \operatorname{End} G/p^j \operatorname{End} R(G) \to \operatorname{End} F(G)$.*
(b) *If f is an idempotent of $\operatorname{End} F(G)$, then there is an idempotent g of $\operatorname{End} G$ with $\phi(g) = f$.*
(c) *The group G is indecomposable if and only if $(\operatorname{End} G)_p$ is a local ring.*

PROOF. (a) Let $\phi(f) \in \operatorname{End} F(G)$ be the endomorphism induced by f, which is well-defined, since $R(G)$ is a fully invariant subgroup of G by Theorem 5.1.6(a). Then $\phi(f) = 0$ if and only if $f : R(G) \to p^j R(G)$. Equivalently, $f \in p^j \operatorname{End} R(G) \subseteq \operatorname{End} G$, since $p^j G \subseteq R(G)$. The proof that ϕ is onto is a direct analogue of the proof of Theorem 5.2.8(a). The crucial observations are that $R(G)$ is a completely decomposable subgroup of G with $p^j G \subseteq R(G)$, and $R(G)$ is a fully invariant subgroup of G.

(b) Suppose f is an idempotent of $\operatorname{End} F(G)$. By (a), there is $h \in \operatorname{End} G$ with $\phi(f) = h$ and $h^2 - h \in p^j \operatorname{End} R(G)$. Now $\operatorname{End} R(G)/N\operatorname{End} R(G)$ is a product of matrix rings over subrings of Q by Proposition 3.1.8.

In fact, $h + p^j \operatorname{End} R(G)$ lifts to an idempotent g of $\operatorname{End} R(G)$. This is so because idempotents of $\Lambda/p^j \Lambda$ lift to idempotents of Λ for Λ a matrix ring over a subring of $\mathbb{Q}$, hence for $\Lambda = \operatorname{End} R(G)/N\operatorname{End} R(G)$. But idempotents of $\operatorname{End} R(G)/N\operatorname{End} R(G)$ lift to idempotents of $\operatorname{End} R(G)$, since $N\operatorname{End} R(G)$ is nilpotent. Since $R(G)$ is a fully invariant subgroup of G with $G/R(G)$ finite, there is an idempotent $g \in \operatorname{End} G$ with $\phi(g) = f$.

(c) If $(\operatorname{End} G)_p$ is a local ring, then 0 and 1 are the only idempotents of $(\operatorname{End} G)_p$. Hence, G is indecomposable at p and so must also be indecomposable as a group. Conversely, suppose that G is indecomposable. Then $\operatorname{End} G$ has only 0 and 1 as idempotents, and so $\operatorname{End} F(G)$ has only 0 and 1 as idempotents by (b). Since $\operatorname{End} F(G)$ is finite, it must be a local ring. By (a), $(\operatorname{End} G)_p/p^j(\operatorname{End} R(G))_p$ is a local ring, whence $(\operatorname{End} G)_p$ is also a local ring, since $p \in J(\operatorname{End} G)_p$.

The near-isomorphism category of torsion-free abelian groups of finite rank has nonunique direct sum decompositions into indecomposables by Example 2.1.11. The near-isomorphism category of almost completely decomposable groups also has nonunique direct sum decompositions by Proposition 5.1.11. However, as a consequence of the next theorem, the near-isomorphism category of p-locally free almost completely decomposable groups G with e_G a power of p is a Krull–Schmidt category.

Corollary 5.4.3 [Faticoni, Schultz 96] *Assume that G and H are p-primary regulating quotient groups. If $G = X_1 \oplus \cdots \oplus X_n$ is nearly isomorphic to $Y_1 \oplus \cdots \oplus Y_n$ with each X_i and Y_i an indecomposable group, then $m = n$ and there is a permutation σ of $\{1, 2, \ldots, n\}$ with X_i nearly isomorphic to $Y_{\sigma(i)}$ for each $1 \leq i \leq n$.*

PROOF. By Corollary 5.1.8(d), it suffices to assume that $G = H$. It is also sufficient to assume that both G and H are p-reduced, hence p-locally free; Exercise 5.1.5. Now apply Theorem 5.4.2(c) and Exercise 1.2.6 to see that there is a permutation σ of $\{1, 2, \ldots, n\}$ with X_i isomorphic at p to $Y_{\sigma(i)}$ for each $1 \leq i \leq n$. Since e_G and e_H are powers of p, it follows from Lemma 5.4.1 that X_i and $Y_{\sigma(i)}$ are nearly isomorphic.

There are examples of almost completely decomposable groups G and H with no rank-1 summands such that $G \oplus H$ has a rank-1 summand; Exercise 5.1.7(b). This cannot happen if G and H are p-primary regulating quotient groups:

Corollary 5.4.4 [Lady 74B] *Suppose G and H are p-primary regulating quotient groups.*

(a) *If G and H have no rank-1 summands, then $G \oplus H$ has no rank-1 summands.*
(b) *Any two direct sum decompositions of G into indecomposable groups have the same number of rank-1 summands of type τ for each τ.*

PROOF. This corollary is a special case of Corollary 5.4.3 and the observation that nearly isomorphic completely decomposable groups are isomorphic.

EXERCISE

1. Assume that $G = A \oplus H$ with A a completely decomposable group. Prove that if B is a regulating subgroup of G, then $B \oplus H$ is a regulating subgroup of H and G/B is isomorphic to $H/H \cap B$. Conclude that if G is a p-primary regulating quotient group, then so is H. What can be said if A is not completely decomposable?

NOTES ON CHAPTER 5

Finite rank completely decomposable groups are classified in [Baer 37]. Examples of pathological direct sum decompositions of almost completely decomposable groups were constructed in [Levi 17] and [Jónsson 57, 59]. For many years, almost completely decomposable groups were considered primarily as a good source of counterexamples; see [Fuchs 73].

Some positive results for almost completely decomposable groups, most of which are given in Section 5.1, first appeared in [Lady 74B]. An equivalence condition on almost completely decomposable groups G and H, $G \oplus C \approx H \oplus C$ for some completely decomposable group C, led to the notion of near-isomorphism for torsion-free abelian groups of finite rank.

Most of Section 5.2 is derived from [Arnold, Dugas 98]. As demonstrated in that section, representations of finite posets can be used to easily construct examples of indecomposable almost completely decomposable groups of arbitrarily large finite rank with types in a fixed typeset. The fact that categories of almost completely decomposable groups with typesets contained in a relatively small lattice of types can have wild modulo p representation type vividly demonstrates the complexity of almost completely decomposable groups.

Almost completely decomposable groups are viewed as extensions of finite torsion groups by completely decomposable groups in [Mader 95]. Near-isomorphism is rediscovered in the sense that an equivalence, called type isomorphism, which arose naturally in this context, is the same as near-isomorphism. Mader's approach, together with the results and techniques of [Burkhardt 84], [Krapf, Mutzbauer 84], and [Schultz 87], has rekindled interest in the subject of almost completely decomposable groups; see [Mutzbauer 99], [Mader, Nonxga 99], [Ould-Beddi, Strüngmann 99], and [Reid 99]. [Mader 00] includes a comprehensive, detailed exposition of the subject of almost completely decomposable groups and an extensive bibliography.

Several classes of almost completely decomposable groups have been classified by invariants up to near isomorphism: uniform block rigid groups [Dugas, Oxford 93]; groups divisible by all but three primes [Cruddis 70]; groups with critical typeset of two elements; and cyclic regulating quotient groups [Mader, Mutzbauer 93], [Mader, Vinsonhaler 95], [Burkhardt, Mutzbauer 96], and [Dittmann, Mader, Mutzbauer 97]. Examples given in these papers demonstrate that classification of almost completely decomposable groups up to isomorphism involves algebraic number-theoretic difficulties that can be avoided by classification up to near isomorphism.

There is a substantial amount of literature on almost completely decomposable groups with small typeset; see [Arnold, Dugas 99]. Results of [Cruddis 70] are extended in [Arnold 73]. In the latter paper, some of the arguments are invalid. Corrected proofs and related results can be found in [Arnold, Dugas 95A], [Lewis 93], and [Mader, Vinsonhaler 94].

Finite rank Butler groups G such that each pure subgroup of G or each torsion-free quotient of G is almost completely decomposable are characterized in terms of the critical typeset of G in [Nonxga, Vinsonhaler 95]. Invariants for almost completely decomposable groups are discussed in [Arnold, Vinsonhaler 93] and [Vinsonhaler 95].

6

Representations over Fields and Exact Sequences

6.1 Projectives, Injectives, and Exact Sequences

The terms *exact sequence*, *kernel*, and *cokernel* are defined for arbitrary categories [Mitchell 65]. In order to avoid lengthy arguments, these terms are defined herein in terms of their descriptions in $\operatorname{rep}(S, k)$ rather than derived.

Let $U = (U_0, U_i : i \in S)$, $V = (V_0, V_i : i \in S)$, and $W = (W_0, W_i : i \in S)$ be objects of $\operatorname{rep}(S, k)$, $f \in \operatorname{Hom}(U, V)$, and $g \in \operatorname{Hom}(V, W)$. The sequence

$$0 \to U \xrightarrow{f} V \xrightarrow{g} W \to 0$$

is an *exact sequence* if the induced sequence

$$0 \to U_i \xrightarrow{f} V_i \xrightarrow{g} W_i \to 0$$

of vector spaces is exact for each i in $S \cup \{0\}$, i.e., for each i, $f : U_i \to V_i$ is one-to-one, $g : V_i \to W_i$ is onto, and $\ker g = \operatorname{image} f$.

A representation morphism $f : U \to V$ is a *monomorphism* if $f : U_0 \to V_0$ is 1-to-1 and an *epimorphism* if the k-linear transformation $f : V_i \to W_i$ is onto for each $i \in S \cup \{0\}$. The kernel of f is

$$\operatorname{Ker} f = (\ker f, (\ker f) \cap U_i : i \in S) \in \operatorname{rep}(S, k),$$

where $\ker f$ is the kernel of the k-linear transformation $f : U_0 \to V_0$. If f is an epimorphism, then $0 \to \operatorname{Ker} f \to U \to V \to 0$ is an exact sequence of representations, since $0 \to (\ker f) \cap U_i \to U_i \to V_i \to 0$ is an exact sequence of vector spaces for each $i \in S \cup \{0\}$.

A representation morphism $f : U \to V$ is a *pure morphism* and $f(U) = (f(U_0), f(U_i) : i \in S)$ is a *pure subrepresentation* of V if $f(U_i) = f(U_0) \cap V_i$ for each $i \in S$. For example, the kernel of a representation morphism is a pure subrepresentation. The terminology arises from the abelian group setting. A pure morphism is called a *proper morphism* in [Simson 92].

If $f : U \to V$ is a pure morphism, then the *cokernel of f* is

$$\text{Coker } f = (V_0/f(U_0), (V_i + f(U_0))/f(U_0) : i \in S) \in \text{rep}(S, k).$$

The sequence $0 \to U \to V \to \text{Coker } f \to 0$ in $\text{rep}(S, k)$ is exact if f is a pure monomorphism, since $0 \to f(U_0) \cap V_i \to V_i \to (V_i + f(U_0))/f(U_0) \to 0$ is an exact sequence of vector spaces for each i.

A representation $U \in \text{rep}(S, k)$ is *projective* if whenever $f : V \to W$ is a representation epimorphism and $g : U \to W$ is a representation morphism with $V, W \in \text{rep}(S, k)$, then there is a morphism $h : U \to V$ with $fh = g$. Dually, a representation U is *injective* if whenever $f : V \to W$ is a pure monomorphism and $g : V \to U$ with $V, W \in \text{rep}(S, k)$, then there is a morphism $h : W \to U$ with $hf = g$.

Recall that $U = (U_0, U_i : i \in S) \in \text{rep}(S, k)$ is a trivial representation if $U_0 \neq 0$ but $U_i = 0$ for each i in S. Up to isomorphism, the only trivial indecomposable representation is $P(0) = (k, P(0)_i : i \in S)$ with $P(0)_i = 0$ for each i in S. Call $U \in \text{rep}(S, k)$ *cotrivial* if $U_i = U_0$ for each i. The only cotrivial indecomposable representation, up to isomorphism, is $I(0) = (k, I(0)_i : i \in S)$ with $I(0)_i = k$ for each i in S.

Lemma 6.1.1 *Let S be a finite poset, k a field, and $\sigma : \text{rep}(S, k) \to \text{rep}(S^{\text{op}}, k)$ the contravariant duality given in Proposition 1.4.7. Then:*

(a) *a sequence $0 \to U \to V \to W \to 0$ is exact in $\text{rep}(S, k)$ if and only if $0 \to \sigma(W) \to \sigma(V) \to \sigma(U) \to 0$ is exact in $\text{rep}(S^{\text{op}}, k)$;*
(b) *a representation U is projective in $\text{rep}(S, k)$ if and only if $\sigma(U)$ is injective in $\text{rep}(S^{\text{op}}, k)$; and*
(c) *U is a trivial representation if and only if $\sigma(U)$ is a cotrivial representation.*

PROOF. Exercise 6.1.1.

Let S be a finite poset and k a field. For $t \in S$, define

$$P(t) = (k, P(t)_i : i \in S) \in \text{rep}(S, k),$$

where $P(t)_j = k$ for each $j \geq t$ in S and $P(t)_j = 0$ if $j \not\geq t$ in S. Dually, define

$$I(t) = (k, I(t)_i : i \in S),$$

where $I(t)_j = k$ for each $j \not\leq t$ in S and $I(t)_j = 0$ if $j \leq t$ in S.

Observe that if $P(t) \in \text{rep}(S, k)$, then $\sigma(P(t)) = I(t) \in \text{rep}(S^{\text{op}}, k)$. For example, let $S = \{1, 2 < 3\}$. Then $P(0) = (k, 0, 0, 0)$, $P(1) = (k, k, 0, 0)$, $P(2) = (k, 0, k, k)$, and $P(3) = (k, 0, 0, k)$. Hence, $S^{\text{op}} = \{1, 3 < 2\}$ and $\sigma(P(0)) = $

$(k, k, k, k) = I(0)$, $\sigma(P(1)) = (k, 0, k, k) = I(1)$, $\sigma(P(2)) = (k, k, 0, 0) = I(2)$, and $\sigma(P(3)) = (k, k, 0, k) = I(3) \in \operatorname{rep}(S^{\mathrm{op}}, k)$.

Lemma 6.1.2 *Let S be a finite poset, k a field, and $U = (U_0, U_i : i \in S) \in$ $\operatorname{rep}(S, k)$. Then:*

(a) *U is projective in $\operatorname{rep}(S, k)$ if and only if U is isomorphic to a finite direct sum of representations $P(0)$ and $P(t)$ for t in S; and*
(b) *U is injective in $\operatorname{rep}(S, k)$ if and only if U is isomorphic to a finite direct sum of representations $I(0)$ and $I(t)$ for t in S.*

PROOF.   The first step is to prove that $P(t)$ is projective for each $t \in S \cup \{0\}$. To this end, assume that $f : V \to W$ is an epimorphism and $g : P(t) \to W$. Since $P(t)_0 = k$ and f is onto, there is a k-linear transformation $h : P(t)_0 \to V_0$ with $fh = g$. Then h is a representation morphism, recalling that if $t \in S$, then $P(t)_j = k$ if $j \geq t$ and $P(t)_j = 0$ otherwise. This shows that each $P(t)$ is projective. It follows that finite direct sums of $P(0)$ and $P(t)$ are projective.

Conversely, let $U = (U_0, U_i : i \in S) \in \operatorname{rep}(S, k)$ be a projective representation. There is an epimorphism $f : P \to U$, where P is a finite direct sum of representations $P(t)$, t in $S \cup \{0\}$. This holds because for a given t, there is a finite direct sum $V(t) = (V_0, V_i : I \in S)$ of copies of $P(t)$ and a representation morphism $V(t) \to U$ such that U_t is contained in the image of V_t. Taking the direct sum of all such $V(t)$'s for $t \in S \cup \{0\}$ gives the desired P.

Since $1_U : U \to U$ and U is projective, there is a morphism $h : U \to P$ with $fh = 1_U$. In particular, U is isomorphic to a representation summand of P. Since P is a finite direct sum of $P(t)$'s for $t \in S \cup \{0\}$ and each $P(t)$ is indecomposable, U is isomorphic to a finite direct sum of copies of $P(0)$ and $P(t)$ for $t \in S$ by Theorem 1.2.2.

(b) Apply (a), Lemma 6.1.1(b), and the observation that $\sigma(P(t)) = I(t)$.

Example 6.1.3 *Suppose S is a chain with n elements and k is a field. If $U \in \operatorname{rep}(S, k)$, then U is both projective and injective.*

PROOF.   The representation U is a finite direct sum of indecomposable representations, and each indecomposable has dimension 1 by Lemma 1.3.3(a). If $U \in \operatorname{rep}(S, k)$ has dimension 1, then $U = P(s) = I(t)$ for some $t < s \in S$ or $U = P(0) = I(n)$. By Lemma 6.1.2, U is both projective and injective, as desired.

An epimorphism $f : P \to U$ in $\operatorname{rep}(S, k)$, with P a projective representation, is a *projective cover* of U if given $g : P' \to U$ an epimorphism with P' projective, then there is an epimorphism $h : P' \to P$ with $fh = g$. In this case, P is isomorphic to a summand of P', since P is projective. Notice that a representation U is projective if and only if the identity homomorphism $U \to U$ is a projective cover of U.

A pure monomorphism $f : U \to I$ in $\operatorname{rep}(S, k)$ with I an injective representation is an *injective envelope* of U if for each pure monomorphism $g : U \to I'$ with I'

injective, there is a pure monomorphism $h : I \to I'$ with $hf = g$. In particular, I is a summand of I'. Moreover, U is injective if and only if the identity is an injective envelope of U. Projective covers and injective envelopes are unique up to isomorphism (Exercise 6.1.10).

Projective covers and injective envelopes of $U \in \mathrm{rep}(S, k)$ can be constructed explicitly. Assume that $U \in \mathrm{rep}(k, S)$ has no trivial summands. For each i in S, let $U_i^\# = \Sigma\{U_j : j < i \text{ in } S\}$. Define

$$P(U) = (P(U)_0, P(U)_i) \text{ with } P(U)_0 = \oplus\{U_i/U_i^\# : i \in S\} \text{ and}$$

$$P(U)_i = \oplus\{U_j/U_j^\# : j \leq i\}$$

for each i in S. If U is trivial, define $P(U) = U$. Then $P(U)$ is a finite direct sum of copies of $P(t)$, $t \in S \cup \{0\}$, whence, by Lemma 6.1.2(a), $P(U)$ is a projective representation in $\mathrm{rep}(S, k)$.

If $U \in \mathrm{rep}(S, k)$ has no cotrivial summands, let $\hat{U}_i = \cap\{U_j : j > i \text{ in } S\}$ agreeing that $\hat{U}_i = U_0$ if there is no j in S with $j > i$. Define

$$I(U) = (I(U)_0, I(U)_i) \text{ with } I(U)_0 = \oplus\{\hat{U}_i/U_i : i \in S\} \text{ and}$$

$$I(U)_i = \oplus\{\hat{U}_j/U_j : i \nleq j\}.$$

If U is cotrivial, define $I(U) = U$. By Lemma 6.1.2(b), $I(U)$ is an injective representation in $\mathrm{rep}(S, k)$ for each U.

Lemma 6.1.4 *Let $U \in \mathrm{rep}(S, k)$ for a finite poset S and field k.*

(a) *There is a projective cover $p_U : P(U) \to U$ and*
(b) *an injective envelope $q_U : U \to I(U)$.*

PROOF. (a) It suffices to assume that U has no trivial summands. For each i, there is a k-linear transformation $h_i : U_i \to U_i/U_i^\#$ given by $h(x) = x + U_i^\#$. Since $\ker h_i = U_i^\#$ is a vector space summand of U_i, there is a k-linear transformation $g_i : U_i/U_i^\# \to U_i \subseteq U_0$ with $h_i g_i = 1$. Define a k-linear transformation $p : P(U)_0 = \oplus\{U_i/U_i^\# : i \in S\} \to U_0$ by $p = \Sigma\{g_i : i \in S\}$. Then p is a representation morphism, since $p(P(U)_i) = p(\oplus\{U_j/U_j^\# : j \leq i\}) = \Sigma\{g_j(U_j/U_j^\#) : j \leq i\} \subseteq \Sigma\{U_j : j \leq i\} \subseteq U_i$ for each $i \in S$.

An induction argument, beginning with a minimal element of S, shows that $p(P(U)_i) = U_i$ for each i in S. In particular, if i is a minimal element of S, then $U_i^\# = 0$, $P(U)_i = U_i$, and $p(P(U)_i) = U_i$. Now assume that i is not a minimal element of S and $p(P(U)_j) = U_j$ for each $j < i$ in S. Then $U_i^\# = \Sigma\{p(U)_j : j < i\}$. Since $U_i = \mathrm{image}\, g_i \oplus U_i^\#$, $p(P(U)_i) = U_i$. But $U_0 = \Sigma\{U_i : i \in S\}$, since U has no trivial summands. Consequently, p is a representation epimorphism.

To see that $p : P(U) \to U$ is a projective cover of U, let $g : P' \to U$ be an epimorphism with P' projective. Since P' is projective and p is an epimorphism, there is $h : P' \to P(U)$ with $ph = g$. An induction argument, beginning with a minimal element of S, shows that $h(P_i') = P(U)_i = U_i$ for each $i \in S$. Hence, h is a representation epimorphism, as desired.

Statement (b) follows from Lemma 6.1.3 and (a).

Projective covers and injective envelopes for representations of antichains are computed explicitly in the next example.

Example 6.1.5 *Let* $S_n = \{1, 2, \ldots, n\}$ *and* $U = (U_0, U_1, \ldots, U_n) \in \mathrm{rep}(S_n, k)$.

(a) *If* U *has no trivial summands, then* $p : P(U) = (U_1 \oplus \cdots \oplus U_n, U_1, \ldots, U_n) \to U$, *with* $p(x_1 \oplus \cdots \oplus x_n) = x_1 + \cdots + x_n$, *is a projective cover of* U.
(b) *If* U *has no cotrivial summands, then* $I(U) = ((U_0/U_1 \oplus \cdots \oplus U_0/U_n), (U_0/U_2 \oplus \cdots \oplus U_0/U_n), \ldots, (U_0/U_1 \oplus \cdots \oplus U_0/U_{n-1}))$ *and* $q : U \to I(U)$ *with* $q(x) = (x + U_1, \ldots, x + U_n)$ *is an injective envelope of* U.

EXERCISES

1. Give a proof of Lemma 6.1.1.

2. Let S be a finite poset and define $T = \{*\} \cup S$, where $* < s$ for each s in S. Use Lemma 6.1.1 and Exercise 1.2.5 to conclude that there is a 1-to-1 correspondence $\mathrm{rep}(S, k) \to \mathrm{rep}(T, k)$.

3. Prove that $\mathrm{rep}(S, k)$ has finite representation type if and only if $\mathrm{rep}(S^{\mathrm{op}}, k)$ has finite representation type.

4. There are two representations U and V of S_5 defined in Exercise 1.2.2 with $\mathrm{End}\, U = \mathrm{End}\, V = C(A, B)$ for given matrices A and B.
    (a) Compute $\sigma(U)$ and $\sigma(V)$, with σ as given in Lemma 6.1.1.
    (b) Find projective covers and injective envelopes for U, V, $\sigma(U)$, and $\sigma(V)$.

5. Prove that if each $U \in \mathrm{rep}(S, k)$ is both projective and injective, then $w(S) = 1$.

6. Confirm the assertions of Lemma 6.1.4(b) and Example 6.1.5.

7. The *Cartan matrix* of a finite poset S is an $|S| \times |S|$ matrix $C(S) = (c_{ij})$, where $c_{ij} = 1$ if $i \leq j$ in S and $c_{ij} = 0$ otherwise. Show that the ith column of $C(S)$ is the transpose of the vector cdn $P(i)$ and that the ith row of $C(S)$ is cdn $I(i)$.

8. Define the *extended Cartan matrix* to be an $(|S| + 1) \times (|S| + 1)$ matrix $C^*(S)$ consisting of $C(S)$ together with last row $(0, 0, \ldots, 0, 1)$ and last column the transpose of $(-1, -1, \ldots, -1, 1)$. Label the elements of S by $\{1, 2, \ldots, n\}$. Given variables $x_1, \ldots, x_{n+1}$, define the *quadratic form* q_S of S by $q_S(x_1, \ldots, x_{n+1}) = (x_1, \ldots, x_{n+1}) C^*(S) (x_1, \ldots, x_{n+1})^{tr}$.
    (a) Show that $q_S(x_1, \ldots, x_{n+1}) = \Sigma\{x_i^2 : 1 \leq i \leq n+1\} - \Sigma\{x_i x_j : i < j \text{ in } S\} + x_{n+1}(\Sigma\{x_i : 1 \leq i \leq n\})$.
    (b) [Ringel 84] In each example below, confirm that $q_S(z) = 0$, where $z = (z_1, \ldots, z_{n+1})$ and $S = \{1, 2, \ldots, n\}$.

(i) $z = (1, 1, 1, 1, 2)$, $S = 1$ 2   3   4

(ii) $z = (1, 1, 1, 1, 1, 1, 3)$, $S = 2$ 4   6
$$
\begin{array}{ccc}
| & | & | \\
1 & 3 & 5
\end{array}
$$

(iii) $z = (2, 1, 1, 1, 1, 1, 1, 4)$, $S = 1$ 4   7
$$
\begin{array}{cc}
| & | \\
3 & 6 \\
| & | \\
2 & 5
\end{array}
$$

(iv) $z = (3, 2, 2, 1, 1, 1, 1, 1, 6)$, $S = 1$ 3   8
$$
\begin{array}{cc}
| & | \\
2 & 7 \\
& | \\
& 6 \\
& | \\
& 5 \\
& | \\
& 4
\end{array}
$$

(v) $z = (1, 2, 2, 1, 1, 1, 1, 1, 5)$, $S = 2$ 4   8
$$
\begin{array}{ccc}
| \backslash & | & | \\
1 \quad 3 & & 7 \\
& & | \\
& & 6 \\
& & | \\
& & 5
\end{array}
$$

(c) Prove that if $q_S(z) > 0$ for each $0 \neq z = (z_1, \ldots, z_n)$ with z_i a nonnegative integer for each i, then S has finite representation type (the converse is also true; [Drozd 74] or [Simson 92]).

9. Prove that $U = (U_0, U_i : i \in S) \in \mathrm{rep}(S, k)$ is a sincere representation if and only if the projective cover $P(U)$ of U contains a copy of $P(i)$ for each $i \in S$.

10. Prove that projective covers and injective envelopes in $\mathrm{rep}(S, k)$ are unique up to isomorphism.

6.2   Coxeter Correspondences

The construction of matrices to find indecomposable representations in $\mathrm{rep}(S, k)$, as described in Section 1.2, is impractical for a general finite poset. In this section, correspondences are defined that can be used to generate indecomposables from a given indecomposable. In fact, if $\mathrm{rep}(S, k)$ has finite representation type, then all indecomposables in $\mathrm{rep}(S, k)$ can be found by applying these correspondences to 1-dimensional projective or injective representations (Theorem 6.2.6).

Let $U = (U_0, U_i : i \in S) \in \mathrm{rep}(S, k)$ and $p_U : P(U) \to U$ be a projective cover of U as defined in Lemma 6.1.4(a) with $P(U)_0 = \oplus\{U_i / U_i^\# : i \in S\}$ and $P(U)_i =$

$\oplus\{U_j/U_j^\#:j\leq i\}$. Define representations $W(U)=(W_0,W_i:i\in S)$, with

$$W_0=P(U)_0=\oplus\{U_i/U_i^\#:i\in S\},\ W_i=\oplus\{U_j/U_j^\#:i\not\leq j\}$$

and

$$L(U)=(L(U)_0,L(U)_i:i\in S),$$

with

$$L(U)_0=U_0,\ L(U)_i=\Sigma\{U_j:i\not\leq j\}.$$

Then

$$C^+(U)=(\ker p_U,(\ker p_U)\cap W_i:i\in S)$$

is a kernel of the representation morphism $p_U:W(U)\to L(U)$. Hence,

$$0\to C^+(U)\to W(U)\to L(U)\to 0$$

is an exact sequence of representations with $W(U)$ an injective representation by Lemma 6.1.2(b).

Let $q_U:U\to I(U)$ be an injective envelope of U as defined in Lemma 6.1.4(b). Define representations

$$W'(U)=(W_0',W'(U)_i),$$

with

$$W_0'=I(U)_0=\oplus\{\hat{U}_i/U_i:i\in S\},\ W'(U)_i=\oplus\{\hat{U}_j:j\leq i\},$$

and

$$K(U)=(K(U)_0,K(U)_i),$$

with

$$K(U)_0=q_U(U_0),\ K(U)_i=q_U(U_0)\cap W'(U)_i.$$

Then

$$C^-(U)=(I(U)_0/q_U(U_0),(W'(U)_i+q_U(U_0))/q_U(U_0):i\in S)$$

is a cokernel of the pure monomorphism $K(U)\to W'(U)$. Hence,

$$0\to K(U)\to W'(U)\to C^-(U)\to 0$$

is an exact sequence of representations with $W'(U)$ a projective representation.

Proposition 6.2.1 *Let S be a finite poset and k a field. Then C^+ and C^- are correspondences from* rep(S,k) *to* rep(S,k). *If $U\in$ rep(S,k), then $C^+U=0$ if and only if U is a projective representation, and $C^-U=0$ if and only if U is an injective representation. Moreover, C^+ and C^- preserve finite direct sums.*

Proof. The representation U is projective if and only if $P(U) = U$, equivalently, the kernel of $p_U: P(U) \to U$ is 0. Thus, $C^+U = 0$ if and only if U is projective. A similar argument shows that U is injective if and only if $C^-U = 0$.

Both $P(U) \oplus P(V)$ and $P(U \oplus V)$ are projective covers of $U \oplus V$. Hence $P(U) \oplus P(V)$ is isomorphic to $P(U \oplus V)$ by Exercise 6.1.10. Now, $C^+(U \oplus V)$ is constructed in terms of $P(U \oplus V)$, while $C^+U \oplus C^+V$ is constructed from $P(U) \oplus P(V)$. It follows that $C^+(U \oplus V)$ is isomorphic to $C^+U \oplus C^+V$. A similar argument, using injective envelopes, shows that $C^-(U \oplus V)$ is isomorphic to $C^-U \oplus C^-V$.

Example 6.1.5 (continued)

(c) *If U has no trivial summands, then*

$$W(U) = (U_1 \oplus \cdots \oplus U_n, U_2 \oplus \cdots \oplus U_n,$$

$$U_1 \oplus U_3 \oplus \cdots \oplus U_n, \ldots, U_1 \oplus \cdots \oplus U_{n-1}),$$

$$L(U) = (U_0, U_2 + \cdots + U_n, \ldots, U_1 + \cdots + U_{n-1}),$$

and

$$C^+U = (\ker p_U, (\ker p_U) \cap (U_2 \oplus \cdots \oplus U_n), \ldots,$$

$$(\ker p_U) \cap (U_1 \oplus \cdots \oplus U_{n-1})).$$

(d) *If U has no cotrivial summands, then*

$$W'(U) = (U_0/U_1 \oplus \cdots \oplus U_0/U_n, U_0/U_1, \ldots, U_0/U_n),$$

$$K(U) = (q_U(U_0), q_U(U_0) \cap (U_0/U_1), \ldots, q_U(U_0) \cap (U_0/U_n)),$$

and

$$C^-U = ((U_0/U_1) \oplus \cdots \oplus (U_0/U_n))/q_U(U_0),$$

$$((U_0/U_i) + q_U(U_0))/q_U(U_0) : 1 \le i \le n).$$

(e) *Suppose* $\mathrm{cdn}\, U = (u_0, u_1, \ldots, u_n)$.
  (i) *If U is not a projective representation, then*

$$\mathrm{cdn}\, C^+U = (u_1 + \cdots + u_n - u_0, u_2 + \cdots + u_n - u_0, \ldots,$$

$$u_1 + \cdots + u_{n-1} - u_0).$$

  (ii) *If U is not an injective representation, then*

$$\mathrm{cdn}\, C^-U = ((n-1)u_0 - u_1 - \cdots - u_n, u_0 - u_1, \ldots, u_0 - u_n).$$

For a positive integer i, define C^{+i} to be the composition of C^+ repeated i times and define C^{-i} to be the composition of C^- repeated i times. Define $C^{+0}U = C^{-0}U = U$. Call U in $\mathrm{rep}(S, k)$ *preprojective* if U is isomorphic to $C^{-i}P$ for some projective P and $i \geq 0$, and *preinjective* if U is isomorphic to $C^{+i}I$ for some injective I and $i \geq 0$. Each projective is preprojective and each injective is preinjective.

Example 6.2.2 *Following is a listing of the indecomposable preprojectives and preinjectives for* rep(S_3, k):

(a) *indecomposable preprojectives:*

$$P(0),\ P(1),\ P(2),\ P(3),$$

$$C^- P(0) = (k \oplus k, k \oplus 0, 0 \oplus k, (1+1)k),$$

$$C^{-2} P(0) = I(0),\ C^- P(1) = I(1),$$

$$C^- P(2) = I(2),\ C^- P(3) = I(3).$$

(b) *indecomposable preinjectives:*

$$I(0),\ I(1),\ I(2),\ I(3)$$

$$C^+ I(0) = C^- P(0),\ C^{+2} I(0) = P(0),\ C^+ I(1) = P(1),$$

$$C^+ I(2) = P(2),\ C^+ I(3) = P(3).$$

PROOF. Recall that $P(0) = (k, 0, 0, 0)$, $P(1) = (k, k, 0, 0)$, $P(2) = (k, 0, k, 0)$, and $P(3) = (k, 0, 0, k)$ are the projectives for rep(S_3, k). Then cdn $P(0) = (1, 0, 0, 0)$, cdn $P(1) = (1, 1, 0, 0)$, cdn $P(2) = (1, 0, 1, 0)$, and cdn $P(3) = (1, 0, 0, 1)$. The injectives for rep(S_3, k) are $I(0) = (k, k, k, k)$, $I(1) = (k, 0, k, k)$, $I(2) = (k, k, 0, k)$, $I(3) = (k, k, k, 0)$ with cdn $I(0) = (1, 1, 1, 1)$, cdn $I(1) = (1, 0, 1, 1)$, cdn $I(2) = (1, 1, 0, 1)$, and cdn $I(3) = (1, 1, 1, 0)$.

(a) By Example 6.1.5(e), cdn $C^- P(0) = (2, 1, 1, 1)$, cdn $C^{-2} P(0) = (1, 1, 1, 1)$ cdn $C^- P(1) = (1, 0, 1, 1)$, cdn $C^- P(2) = (1, 1, 0, 1)$, and cdn $C^- P(3) = (1, 1, 1, 0)$. Since 1-dimensional representations are determined up to isomorphism by their coordinate vector, it remains only to determine $C^- P(0)$.

By Example 6.1.5(d), $W'(P(0)) = (k \oplus k \oplus k, k \oplus 0 \oplus 0, 0 \oplus k \oplus 0, 0 \oplus 0 \oplus k)$, $K(P(0)) = ((1+1+1)k, 0, 0, 0)$, and $C^- P(0) = (V_0, V_1, V_2, V_3)$ with

$$V_0 = (k \oplus k \oplus k)/(1+1+1)k,$$

$$V_1 = ((k \oplus 0 \oplus 0) + (1+1+1)k)/(1+1+1)k,$$

$$V_2 = ((0 \oplus k \oplus 0) + (1+1+1)k)/(1+1+1)k,$$

$$V_3 = ((0 \oplus 0 \oplus k) + (1+1+1)k)/(1+1+1)k.$$

It follows that $C^- P(0)$ is isomorphic to $(k \oplus k, k \oplus 0, 0 \oplus k, (1+1)k)$.

(b) By Example 6.1.5(e), cdn $C^+ I(0) = (2, 1, 1, 1)$, cdn $C^{+2} I(0) = (1, 0, 0, 0)$, cdn $C^+ I(1) = (1, 1, 0, 0)$, cdn $C^+ I(2) = (1, 0, 1, 0)$ and cdn $C^+ I(3) = (1, 0, 0, 1)$. Once again, only $C^+ I(0)$ needs to be computed. By Example 6.1.5(c),

$W(I(0)) = (k \oplus k \oplus k, 0 \oplus k \oplus k, k \oplus 0 \oplus k, 0 \oplus 0 \oplus k), L(I(0)) = (k, k, k, k),$
and $C^+I(0) = (k \oplus k, k \oplus 0, 0 \oplus k, (1+1)k)$, as desired.

There are several intriguing observations about the list in Example 6.2.2. The first is that C^+ and C^- applied to an indecomposable injective representation of S_3, respectively an indecomposable projective representation, resulted in an indecomposable representation. This property is, in fact, true for each finite poset and each indecomposable representation by Corollary 6.2.5(d). Secondly, each indecomposable in $\text{rep}(S_3, k)$ is both preprojective and preinjective. This property holds for each S such that $\text{rep}(S, k)$ has finite representation type by Theorem 6.2.6. Finally, if $U \in \text{rep}(S_3, k)$ and U is indecomposable, then U is isomorphic to C^-C^+U if $C^+U \neq 0$ and U is isomorphic to C^+C^-U if $C^-U \neq 0$. This property holds for each finite poset S by Corollary 6.2.4(c) and (d).

Fundamental properties of the correspondences C^+ and C^- are derived in Corollary 6.2.4 by showing that the duality σ of Lemma 6.1.1 is a factor of both C^+ and C^-.

Proposition 6.2.3 *If $U \in \text{rep}(S, k)$, S a finite poset, and k a field, then*

(a) $\sigma C^+(U) = C^- \sigma(U)$;
(b) *if $\rho = \sigma C^+$ and U has no projective summands, then $\rho^2(U)$ is naturally isomorphic to U;*
(c) $0 \to C^+U \to W(U)$ *is an injective envelope of C^+U; and*
(d) $W'(U) \to C^-U \to 0$ *is a projective cover of C^-U.*

PROOF. (a) There is a projective cover $0 \to M(U) \to P(U) \to U \to 0$ of U in $\text{rep}(S, k)$ with $P(U)_0 = \oplus \{U_i/U_i^\# : i \in S\}$, and $P(U)_i = \oplus \{U_j/U_j^\# : j \leq i\}$. Then $0 \to C^+(U) \to W(U) \to L(U) \to 0$ is an exact sequence of representations with

$$W(U)_0 = P(U)_0 = \oplus\{U_i/U_i^\# : i \in S\}, \ W(U)_i = \oplus\{U_j/U_j^\# : i \nleq j\},$$

and

$$L(U)_0 = U_0, \ L(U)_i = \Sigma\{U_j : i \nleq j\}.$$

Hence, $0 \to \sigma L(U) \to \sigma W(U) \to \sigma C^+(U) \to 0$ is an exact sequence of representations with

$$\sigma W(U)_0 = \sigma P(U)_0 = \oplus\{\sigma(U_i/U_i^\#) : i \in S\}, \ \sigma W(U)_i = \oplus\{\sigma(U_j/U_j^\#) : j \leq i\},$$

and

$$\sigma L(U)_0 = \sigma U_0, \ \sigma L(U)_i = \sigma(\Sigma\{U_j : i \nleq j\}).$$

On the other hand, $0 \to \sigma U \to \sigma P(U) \to \sigma M(U) \to 0$ is an injective envelope of σU. Then $0 \to K(\sigma U) \to W'(\sigma U) \to C^-(\sigma U) \to 0$ is an exact sequence of representations with

$$W'(\sigma U)_0 = \sigma P(U)_0 = \oplus\{\sigma(U_i/U_i^\#) : i \in S\},$$
$$W'(\sigma U)_i = \oplus\{\sigma(U_j/U_j^\#) : j \leq i\},$$

and

$$K(\sigma U)_0 = \sigma U_0, \ K(U)_i = \sigma(\Sigma\{U_j : i \not\leq j\}).$$

Thus, $\sigma W(U) = W'(\sigma U)$ and $\sigma L(U)_i = K(U)_i$ for each $i \in S$. This shows that $\sigma C^+(U) = C^-\sigma U$, as desired.

(b) There is an exact sequence of representations

$$0 \to C^+(U) \to W(U) \to L(U) \to 0,$$

recalling that $W(U)$ is injective with $W(U)_0 = P(U)_0$ and $L(U)_0 = U_0$.

Applying σ yields an exact sequence of representations

$$0 \to \sigma L(U) \to \sigma W(U) \to \rho(U) = \sigma C^+(U) \to 0$$

with $\sigma W(U)$ projective. Now, $P(\rho(U))$ is a projective cover of $\rho(U)$, so that $\sigma W(U) = P(\rho(U)) \oplus P$ for some projective representation P. The definition of $C^+(\rho(U))$ yields an exact sequence

$$0 \to C^+(\rho(U)) \oplus I \to W(\rho(U)) \oplus I \to L(\rho(U)) \to 0$$

for some injective I with $W(\rho(U))_0 \oplus I_0 = P(\rho(U))_0 \oplus P_0 = \sigma W(U)_0 = \sigma P(U)_0$. Hence,

$$0 \to \sigma L(\rho(U)) \to \sigma(W(\rho(U)) \oplus I) \to \sigma C^+\rho(U) \oplus \sigma(I) = \rho^2(U) \oplus \sigma(I) \to 0$$

is an exact sequence of representations with $\sigma(W(\rho(U)) \oplus I)$ projective and $\sigma(W(\rho(U)) \oplus I)_0 = \sigma^2 P(U)_0$.

Thus, $\sigma^2 : P(U)_0 \to \sigma(W(\rho(U)) \oplus I)_0$ is a vector space isomorphism that induces a representation isomorphism $\alpha : U \to \rho^2(U) \oplus \sigma(I)$ with $\sigma(I)$ projective. Since U has no projective summands, $\sigma(I) = 0$, whence $I = 0$ and $P = 0$. Hence, U is isomorphic to $\rho^2(U)$. It is straightforward to see that this isomorphism is natural.

(c) As in the proof of (b), $\sigma(W(U)) \to \sigma C^+U$ is a projective cover of σC^+U. By Lemma 6.1.1, $C^+(U) \to W(U)$ must be an injective envelope of C^+U.

(d) Since, $C^-\sigma U = \sigma C^+U$, it follows that $W'(U) \to C^-U$ is a projective cover of C^-U.

Corollary 6.2.4 *Let $\rho = \sigma C^+$ and $U \in \mathrm{rep}(S, k)$. Then:*

(a) *if U has no trivial summands, then C^+U is isomorphic to $\sigma\rho(U)$;*
(b) *if U has no cotrivial summands, then C^-U is isomorphic to $\rho\sigma(U)$;*
(c) *if U has no projective summands, then C^-C^+U is isomorphic to U;*
(d) *if U has no injective summands, then C^+C^-U is isomorphic to U;*

(e) *U is preprojective if and only if $C^{+i}U = 0$ for some $i \geq 1$; and*
(f) *U is preinjective if and only if $C^{-i}U = 0$ for some $i \geq 1$.*

PROOF. (a) As a consequence of Lemma 6.1.1, $\sigma\rho(U) = \sigma^2 C^+U$ is isomorphic to C^+U for each $U \in \text{rep}(S, k)$. As for (b), $\rho\sigma(U) = \sigma C^+\sigma(U) = C^-(\sigma^2(U))$ by Proposition 6.2.3(a), and $C^-(\sigma^2(U))$ is isomorphic to $C^-(U)$ by Lemma 6.1.1.

(c), (d) Notice that $C^-C^+U = (C^-\sigma)(\sigma C^+)(U) = \rho^2(U)$ is isomorphic to U as a consequence of Lemma 6.1.1 and Proposition 6.2.3(b).

Assertions (e) and (f) are consequences of Proposition 6.2.1 and (c) and (d), respectively.

Corollary 6.2.5 *Suppose $U, V \in \text{rep}(S, k)$, S is a finite poset, and k is a field.*

(a) *If U has no injective summands and V has no projective summands, then $\text{Hom}(U, C^+V)$ is isomorphic to $\text{Hom}(C^-U, V)$.*
(b) *If U has no projective summands, then $\text{End}\, U$ is isomorphic to $\text{End}\, C^+U$.*
(c) *If U has no injective summands, then $\text{End}\, U$ is isomorphic to $\text{End}\, C^-U$.*
(d) *If U is indecomposable, then C^+U and C^-U are indecomposable.*
(e) *If U is either preprojective or preinjective, then $\text{End}\, U$ is isomorphic to k.*

PROOF. (a) It is sufficient to prove that $\text{Hom}(U, V)$ is isomorphic to $\text{Hom}(\rho(V), \rho(U))$. To see this, observe that $\text{Hom}(U, C^+V) = \text{Hom}(U, \sigma\rho(V))$ is isomorphic to $\text{Hom}(\rho(V), \sigma(U))$ by Lemma 6.1.1. Then $\text{Hom}(\rho(V), \sigma(U))$ is isomorphic to $\text{Hom}(\rho\sigma(U), \rho^2(V)) = \text{Hom}(C^-U, V)$ via Proposition 6.2.3.

Let $p_U : P(U) \to U \to 0$ be the projective cover of U given in Lemma 6.1.4 and $f : U \to V$. By the construction of $P(U)$, $W(U)$, and $L(U)$, f induces $g : W(U) \to W(V)$ and $h : L(U) \to L(V)$ with $hp_U = p_V g$, where $p_U : W(U) \to L(U)$. Since $C^+U = \text{Ker}\, p_U$, f induces $f' : C^+U \to C^+V$. This yields a homomorphism $\alpha : \text{Hom}(U, V) \to \text{Hom}(C^+U, C^+V)$ given by $\alpha(f) = f'$. By Lemma 6.1.1, $\sigma : \text{Hom}(C^+U, C^+V) \to \text{Hom}(\sigma C^+V, \sigma C^+U)$ is an isomorphism. Since $\rho = \sigma C^+$, $\sigma\alpha : \text{Hom}(U, V) \to \text{Hom}(\rho(V), \rho(U))$ is a homomorphism. Similarly, there is a homomorphism $\text{Hom}(\rho(V), \rho(U)) \to \text{Hom}(\rho^2 U, \rho^2 V)$. It follows from Proposition 6.2.3(b) that $\text{Hom}(U, V)$ is isomorphic to $\text{Hom}(\rho(V), \rho(U))$.

(b), (c) Now, $\text{Hom}(U, U)$ is isomorphic to $\text{Hom}(U, C^+C^-U)$ by Corollary 6.2.4(d), and by (a), $\text{Hom}(U, C^+C^-U)$ is isomorphic to $\text{Hom}(C^-U, C^-U)$. On the other hand, $\text{Hom}(U, U)$ is isomorphic to $\text{Hom}(C^-C^+U, U)$, by Corollary 6.2.4(c), which is isomorphic to $\text{Hom}(C^+U, C^+U)$ by (a).

Statement (d) is an immediate consequence of (b) and (c), since a representation U is indecomposable if and only if 0 and 1 are the only idempotents of $\text{End}\, U$.

(e) Assume that U is an indecomposable preprojective representation. Then $U = C^{-i}P(t)$ for some $t \in S \cup \{0\}$. By (a), $\text{End}\, U = \text{End}\, C^{+i}C^{-i}P(t) = \text{End}\, P(t)$. But $\text{End}\, P(t) = k$, since $P(t)$ is a 1-dimensional representation. Similarly, if U is an indecomposable preinjective representation, then $\text{End}\, U = \text{End}\, I(t) = k$ for some $t \in S \cup \{0\}$.

Theorem 6.2.6 *Let S be a finite poset and k a field. The following are equivalent:*

(a) $\mathrm{rep}(S, k)$ *has finite representation type;*
(b) *If U is an indecomposable in* $\mathrm{rep}(S, k)$, *then U is preinjective;*
(c) *If U is an indecomposable in* $\mathrm{rep}(S, k)$, *then U is preprojective;*
(d) *End U is isomorphic to k for each indecomposable U in* $\mathrm{rep}(S, k)$.

In this case, if U is an indecomposable in $\mathrm{rep}(S, k)$, *then U is determined up to isomorphism by* cdn U.

PROOF. (a) $\Rightarrow$ (b) [Ringel, 80].

(b) $\Rightarrow$ (d) follows from Corollary 6.2.5(e).

(d) $\Rightarrow$ (a) If $\mathrm{rep}(S, k)$ has infinite represention type, then S contains one of the critical posets $T = S_4, (2, 2, 2), (1, 3, 3), (1, 2, 5)$, or $(N, 4)$ as a subposet by Theorem 1.3.6. Choose an indecomposable U in $\mathrm{rep}(T, k)$ with End U not isomorphic to k, as given in the proof of Theorem 1.3.6. By Proposition 1.3.1, U gives rise to an indecomposable V in $\mathrm{rep}(S, k)$ with End V not isomorphic to k.

(c) $\Rightarrow$ (d) is a consequence of Corollary 6.2.5(e).

(a) $\Rightarrow$ (c) follows from (a) $\Rightarrow$ (b), Corollary 6.2.4, and Proposition 1.4.7.

Example 6.2.7 *The category* $\mathrm{rep}(S_4, k)$ *has tame representation type. Following is a complete list of elements of* $\mathrm{Ind}(S_4, k)$:

(a) *indecomposable preprojectives U; all have endomorphism ring k and are classified by* cdn U:

(i) $\mathrm{cdn}\, C^{-n} P(0) = (2n + 1, n, n, n, n)$,

$$C^{-n} P(0) = (k^n \oplus k^n \oplus k, k^n \oplus 0, 0 \oplus k^n, u(k^n), v(k^n)), \text{ where}$$

$$u(x_1, \ldots, x_n) = (0, x_1, \ldots, x_n, x_n, \ldots, x_1) \text{ and}$$

$$v(x_1, \ldots, x_n) = (x_1, \ldots, x_n, x_n, \ldots, x_1, 0).$$

(ii) $\mathrm{cdn}\, C^{-2n} P(1) = (2n - 1, n, n - 1, n - 1, n - 1)$,

$$C^{-2n} P(1) = (k^n \oplus k^{n-1}, k^n \oplus 0, 0 \oplus k^{n-1}, u'(k^{n-1}), v'(k^{n-1})), \text{ where}$$

$$u'(x_1, \ldots, x_{n-1}) = (x_1, \ldots, x_{n-1}, x_1, x_2, \ldots, x_{n-1}, 0) \text{ and}$$

$$v'(x_1, \ldots, x_{n-1}) = (x_1, \ldots, x_{n-1}, 0, x_1, \ldots, x_{n-1}).$$

(iii) $\mathrm{cdn}\, C^{-2n} P(2) = (2n - 1, n - 1, n, n - 1, n - 1)$,

$$C^{-2n} P(2) = (k^n \oplus k^{n-1}, 0 \oplus k^{n-1}, k^n \oplus 0, u'(k^{n-1}), v'(k^{n-1})).$$

(iv) $\mathrm{cdn}\, C^{-2n} P(3) = (2n - 1, n - 1, n - 1, n, n - 1)$,

$$C^{-2n} P(3) = (k^n \oplus k^{n-1}, 0 \oplus k^{n-1}, u'(k^{n-1}), k^n \oplus 0, v'(k^{n-1})).$$

(v) cdn $C^{-2n}P(4) = (2n-1, n-1, n-1, n-1, n)$,

$$C^{-2n}P(4) = (k^n \oplus k^{n-1}, 0 \oplus k^{n-1}, u'(k^{n-1}), v'(k^{n-1}), k^n \oplus 0).$$

(vi) cdn $C^{-(2n-1)}P(1) = (2n, n-1, n, n, n)$,

$$C^{-2n-1}P(1) = (k^n \oplus k^n, w(k^{n-1}), k^n \oplus 0, 0 \oplus k^n, (1+1)(k^n)),$$

$$w(x_1, \ldots, x_{n-1}) = (0, x_1, \ldots, x_{n-1}, x_1, \ldots, x_{n-1}, 0).$$

(vii) cdn $C^{-(2n-1)}P(2) = (2n, n, n-1, n, n)$,

$$C^{-(2n-1)}P(2) = (k^n \oplus k^n, k^n \oplus 0, w(k^{n-1}), 0, k^n, (1+1)(k^n)).$$

(viii) cdn $C^{-(2n-1)}P(3) = (2n, n, n, n-1, n)$,

$$C^{-(2n-1)}P(3) = (k^n \oplus k^n, k^n \oplus 0, 0 \oplus k^n, w(k^{n-1}), (1+1)(k^n)).$$

(ix) cdn $C^{-(2n-1)}P(4) = (2n, n, n, n, n-1)$,

$$C^{-(2n-1)}P(4) = (k^n \oplus k^n, k^n \oplus 0, 0 \oplus k^n, (1+1)(k^n), w(k^{n-1})).$$

(b) *indecomposable preinjectives U; all have endomorphism ring k and are classified by* cdn U.

   (i)–(ix) Apply σ to the representations in (a).

(c) cdn $U_A = (2n, n, n, n, n)$

   $U_A = (k^n \oplus k^n, k^n \oplus 0, 0 \oplus k^n, (1+1)k^n, (1+A)k^n)$, End $U_A \approx k[x]/$ $\langle g(x)^e \rangle$, A an $n \times n$ k-matrix with minimal polynomial $g(x)^e$, $g(x)$ *irreducible.*

(d) cdn $U = (2n+1, n, n+1, n+1, n)$

   $U = (k^n \oplus k^{n+1}, k^n \oplus 0, 0 \oplus k^{n+1}, a(k^{n+1}), b(k^n))$, End $U \approx k[x]/\langle x^n \rangle$, *where* $a(x_1, \ldots, x_{n+1}) = (x_1, \ldots, x_n, x_1, \ldots, x_{n+1})$ *and* $b(x_1, \ldots, x_n) = (x_1, \ldots, x_n, 0, x_1, \ldots, x_n)$ *and five other similar representations obtained by permuting the subspaces, U_1, U_2, U_3, U_4 corresponding to permutations of* cdn U.

PROOF. [Gelfand, Ponomarev 70] or [Brenner 74B].

EXERCISES

1. Prove that if S is a forest, then $C^+, C^- : \text{rep}(S, k) \to \text{rep}(S, k)$ are functors. What is the situation if S is not a forest?

2. Confirm the assertions of Example 6.1.5(c), (d), and (e).

3. Explicitly construct the representations in Example 6.2.7(b) and (d).

4. Suppose $U \in \text{rep}(S, k)$ with $\text{cdn}\, U = (u_0, u_i : i \in S)$.
   (a) Prove that if $\text{cdn}\,\sigma(U) = (v_0, v_i : i \in S)$, then $v_0 = u_0$ and $\Sigma\{u_j : j \leq i\} + \Sigma\{v_j : j > i\} = u_0$ for each $i \in S$.
   (b) Prove that if $\text{cdn}\,\rho(U) = (w_0, w_i : i \in S)$, then $w_0 = \Sigma\{(u_i - u_0) : i \in S\}$ and $w_i = u_i$ for each $i \in S$.

5. Let $X, Y \in \text{rep}(S, k)$ be 1-dimensional representations. Prove that for each positive integer n:
   (a) $C^{+n}X$ is isomorphic to $C^{+n}Y$ if and only if $\text{cdn}\, C^{+n}X = \text{cdn}\, C^{+n}Y$.
   (b) $C^{-n}X$ is isomorphic to $C^{-n}Y$ if and only if $\text{cdn}\, C^{-n}X = \text{cdn}\, C^{-n}Y$.

6.3 Almost Split Sequences

Almost split sequences are one of the most important tools for classification of finitely generated modules over finite-dimensional k-algebras with tame representation type. The most important theorem about almost split sequences in $\text{rep}(S, k)$ is that they exist (Theorem 6.3.4). Even more, the left-hand term of an almost split sequence in $\text{rep}(S, k)$ determines the right-hand term, and the right-hand term determines the left-hand term. Almost split sequences in $\text{rep}(S, k)$ can be constructed explicitly (Corollary 6.3.5).

A representation morphism $f : U \to V$ in $\text{rep}(S, k)$ is a *splitting monomorphism* if there is $g : V \to U$ with $gf = 1_U$ and a *splitting epimorphism* if there is $g : V \to U$ with $fg = 1_V$. An exact sequence

$$E : 0 \to U \xrightarrow{f} V \xrightarrow{g} W \to 0$$

of representations in $\text{rep}(S, k)$ is *split exact* if there is $h : V \to U$ with $hf = 1_U$, the identity on U. Equivalently, there is $h : W \to V$ with $gh = 1_W$. Then E is an *almost split* sequence if:

(i) E is not a split exact sequence;
(ii) if $h : Y \to W$ is not a splitting epimorphism, then there is $\alpha : Y \to V$ with $g\alpha = h$; and
(iii) if $h : U \to Y$ is not a splitting monomorphism, then there is $\alpha : V \to Y$ with $\alpha f = h$.

The sequence E is *left almost split* if (i) and (iii) hold and *right almost split* if (i) and (ii) hold. Given exact sequences

$$E : 0 \to U \xrightarrow{f} V \xrightarrow{g} W \to 0,$$
$$E' : 0 \to U' \xrightarrow{f'} V' \xrightarrow{g'} W \to 0,$$

call $(\alpha, \beta, \gamma) : E \to E'$ a *homomorphism of exact sequences* if $\alpha : U \to U'$, $\beta : V \to V'$, and $\gamma : W \to W'$ are representation morphisms with $\beta f = f'\alpha$, and $\gamma g = g'\beta$. The homomorphism (α, β, γ) is an *isomorphism* if α, β, and γ are representation isomorphisms.

As a consequence of the following lemma, almost split sequences are unique up to isomorphism.

Lemma 6.3.1 *Suppose S is a finite poset, k is a field, and*

$$E: 0 \to U \xrightarrow{f} V \xrightarrow{g} W \to 0$$

is an almost split sequence in $\operatorname{rep}(S, k)$

(a) *Both* $\operatorname{End} U$ *and* $\operatorname{End} W$ *are local rings. In particular, U and W are indecomposable representations.*

(b) *If*

$$E': 0 \to U' \xrightarrow{f'} V' \xrightarrow{g'} W \to 0$$

is a right almost split sequence with U' indecomposable, then there is an isomorphism $(\alpha, \alpha, 1): E' \to E$.

(c) *If*

$$E': 0 \to U \xrightarrow{f'} V' \xrightarrow{g'} W' \to 0$$

is a left almost split sequence with W' indecomposable, then there is an isomorphism $(1, \alpha, \alpha): E' \to E$.

PROOF. (a) Suppose h and h' are nonunits in $\operatorname{End} W$. Since $\operatorname{End} W$ is a finite-dimensional k-algebra, h and h' are not splitting epimorphisms. Choose β, β': $W \to V$ with $g\beta = h$ and $g\beta' = h'$. Then $g(\beta + \beta') = h + h' = u$ is not a unit of $\operatorname{End} W$; otherwise, $u^{-1}g(\beta + \beta') = 1$, contradicting the assumption that E is not split exact.

(b) Since E' is not split exact, g' is not a splitting epimorphism. Choose $\alpha : V' \to V$ with $g\alpha = g'$. This choice is possible, since E is almost split. But E' is right almost split, so that there is $\beta : V \to V'$ with $g'\beta = g$. Since $g'\beta\alpha = g'1_W$, there is a homomorphism $(\gamma, \beta\alpha, 1): E' \to E'$ with $\beta\alpha f' = f'\gamma$.

If γ is a unit of $\operatorname{End} U'$, then $\alpha : f'(U') \to f'(U')$ is a unit, since U' is indecomposable. In this case, $(\alpha, \alpha, 1): E' \to E$ is the desired isomorphism. Now assume that γ is not a unit of $\operatorname{End} U'$. Since U' is indecomposable, γ is nilpotent, say $\gamma^n = 0$ for some n. Then $g'(\beta\alpha)^n = g'$, an epimorphism from V' to W, and $(\beta\alpha)^n$ induces $h' : W \to V'$, since $\gamma^n = 0$. As $g'(\beta\alpha)^n = g'1_{W'}$ and g' is onto, it follows that $g'h'$ is a unit of $\operatorname{End} W$, a contradiction to the fact that g' is not a splitting epimorphism.

(c) The proof is dual to that of (b).

Almost split sequences are minimal with respect to being a nonsplit exact sequence, provided that the left- and right-hand terms of the sequence are known.

Lemma 6.3.2 *Assume that S is a finite poset, k is a field, and that there is an almost split sequence*

$$E: 0 \to U \xrightarrow{f} V \xrightarrow{g} W \to 0.$$

Let

$$E': 0 \to U \xrightarrow{f'} V' \xrightarrow{g'} W \to 0$$

be a nonsplit exact sequence in $\mathrm{rep}(S, k)$. *The following statements are equivalent:*

(a) E' *is a left almost split sequence.*
(b) *If U' is a nonzero pure subrepresentation of U, then $0 \to U/U' \to V'/f'(U') \to W \to 0$ is a split exact sequence.*
(c) E' *is an almost split sequence.*
(d) *If $W' \neq W$ is a pure subrepresentation of W, then the sequence $0 \to U \to (g')^{-1}(W') \to W' \to 0$ is split exact.*
(e) E' *is a right almost split sequence.*

PROOF. (a) $\Rightarrow$ (b) Since U' is nonzero, $h : U \to U/U'$ is not a splitting monomorphism. By (a), there is $\alpha : V' \to U/U'$ with $\alpha f' = h$. Now, $\alpha f'(U') \subseteq \ker h = U'$, so $\alpha : V'/f'(U') \to U/U'$ with $\alpha f' = 1$, as desired.

(b) $\Rightarrow$ (c) Notice that $g' : V' \to W$ is not a splitting epimorphism, since E' is not a split exact sequence. Since E is almost split, there is $\alpha : V' \to V$ with $g\alpha = g'$. If $\alpha' = f^{-1}\alpha f'$ is a unit of $\mathrm{End}\, U$, then $(\alpha', \alpha, 1): E' \to E$ is an isomorphism, whence E' is an almost split sequence. If α' is not a unit, then $\ker \alpha'$ is a pure nonzero subrepresentation of U. By (b), there is a split exact sequence $0 \to U/\ker \alpha' \to V'/f'(\ker \alpha) \to W \to 0$. Since α sends $V'/\ker \alpha'$ isomorphically to a subrepresentation of V and $g\alpha = g'$, this split exact sequence induces a splitting of E', a contradiction.

(c) $\Rightarrow$ (a) is clear. The equivalence of (c), (d), and (e) is proved by a dual argument.

Corollary 6.3.3 *Suppose*

$$E: 0 \to U \xrightarrow{f} V \xrightarrow{g} W \to 0$$

is an almost split sequence in $\mathrm{rep}(S, k)$ *and either* $\mathrm{End}\, U$ *or* $\mathrm{End}\, W$ *is a division algebra. If*

$$E': 0 \to U \xrightarrow{f'} V' \xrightarrow{g'} W \to 0$$

is an exact sequence that is not split exact, then E' is an almost split sequence.

PROOF. Assume that $\mathrm{End}\, W$ is a division algebra. Then $f' : U \to V'$ is not a splitting monomorphism, since E' is not a split exact sequence. Since E is almost split, there is $\alpha : V \to V'$ with $\alpha f = f'$. Then α induces $\beta : W \to W$ with $\beta g = g'\alpha$. Moreover, $0 \neq \beta$, because E is not split exact. Since End W is a division algebra, β is an automorphism of W. Thus, $(1, \alpha, \beta) : E \to E'$ is an isomorphism, and E' is an almost split sequence. The proof for the case that End U is a division algebra is dual.

The proof of the following theorem is beyond the scope of this book.

Theorem 6.3.4 [Bautista, Martinez 79] *Let S be a finite poset and k a field.*

(a) *If W is an indecomposable in* rep(S, k) *that is not projective, then there is an almost split sequence* $0 \to C^+W \to V \to W \to 0$.
(b) *If U is an indecomposable in* rep(S, k) *that is not injective, then there is an almost split sequence* $0 \to U \to V \to C^-U \to 0$.

Given that almost split sequences exist, they can be constructed as follows:

Corollary 6.3.5 *Suppose S is a finite poset, k is a field, and $U \in$ rep(S, k) is a nonprojective indecomposable representation. Let $p_U : P(U) \to U$ be a projective cover of U and*

$$0 \to C^+U \to W(U) \xrightarrow{p} L(U) \to 0$$

the resulting injective envelope of C^+U.

(a) *There is a morphism $i : U \to L(U)$ induced by inclusion of U_0 in $L(U)_0$.*
(b) *Define $B = (B_0, B_i \in S)$ with $B_0 = \{(x, y) \in L(U) \oplus W(U) : ip_U(x) = i(y)\}$ and $B_i = B_0 \cap ((C^+U)_i \oplus L_i)$ for each $i \in S$. The sequence $0 \to C^+U \to B \to U \to 0$ is an almost split sequence in* rep(S, k).

PROOF. Exercise 6.3.4.

Let $(X_1, \ldots, X_n)$ be an n-tuple of 1-dimensional representations in rep(S, k). Choose a representation embedding $f_i : X_i \to I(0)$ and define $U(X_1, \ldots, X_n) \in$ rep(S, k) to be the kernel of $\oplus f_i : \oplus\{X_i : i \in S\} \to I(0)$. Representations $U(X_i, \ldots, X_n)$ are the representation analogue of bracket groups, as defined in Chapter 7.3.

Example 6.3.6 *Suppose $X \in$ rep(S, k) is nonprojective with dimension 1. Then $C^+X = U(I(i) : X_i \neq 0)$ for 1-dimensional injective representations $I(i)$, and the sequence*

$$0 \to U(I(i) : X_i \neq 0) \xrightarrow{f} U(X, I(i) : X_i \neq 0) \xrightarrow{g} X \to 0$$

with $f(x_1, \ldots, x_n) = (0, x_1, \ldots, x_n)$, $g(x, x_1, \ldots, x_n) = x$ is an almost split sequence.

PROOF. Up to isomorphism, it suffices to assume $X \subseteq I(0)$. There is a projective cover

$$p_X : P(X) = \oplus\{P(i) : X_i \neq 0\} \to X \text{ with } p_X(\oplus_i y_i) = \Sigma_i y_i.$$

Then $0 \to C^+X \to W(X) \to L(X) \to 0$ is an injective envelope of C^+U with $W(X) = \oplus\{I(i) : X_i \neq 0\}$ an injective representation and $X \subseteq L(X)$ a 1-dimensional representation. Hence, $C^+X = U(I(i) : (X)_i \neq 0)$, being the representation kernel of $p_X : W(X) \to L(X)$. Notice that $U(X, I(i) : (X)_i \neq 0) = B$, as defined in Corollary 6.3.5.

EXERCISES

1. A representation morphism $f : U \to V$ in rep(S, k) is *irreducible* if f is neither a split monomorphism nor a split epimorphism and if $f = gh$ for some $h : U \to W$ and $g : W \to V$, then either h is a splitting monomorphism or g is a splitting epimorphism. Prove that if

$$0 \to U \xrightarrow{f} V \xrightarrow{g} W \to 0$$

   is an almost split sequence in rep(S, k) and $V = V_1 \oplus \cdots \oplus V_n$ with each V_i an indecomposable representation, injections i_j, and projections π_j, then $\pi_j f$ and $g i_j$ are irreducible morphisms for each j.

2. Prove that if $f : U \to V$ is irreducible, then $f \neq \Sigma_j h_j g_j$ whenever Z_j is an indecomposable representation and $g_j : U \to Z_j$, $h_j : Z_j \to V$ are not isomorphisms.

3. Prove Lemma 6.3.1(c) and the equivalence of (c), (d), and (e) in Lemma 6.3.2.

4. Provide a proof for Corollary 6.3.5.

6.4   A Torsion Theory and Localizations

Given a finite poset S with unique maximal element ∞ and discrete valuation ring R, there is an associated category of representations of a quiver over R and a torsion theory for which the poset representations of S are the torsion-free elements (Corollary 6.4.6).

A poset S can be viewed as a *quiver*, a directed graph. In this case, vertices are elements of S, and there is a directed edge $i \to j$ if and only if $i \leq j$ in S. Representations of quivers and finitely generated modules over finite-dimensional k-algebras, k a field, are investigated thoroughly in [Ringel 84] and [Simson 92]. This chapter is devoted to a brief discussion of the category of quiver representations of S over R with relations induced by the poset S and its connection with rep(S, R).

Define a category Qrep(S, R) with objects $M = (M_i : f_{ij} : i \leq j)$ where each M_i is a finitely generated free R-module, $f_{ij} : M_i \to M_j$ is an R-homomorphism with pure image whenever $i \leq j$ in S, f_{ii} is the identity of M_i for each $i \in S$, and

$f_{jk} f_{ij} = f_{ik}$ whenever $i \leq j \leq k$ in S. A morphism from M to M' is $g = \{g_i : i \in S\}$ with each $g_i : M_i \to M'_i$ an R-homomorphism and $g_i f_{ij} = f'_{ij} g_i$ whenever $i \leq j$ in S.

Alternatively, $Q\mathrm{rep}(S, R)$ can be thought of as a category of modules over the R-incidence algebra of S that are finitely generated and free as R-modules; Exercise 6.4.1.

Proposition 6.4.1 *Let S be a finite poset with unique maximal element ∞ and R a discrete valuation ring. Then $Q\mathrm{rep}(S, R)$ is an additive category closed under direct sums, idempotents split in $Q\mathrm{rep}(S, R)$, and $\mathrm{rep}(S, R)$ is a full subcategory of $Q\mathrm{rep}(S, R)$.*

PROOF. It is straightforward to confirm that $Q\mathrm{rep}(S, R)$ is a category and the set of morphisms from M to M' is an abelian group for each M, M' in $Q\mathrm{rep}(S, R)$. If $M = (M_i, f_{ij} : i \leq j)$ and $M' = (M'_i, f'_{ij} : i \leq j)$ are in $Q\mathrm{rep}(S, R)$, then $M \oplus M' = (M_i \oplus M'_i, f_{ij} \oplus f'_{ij} : i \leq j)$ is a direct sum of M and M' in $Q\mathrm{rep}(S, R)$.

If $e = \{e_i : i \in S\}$ is an idempotent in $\mathrm{End}\, M$, the endomorphism ring of M in $Q\mathrm{rep}(S, R)$, then $e_i^2 = e_i$ for each i. Hence, $M = K \oplus L$ with $K = (e_i(M_i), e_j f_{ij} e_i : i \leq j)$ and $K \in Q\mathrm{rep}(S, R)$, since $e_j f_{ij} e_i$ has pure image in $e_j(M_j)$ and $e_k f_{jk} e_j f_{ij} e_i = e_k f_{ik} e_i$ whenever $i \leq j \leq k$ in S.

Let $U = (U_\infty, U_i : i \in S) \in \mathrm{rep}(S, R)$. Then $M = (U_i, f_{ij} : i \leq j) \in Q\mathrm{rep}(S, R)$, where $f_{ij} : U_i \to U_j$ is inclusion for $i \leq j$ in S. Each f_{ij} has pure image, since U_i is pure in U_∞, hence in U_j, whenever $i \leq j$, and $f_{jk} f_{ij} = f_{ik}$ whenever $i \leq j \leq k$ in S. The correspondence from $\mathrm{Hom}(U, U')$ in $\mathrm{rep}(S, R)$ to $\mathrm{Hom}(M, M')$ in $Q\mathrm{rep}(S, R)$ given by $f \to \{f_i = f : i \in S\}$ is an isomorphism for each $U, U' \in \mathrm{rep}(S, R)$. It follows that $\mathrm{rep}(S, R)$ is a full subcategory of $Q\mathrm{rep}(S, R)$.

If $N = (N_i, g_{ij})$ and (M_i, f_{ij}) are in $Q\mathrm{rep}(S, R)$, then N is a *subrepresentation* of M if N_i is a submodule of M_i for each i and each g_{ij} is the restriction of f_{ij} to N_i. Given $M = (M_i, f_{ij} : i \leq j) \in Q\mathrm{rep}(S, R)$, define $T(M) = (T(M)_i, g_{ij} : i \leq j)$ by $T(M)_\infty = 0$, $T(M)_i = \ker f_{i\infty}$ if $i \in S \backslash \{\infty\}$, and $g_{ij} : T(M)_i \to T(M)_j$ the restriction of f_{ij} to $T(M)_i$. Then g_{ij} is well-defined, and the image of each g_{ij} is pure, since $f_{j\infty} f_{ij} = f_{i\infty}$ and the image of $f_{i\infty}$ is pure. Notice that $T(M)$ is a subrepresentation of M and each $T(M)_i$ is a free R-module, being a submodule of the free R-module M_i. Moreover, $M_i / T(M)_i$ is a finitely generated torsion-free R-module, hence free.

Proposition 6.4.2 *Let S be a finite poset with unique maximal element ∞, R a discrete valuation ring, and $M, N \in Q\mathrm{rep}(S, R)$. Then T is a functor from $Q\mathrm{rep}(S, R)$ to $Q\mathrm{rep}(S, R)$ such that:*

(a) *$T(M)$ is a subrepresentation of M that preserves monomorphisms;*
(b) *$T(M \oplus N) = T(M) \oplus T(N)$;*
(c) *$T(T(M)) = T(M)$;*
(d) *$T(M) = 0$ if and only if $f_{i\infty}$ is a monomorphism for each $i \in S$; and*
(e) *$T(M/T(M)) = 0$.*

PROOF. (a) If $f : M \rightarrow N$, then $T(f) = f : T(M) \rightarrow T(N)$. Hence, $T : Q\mathrm{rep}(S, R) \rightarrow Q\mathrm{rep}(S, R)$ is a functor and $T(f)$ is a monomorphism if f is a monomorphism.

(b) Apply the definitions of $T(M)$ and direct sums.

(c) By definition, $T(M) = (T(M)_i, g_{ij} : i \leq j)$ with $T(M)_\infty = 0$, $T(M)_i = \ker f_{i\infty}$ if $i \in S \backslash \{\infty\}$, and $g_{ij} : T(M)_i \rightarrow T(M)_j$, the restriction of f_{ij} to $T(M)_i$. Then $0 = g_{i\infty} : T(M)_i = \ker f_{i\infty} \rightarrow T(M)_\infty = 0$, whence $\ker g_{i\infty} = T(M)_i$ for each i. This shows that $T(T(M)) = T(M)$.

Statement (d) follows immediately from the definition of $T(M)$.

(e) Write $M/T(M) = (M_i/T(M)_i, h_{ij})$, where $h_{ij} : M_i/T(M)_i \rightarrow M_j/T(M)_j$ is induced by f_{ij}. Since $h_{i\infty} : M_i/T(M)_i \rightarrow M_\infty/T(M)_\infty = M_\infty$ is a monomorphism, $\ker h_{i\infty} = 0$ for each i. Now apply (d).

Call $M \in Q\mathrm{rep}(S, R)$ *torsion* if $T(M) = M$ and *torsion-free* if $T(M) = 0$. Then $M = (M_i, f_{ij})$ is torsion if and only if $f_{i\infty} = 0$ for each $i \in S$.

A nonempty subset P of S is an *up-set in S* if given $i \in P$ and $i < j$ in S, then $j \in P$. An up-set is called a *filter* by some authors. Let $M = (M_i, f_{ij}) \in Q\mathrm{rep}(S, R)$ and define $M_P = ((M_P)_i, g_{ij})$, where $(M_P)_i = M_\infty$ if $i \in P$, $(M_P)_i = M_i$ if $i \notin P$, $g_{ij} = 1$ if $i \leq j$ in S with $i \in P$, and $g_{ij} = f_{ij}$ if $i \leq j$ in S with $j \notin P$. Since P is an up-set, M_P is a well-defined element of $Q\mathrm{rep}(S, R)$.

Call $M = (M_i, f_{ij}) \in Q\mathrm{rep}(S, R)$ *cotrivial* if each $M_i = M_\infty$ and f_{ij} is the identity for each $i \leq j$ in S. If S has a unique minimal element and M is cotrivial, then M is both injective and projective. The constructions of the next two propositions resemble localizations of modules.

Proposition 6.4.3 *Suppose S is a finite poset with unique maximal element, R is a discrete valuation ring, $M \in Q\mathrm{rep}(S, R)$, and P is an up-set in S. Then*

(a) $M \subseteq M_P, (M_P)_P = M_P$, and $T(M_P) = T(M)_P$;
(b) *if M is torsion-free, then M_P is torsion-free, and if M is torsion, then M_P is torsion; and*
(c) *if $P = S$, then M_P is cotrivial.*

PROOF. Immediate consequences of the definitions.

There is a localization "dual" to that of M_P. If P is an up-set in S and $M \in Q\mathrm{Rep}(S, R)$, define $M^P = ((M^P)_i, g_{ij})$, where $(M^P)_i = M_i$ if $i \in P$; $(M^P)_i = 0$ if $i \notin P$; $g_{ij} = f_{ij}$ if $i \leq j, i \in P$; and $g_{ij} = 0$ if $i \leq j, j \notin P$. Since P is an up-set, M^P is a well-defined element of $Q\mathrm{Rep}(S, R)$.

Proposition 6.4.4 *Suppose S is a finite poset with unique maximal element, R is a discrete valuation ring, $M = (M_i, f_{ij}) \in Q\mathrm{Rep}(S, R)$, and P is an up-set in S. Then*

(a) $M^P \subseteq M, (M^P)^P = M^P$, and $T(M^P) = T(M)^P$;
(b) $(M_P)^P = (M^P)_P = (N_i, g_{ij})$ *with $N_i = M_\infty$ for $i \in P$, $N_i = 0$ if $i \notin P$, and $g_{ij} = 1$ for each i;*

(c) *If M is torsion-free, then M^P is torsion-free, and if M is torsion, then M^P is torsion;*

(d) $M = \Sigma\{M^P : P \text{ up-set}\}$*; and*

(e) *there is a subset X of the set of up-sets in S with $M = \oplus\{M^P : P \in X\}$ for each torsion M in $Q\mathrm{rep}(S, R)$ if and only if $S\backslash\{\infty\}$ is the disjoint union of $\{P\backslash\{\infty\} : P \in X\}$.*

PROOF. Statements (a), (b), and (c) are immediate consequences of the definitions.
As for (d), if $i \in S$, then i is an element of the up-set $P_i = \{s \in S : s \geq i\}$. Hence, if $M = (M_i, f_{ij}) \in Q\mathrm{rep}(S, R)$, then $M = \Sigma\{M^P : P \text{ up-set}\}$, since $M_i = (M^P)_i$ for $P = P_i$.

Assertion (e) follows from the arguments in (d), observing that if M is torsion, then $M_\infty = 0$. $\blacksquare$

The next corollary is a reduction step in a characterization of those finite posets S with unique maximal element such that $T(M)$ is a summand of M for each $M \in Q\mathrm{rep}(S, R)$. The complete characterization is left as an exercise (Exercise 6.4.2).

Corollary 6.4.5 *Assume that S is a finite poset with unique maximal element ∞, R is a discrete valuation ring, and there is a subset X of the set of up-sets in S with $S\backslash\{\infty\}$ the disjoint union of $\{P\backslash\{\infty\} : P \in X\}$. If $M \in Q\mathrm{rep}(S, R)$, then $T(M)$ is a summand of M if and only if $T(M^P)$ is a summand of M^P for each $P \in X$.*

PROOF. If $T(M)$ is a summand of M and P is an up-set, then $T(M)^P$ is a summand of M^P. But $T(M)^P = T(M^P)$ by Proposition 6.4.4(a). Conversely, suppose that $f^P : M^P \to T(M)^P$ is a splitting of the inclusion of $T(M^P) = T(M)^P$ in M^P for each $P \in X$. By Proposition 6.4.4(e), $T(M) = \oplus\{T(M)^P : P \in X\}$ and $M_i = \oplus\{(M^P)_i : P \in X\}$ for each $i \in S\backslash\{\infty\}$. Hence, $f_i = \oplus_P f^P : M_i \to T(M)_i$ is a splitting for each $i \in S$. Define $f_\infty = 0$. Then $f = \{f_i : i \in S\} : M \to T(M)$ is a splitting of the inclusion of $T(M)$ in M, since $T(M)_\infty = 0$. $\blacksquare$

For $s \in S$, $P(s) = (P(s)_i, f_{ij}) \in Q\mathrm{rep}(S, R)$ is defined by $P(s)_i = R$ if $s \leq i$, $P(s)_i = 0$ if s is not less than or equal to i, $f_{ij} = 1$ if $s \leq i$, and $f_{ij} = 0$ otherwise. Define $P(0) = (P(0)_i, f_{ij}) \in Q\mathrm{rep}(S, R)$ by letting $P(0)_i = R$ and $f_{ij} = 1$ for each $i \in S$ and $i \leq j$ in S. Observe that each $P(s)$ is torsion-free.

The following result is given in [Butler 87], [Bautista, Martinez 79], and [Buenermann 81, 83] for the case that R is a field.

Corollary 6.4.6

(a) *Projectives in $Q\mathrm{rep}(S, R)$ are, up to isomorphism, summands of direct sums of representations of the form $P(s)$.*

(b) *Torsion-free elements of $Q\mathrm{rep}(S, R)$ are, up to isomorphism, the elements of $\mathrm{rep}(S, R)$.*

(c) *An $M \in Q\mathrm{rep}(S, R)$ is torsion-free if and only if M is a subrepresentation of a projective representation of $Q\mathrm{rep}(S, R)$.*

PROOF. (a) Each $P(s)$, hence the direct sum of $P(s)$'s, is readily seen to be projective. Moreover, if $M \in Q\mathrm{rep}(S, R)$, then there is $N \in Q\mathrm{rep}(S, R)$ with N isomorphic to a direct sum of $P(s)$'s and an onto morphism $N \to M$. Specifically, $N = \oplus\{N(s) : s \in S\}$, where $N(s) = (N(s)_i, f_{ij})$ with $N(s)_i = M_s$ if $i \geq s$, $N(s)_i = 0$ if $i \not\geq s$, $f_{ij} = 1$ if $s \leq i \leq j$, and $f_{ij} = 0$ otherwise. Notice that each $N(s)$ is isomorphic to a direct sum of copies of $P(s)$. If M is projective, then M is a summand of N.

(b) If $U \in \mathrm{rep}(S, R)$, then $f_{i\infty} : U_i \to U_\infty$ is inclusion, so that $T(U)_i = \ker f_{i\infty} = 0$ for each i. Thus, $T(U) = 0$. Conversely, if $M = (M_i, f_{ij} : i \leq j) \in Q\mathrm{rep}(S, R)$ is torsion-free, then each $f_{i\infty}$ is a monomorphism by Proposition 6.4.2(d). But $f_{j\infty} f_{ij} = f_{i\infty}$ if $i \leq j$ in S, whence each f_{ij} is a monomorphism. Hence, M is isomorphic to $U = (M_\infty, f_{i\infty}(M_i) : i \in S\backslash\{\infty\}) \in \mathrm{rep}(S, R)$, observing that each $f_{i\infty}$ has pure image in M_∞.

(c) In view of (a), Proposition 6.4.2(b), and the fact that each $P(s)$ is torsion-free, a projective representation is torsion-free. Since T preserves monomorphisms, a subrepresentation of a projective representation is also torsion-free. Conversely, if $M = (M_0, M_i) \in \mathrm{rep}(S, R)$, then, by Proposition 6.4.3, M is a subrepresentation of a cotrivial projective representation.

EXERCISES

1. Let S be a finite poset and R a discrete valuation ring. Define the *incidence algebra* RS to be the set of matrices $M = (a_{ij})$ such that $a_{ij} \in R$ if $i \leq j$ in S and $a_{ij} = 0$ otherwise.
   (a) Prove that RS is a ring with identity.
   (b) Prove that there is a fully faithful embedding $Q\mathrm{rep}(S, R) \to \mathrm{mod}\, RS$, the category of finitely generated RS-modules. Describe the image of this embedding.

2. Let S be a finite poset with unique maximal element ∞ and R a discrete valuation ring. Prove that $T(M)$ is a summand of M for each $M \in Q\mathrm{rep}(S, R)$ if and only if $S\backslash\{\infty\}$ is a forest.

NOTES ON CHAPTER 6

Descriptions of projectives, injectives, projective covers, and injective envelopes in $\mathrm{rep}(S, k)$ are given in [Gabriel 73A]. Coxeter correspondences on $\mathrm{rep}(S, k)$ are commonly defined by regarding $\mathrm{rep}(S, k)$ as a subcategory of the category $\mathrm{mod}\, kS$ of finitely generated modules over the incidence algebra kS of S and using the Auslander–Reiten correspondences DTr and TrD for these modules [Bautista, Martinez 79], [Drozd 74], and [Bünermann 81, 83]. In this chapter, the definition and development of Coxeter correspondences, in the spirit of [Gelfand, Ponomarev 70] for the case that S is an antichain, are entirely within the category rep(S, k) and without reference to any module category. Hence, they are immediately available for quasi-homomorphism categories of finite rank Butler groups, as briefly discussed in Section 7.1.

The existence of almost split sequences in rep(S, k) is derived in [Auslander, Smálo 81], as well as [Bautista, Martinez 79] and [Bünermann 81, 83] using the torsion theory of Section 6.4, from the existence of almost split sequences in mod kS [Auslander, Reiten 75]. There does not seem to be a proof of their existence entirely within the category rep(S, k).

All indecomposable representations for the critical posets S_4, $(2, 2, 2)$, $(1, 3, 3)$, $(N, 4)$, and $(1, 2, 5)$ are given in [Ringel 84] via a description of the Auslander–Reiten quiver. The list is not quite as explicit as that of Example 6.2.8 for S_4.

An intriguing aspect that is not discussed herein is the characterization of the representation type of rep(S, k) in terms of an integral quadratic form q_S, see Exercise 6.1.8. If rep(S, k) has finite representation type, then there is a one-to-one correspondence from the vectors cdn U, $U \in$ Ind(S, k), to the roots of q_S [Drozd 74]. Moreover, the critical posets can be associated with certain Dynkin diagrams. See [Ringel 84] and [Simson 92] for comprehensive discussions of the role of quadratic forms in representations of partially ordered sets and quivers and characterizations of the representation type of rep(S, k) in terms of properties of q_S.

Techniques from the theory of torsion-free abelian groups, particularly the analogue of rank-1 groups and balanced exact sequences as discussed in Chapter 7, have been carried over to rep(S, k) for a finite poset S and field k in [Arnold, Vinsonhaler 92] and [Nonxga, Vinsonhaler 97].

The representation type of the category Qrep(S, k) for a finite poset S and a field k is completely described in [Loupias 75].

7

Finite Rank Butler Groups

7.1 Projectives, Injectives, and Exact Sequences

Free groups are the projectives and divisible groups are the injectives for the category of torsion-free abelian groups [Hungerford 74]. Changing the category and the class of defining homomorphisms can change the projectives and injectives. Properties of Coxeter correspondences and almost split sequences are applied to categories of finite rank Butler groups in this section.

A group G in $B(T)$ is *projective in* $B(T)$ if whenever $f : H \rightarrow K$ is an epimorphism of groups and $g : G \rightarrow K$, then there is $h : G \rightarrow H$ with $fh = g$. The group G is *injective in* $B(T)$ if for $f : H \rightarrow K$ a monomorphism of groups in $B(T)$ with pure image and $g : H \rightarrow G$, there is $h : K \rightarrow G$ with $hf = g$.

Proposition 7.1.1 *Let T be a finite lattice of types.*

(a) *The projectives in* $B(T)$ *are* $\tau(0)$-*homogeneous completely decomposable groups,* $\tau(0)$ *the smallest element of* T.
(b) *The injectives in* $B(T)$ *are* $\tau(\infty)$-*homogeneous completely decomposable groups,* $\tau(\infty)$ *the largest element of* T.

PROOF. (a) Suppose $f : H \rightarrow K$ is an epimorphism of groups in $B(T)$, X is a rank-1 group with type $\tau(0)$, and $0 \neq g : X \rightarrow K$. Let $0 \neq x \in X$ and $0 \neq y \in H$ with $f(y) = g(x)$. If Y is the pure rank-1 subgroup of H generated by y, then $Y \in B(T)$, and so $\tau(0) \leq \text{type } Y$. Since type $X = \tau(0)$, there is some $h : X \rightarrow H$ with $fh = g$. A standard argument using injections and projections for finite direct sums [Hungerford 74] shows that a $\tau(0)$-homogeneous completely decomposable group is projective in $B(T)$.

Conversely, let G be projective in $B(T)$. Since $G \in B(T)$, there is an epimorphism $f : C \to G$, with C a completely decomposable group in $B(T)$, by Lemma 3.3.1. Because G is projective, there is $h : G \to C$ with $hf = 1_G$. Therefore, G is a summand of C, hence completely decomposable by Theorem 3.1.7(a).

It remains to show that if X is a rank-1 summand of C with type $X > \tau(0)$, then X is not projective in $B(T)$. To see this, let Y be a rank-1 group in $B(T)$ with type $Y = \tau(0) <$ type X. Choose $0 \neq x \in X$ and $0 \neq y \in Y$. Since Y is reduced, there is an integer n such that $na = x$ has no solution $a \in Y \oplus X$. Define G to be the pure subgroup of $Y \oplus X$ generated by $ny + x$. Then type $G =$ type Y and type $(Y \oplus X)/G =$ type $X = \tau$.

Now assume, by way of contradiction, that X is projective in $B(T)$. Since $(Y \oplus X)/G$ is isomorphic to X, G is a summand of $H = Y \oplus X$, say $H = G \oplus D$. Then type $D = \tau$, from which it follows that $H(\tau) = X \oplus D$. But $y \notin X \oplus D$, since $na = x$ has no solution $a \in Y \oplus X$, a contradiction.

(b) Let X be a rank-1 group with type $X = \tau(\infty)$, G and $H \in B(T)$, $f : G \to H$ a monomorphism with pure image, and $0 \neq g : G \to X$. Then $\mathbb{Q}X = \mathbb{Q} \otimes_{\mathbb{Z}} X$ is an injective abelian group, so there is $h : H \to \mathbb{Q}X$ with $hf = g$. Now, type $h(H) \in$ cotypeset $H \subseteq T$, so that type $h(H) \leq$ type X, the largest element of T. Consequently, there is $h : H \to X$ with $hf = g$. This shows that X, hence a $\tau(\infty)$-homogeneous completely decomposable group, is injective in $B(T)$.

Conversely, suppose $G \in B(T)$ is injective in $B(T)$. Then G is a pure subgroup of a completely decomposable group C in $B(T)$ by Lemma 3.3.1. Since G is injective in $B(T)$, G is a summand of C, hence completely decomposable by Theorem 3.1.7.

It now suffices to show that if Y is a rank-1 summand of G with type $Y <$ type $X = \tau(\infty)$, then Y is not injective in $B(T)$. In this case, G must be a $\tau(\infty)$-homogeneous completely decomposable group. A construction like that of (a) shows that $Y \oplus X$ has a pure rank-1 subgroup with type equal to type Y that is not a summand. Hence, Y cannot be injective in $B(T)$.

A finite rank Butler group G is *balanced projective* if whenever $f : H \to K$ is a balanced epimorphism of finite rank Butler groups H and K and $g : G \to K$, then there is $h : G \to H$ with $fh = g$. The group G is *cobalanced injective* if for H and K finite rank Butler groups, $f : H \to K$ a cobalanced monomorphism, and $g : H \to G$, there is $h : K \to G$ with $hf = g$.

Proposition 7.1.2 *For finite rank Butler groups, completely decomposable groups are the balanced projective groups and the cobalanced injective groups.*

PROOF. Suppose $f : H \to K$ is a balanced epimorphism, X is a rank-1 group with type $X = \tau$, and $0 \neq g : X \to K$. There is a pure rank-1 subgroup Y of H with type $Y \geq \tau$ and $f(Y) = g(X)$. But type $Y \leq$ type $f(Y) =$ type $X = \tau$, whence type $Y = \tau =$ type X. It follows that there is an isomorphism $h : X \to H$ with $fh = g$, and so X is balanced projective. Consequently, completely decomposable groups are balanced projective.

Conversely, if G is a balanced projective finite rank Butler group, then there is a completely decomposable group C and a balanced epimorphism $f : C \to G$, by Lemma 3.2.3(a). Since G is balanced projective, G is isomorphic to a summand of C, hence completely decomposable. This shows that completely decomposable groups are the balanced projectives.

To see that a completely decomposable group is cobalanced injective, assume that $f : H \to K$ is a cobalanced monomorphism and X is a rank-1 group with type $X = \tau$. It suffices to prove that the induced homomorphism $\mathrm{Hom}(K, X) \to \mathrm{Hom}(H, X)$, given by $h \mapsto hf$, is onto. In this case, X, hence a completely decomposable group, is cobalanced injective.

It is also sufficient to assume $H[\tau] = K[\tau] = 0$. This is so because $\mathrm{Hom}(H, X) = \mathrm{Hom}(H/H[\tau], X)$ by the definition of $G[\tau]$ and $(H/H[\tau])[\tau] = 0$ by Lemma 3.1.9(a). Similarly, $\mathrm{Hom}(K, X) = \mathrm{Hom}(K/K[\tau], X)$ and $(K/K[\tau])[\tau] = 0$.

To complete the proof that X is cobalanced injective, let $g : H \to X$ and $Y = g(H)$ with type $Y = \sigma \leq \tau$. Then $g : H/H[\sigma] \to Y$ is onto and $f : H/H[\sigma] \to K/K[\sigma]$ is a pure monomorphism, since $f : H \to K$ is cobalanced. Define $G = (Y \oplus (K/K[\sigma]))/M$, M the pure subgroup of $Y \oplus (K/K[\sigma])$ generated by $\{(g(a), -f(a)) : a \in H/H[\sigma]\}$. Then $K/K[\sigma] \to G$ is onto, from which it follows that $G[\sigma] = 0$, since $(K/K[\sigma])[\sigma] = 0$ by Lemma 3.1.9(a). As Y is a pure subgroup of G with $Y(\sigma) = Y$ and $G[\sigma] = 0$, Y is a summand of G by Proposition 3.1.15(b). Since Y is a summand of G, there is $h' : G \to Y$ with the restriction of h' to Y the identity endomorphism of Y. Define $h : K \to Y$ by $h(k) = h'((k + K[\sigma]) + M) \in h'(G) = Y$. If $a \in H$, then $hf(a) = h'((f(a) + K[\sigma]) + M) = h'(g(a) + M) = g(a)$, and so $hf = g$, as desired.

Conversely, assume that G is cobalanced injective. Then G is a cobalanced subgroup of a completely decomposable group C by Lemma 3.2.3(b). Hence, G is a summand of C, so that G is completely decomposable, as desired.

A sequence $0 \to G \xrightarrow{f} L \xrightarrow{g} K \to 0$ of groups in $B(T)_{\mathbb{Q}}$ is *exact* if f is a monomorphism, $(\ker g + \mathrm{image}\ f)/(\ker g \cap \mathrm{image} f)$ is finite, and $(\mathrm{image}\ g + K)/(\mathrm{image}\ g) \cap K$ is finite. In this case, $0 \to \mathbb{Q}G \to \mathbb{Q}L \to \mathbb{Q}K \to 0$ is an exact sequence of $\mathbb{Q}$-vector spaces.

Proposition 7.1.3 *Let $H : B(T)_{\mathbb{Q}} \to \mathrm{rep}(S_T, \mathbb{Q})$ be the category equivalence given in Theorem 3.3.2 with $H(G) = (\mathbb{Q}G, \mathbb{Q}G(\tau) : \tau \in S_T)$. Then $0 \to G \to L \to K \to 0$ is an exact sequence of groups in $B(T)_{\mathbb{Q}}$ if and only if $0 \to H(G) \to H(L) \to H(K) \to 0$ is an exact sequence in $\mathrm{rep}(S, k)$.*

PROOF. Assume that $0 \to G \xrightarrow{f} L \xrightarrow{g} K \to 0$ is an exact sequence in $B(T)_{\mathbb{Q}}$. Then $0 \to \mathbb{Q}G \to \mathbb{Q}L \to \mathbb{Q}K \to 0$ is an exact sequence of vector spaces. It is sufficent to prove that if $\tau \in S_T$, then $\mathbb{Q}G(\tau) \to \mathbb{Q}K(\tau) \to 0$ is onto. In this case, $\mathbb{Q}G(\tau) = \mathbb{Q}G \cap \mathbb{Q}L(\tau)$ and $0 \to \mathbb{Q}G(\tau) \to \mathbb{Q}L(\tau) \to \mathbb{Q}K(\tau) \to 0$ is an exact sequence for each $\tau \in S_T$, as desired.

Let X be a pure rank-1 subgroup of K with type $X \geq \tau \in S_T$ and $A = g^{-1}(X)$ the preimage of X in L. Then $A \in B(T)$, being quasi-isomorphic to a pure subgroup of $L \in B(T)$. Hence, $A = Y_1 + \cdots + Y_n$ for pure rank-1 subgroups Y_i of A with $g(Y_i) \neq 0$, type $Y_i = \sigma_i$, and type $X = \cup\{\sigma_i : 1 \geq i \geq n\} \geq \tau$. Since the lattice of all types is distributive, $\tau = \cup\{\tau \cap \sigma_i : 1 \geq i \geq n\}$. But τ is join irreducible, whence $\tau = \tau \cap \sigma_i$ and $\sigma_i \geq \tau$ for some i. Thus, $\mathbb{Q}X = \mathbb{Q}g(Y_i)$ with $Y_i \in L(\tau)$, as desired.

As for the converse, since H is a category equivalence, there is a category equivalence $H' : \mathrm{rep}(S_T, k) \to B(T)_\mathbb{Q}$ with HH' naturally equivalent to the identity functor on $\mathrm{rep}(S_T, k)$ and $H'H$ naturally equivalent to the identity functor on $B(T)_\mathbb{Q}$. It follows that if $0 \to H(G) \to H(L) \to H(K) \to 0$ is an exact sequence in $\mathrm{rep}(S, k)$, then $0 \to G \to L \to K \to 0$ is an exact sequence in $B(T)_\mathbb{Q}$.

An element τ of a finite lattice of types T is *meet-irreducible* if whenever $\tau = \delta \cap \sigma$ with $\delta, \sigma \in T$, then $\tau = \delta$ or $\tau = \sigma$.

Corollary 7.1.4 *Let T be a finite lattice of types.*

(a) *Projectives in $B(T)_\mathbb{Q}$ are direct sums of rank-1 groups X with type X a join-irreducible type in T.*

(b) *Injectives in $B(T)_\mathbb{Q}$ are direct sums of rank-1 groups X with type X a meet-irreducible type in T.*

(c) *Each $G \in B(T)$ has an projective cover $p_G : P(G) \to G$ in $B(T)_\mathbb{Q}$, where $P(G) = \oplus\{G(\tau)/G^\#(\tau) : \tau \in S_T\}$.*

(d) *Each $G \in B(T)_\mathbb{Q}$ has an injective envelope $i_G : G \to I(G)$ in $B(T)_\mathbb{Q}$, where $I(G) = \oplus\{G^*[\tau]/G[\tau] : \tau \in S_T\}$.*

PROOF. Statements (a) and (b) follow from Proposition 7.1.3 and Lemma 6.1.2, while (c) and (d) are consequences of Lemma 6.1.4 and Proposition 3.2.8.

The duality σ of Lemma 6.1.1 induces a duality for categories of Butler groups.

Theorem 7.1.5 *Let T' and T' be finite lattices of types and $\alpha : T' \to T$ a lattice anti-isomorphism. Then:*

(a) *there is a contravariant category equivalence $D_\alpha : B(T')_\mathbb{Q} \to B(T)_\mathbb{Q}$ such that $D_{\alpha-1}D_\alpha$ is naturally equivalent to the identity functor on $B(T')_\mathbb{Q}$;*

(b) *$D_\alpha(G(\tau))$ is quasi-isomorphic to $D_\alpha(G)/D_\alpha(G)[\alpha(\tau)]$ for each $\tau \in T'$;*

(c) *D_α sends balanced sequences in $B(T')_\mathbb{Q}$ to cobalanced sequences in $B(T)_\mathbb{Q}$; and*

(d) *D_α sends cobalanced sequences in $B(T')_\mathbb{Q}$ to balanced sequences in $B(T)_\mathbb{Q}$.*

PROOF. (a) Notice that α induces an anti-isomorphism $\alpha : S_{T'} \to S^T$, S^T the opposite of the poset of meet irreducible elements of T. Now, $H : B(T')_\mathbb{Q} \to \mathrm{rep}(S_{T'}, k)$ with $H(G) = (\mathbb{Q}G, \mathbb{Q}G(\tau) : \tau \in S_T)$ is an exact category equivalence by Proposition 7.1.3, and $\sigma : \mathrm{rep}(S_T, k) \to \mathrm{rep}(S^{T'}, k)$ is an exact contravariant category equivalence by Lemma 6.1.1.

Define a correspondence $F : B(T)_{\mathbb{Q}} \to \text{rep}(S^T, k)$ by $F(G) = (\mathbb{Q}G, \mathbb{Q}G[\tau]:$ $\tau \in S^T)$. Then F is a functor, as a consequence of Lemma 3.1.9(b). Now, $\mathbb{Q}G[\tau] =$ $\Sigma\{\mathbb{Q}G(\sigma): \sigma \not\le \tau\}$ and $\mathbb{Q}G(\tau) = \cap\{\mathbb{Q}G[\sigma]: \sigma \not\ge \tau\}$ for each type τ by Proposition 3.2.8. Since H is a category equivalence, a representation U in $\text{rep}(S_T, \mathbb{Q})$ is completely determined by $(\mathbb{Q}G, \mathbb{Q}G(\tau): \tau \in T)$. Hence, U is also completely determined by $(\mathbb{Q}G, \mathbb{Q}G[\tau]: \tau \in T)$. In fact, $(\mathbb{Q}G, \mathbb{Q}G[\tau]: \tau \in T)$ is completely determined by $(\mathbb{Q}G, \mathbb{Q}G[\tau]: \tau \in S^T)$, since $\mathbb{Q}G[\sigma \cap \tau] = \mathbb{Q}G[\sigma] + \mathbb{Q}G[\tau]$ by Proposition 3.2.8(d). It now follows that F is an exact category equivalence, in particular, H and F are manifestations of the same functor from $B(T)_{\mathbb{Q}}$ to $\text{rep}(T, \mathbb{Q})$.

Define D_α by $F^{-1}\sigma H : B(T')_{\mathbb{Q}} \to B(T)_{\mathbb{Q}}$. Then D_α is an exact contravariant equivalence. Since H and F are manifestations of the same functor, $D_{\alpha-1} = H^{-1}\sigma F$, whence $D_{\alpha-1}D_\alpha = (H^{-1}\sigma F)(F^{-1}\sigma H) = H^{-1}\sigma^2 H$ is naturally equivalent to the identity functor.

Assertions (b), (c), and (d) are left as exercises (Exercise 7.1.1).

Corollary 7.1.6 *Suppose T is a finite lattice of types and $G, H \in B(T)_{\mathbb{Q}}$. There are correspondences $C^+, C^- : B(T)_{\mathbb{Q}} \to B(T)_{\mathbb{Q}}$ such that:*

(a) *$C^+G = 0$ if and only if G is projective in $B(T)_{\mathbb{Q}}$ and $C^-G = 0$ if and only if G is injective in $B(T)_{\mathbb{Q}}$.*
(b) *If G has no injective summands and H has no projective summands, then $\mathbb{Q}\text{Hom}(G, C^+H)$ is isomorphic to $\mathbb{Q}\text{Hom}(C^-G, H)$.*
(c) *If G has no projective summands, then $\mathbb{Q}\text{End}\, G$ is isomorphic to $\mathbb{Q}\text{End}\, C^+G$, and if G has no injective summands, then $\mathbb{Q}\text{End}\, G$ is isomorphic to $\mathbb{Q}\text{End}\, C^-G$.*
(d) *If $G \in B(T)_{\mathbb{Q}}$ is a strongly indecomposable group, then C^+G and C^-G are strongly indecomposable groups.*

PROOF.   Apply Proposition 7.1.3 and Corollary 6.2.5.

Call $G \in B(T)_{\mathbb{Q}}$ *preprojective* if $C^{+i}G = 0$ for some $i \ge 1$ and *preinjective* if $C^{-i}G = 0$ for some $i \ge 1$. Let cdn $G = \{\text{rank}\, G, \text{rank}\, G(\tau): \tau \in S_T\}$.

Corollary 7.1.7 *Let T be a finite lattice of types. The following statements are equivalent:*

(a) *$B(T)_{\mathbb{Q}}$ has finite representation type.*
(b) *If G is a strongly indecomposable group in $B(T)_{\mathbb{Q}}$, then G is preinjective.*
(c) *If G is a strongly indecomposable group in $B(T)_{\mathbb{Q}}$, then G is preprojective.*
(d) *The quasi-endomorphism ring of G is isomorphic to $\mathbb{Q}$ for each strongly indecomposable $G \in B(T)_{\mathbb{Q}}$.*

In this case, if G is an strongly indecomposable group in $B(T)_{\mathbb{Q}}$, then G is determined up to isomorphism by cdn G.

PROOF. Theorem 6.2.6 and Proposition 7.1.3.

> **Corollary 7.1.8** *Let T be a finite lattice of types.*
>
> (a) *If $G \in B(T)_\mathbb{Q}$ is a strongly indecomposable group in $B(T)_\mathbb{Q}$ that is not projective, then there is an almost split sequence $0 \to C^+G \to B \to G \to 0$ in $B(T)_\mathbb{Q}$.*
> (b) *If G is a strongly indecomposable group in $B(T)_\mathbb{Q}$ that is not injective, then there is an almost split sequence $0 \to G \to B \to C^-G \to 0$ in $B(T)_\mathbb{Q}$.*

PROOF. Proposition 7.1.3 and Theorem 6.3.4

> **Open Question:** *Given a finite lattice T of types, is there a category $\mathbf{X}$ of abelian groups containing $B(T)_\mathbb{Q}$ and a category equivalence $\mathbf{X} \to \mathbb{Q}\mathrm{rep}(S_T, \mathbb{Q})$ that restricts to the category equivalence $B(T)_\mathbb{Q} \to \mathrm{rep}(S, \mathbb{Q})$ of Theorem 3.3.2? What is the situation for $B(T)_p$ and $\mathbb{Q}\mathrm{rep}(S_T, Z_p)$?*

EXERCISES

1. Complete the proof of Theorem 7.1.5.

2. Give explicit constructions for C^+G and C^-G, $G \in B(T)_\mathbb{Q}$. Give examples of preprojective groups that are not projective and preinjective groups that are not injective.

3. Given X a rank-1 group in $B(T)_\mathbb{Q}$:
   (a) Compute C^+X and C^-X.
   (b) Find an almost split sequence $0 \to C^+X \to B \to X \to 0$.
   (c) Find an almost split sequence $0 \to X \to B \to C^-X \to 0$.

4. For $G \in B(T)_\mathbb{Q}$ and M a subset of T, define $G(M)$ to be the pure subgroup of G generated by $\{G(\tau) : \tau \in M\}$ and $G[M] = \cap\{G[\tau] : \tau \in M\}$. Let $0 \to G \to H \to K \to 0$ be an almost split sequence in $B(T)_\mathbb{Q}$. Prove the following:
   (a) If $G(M) \neq G$, then $0 \to G(M) \to H(M) \to K(M) \to 0$ is a split exact sequence in $B(T)_\mathbb{Q}$.
   (b) If $G[M] \neq 0$, then $0 \to G/G[M] \to H/H[M] \to K/K[M] \to 0$ is a split exact sequence in $B(T)_\mathbb{Q}$.

7.2   Endomorphism Rings

If R is a ring with additive group a reduced torsion-free abelian group, then R is the endomorphism ring of G for some reduced torsion-free abelian group G of finite rank by Theorem 2.4.6. However, the G constructed in that theorem is not, in general, a Butler group. Endomorphism rings of finite rank Butler groups are described in this section. Given a ring R, R^+ denotes the additive group of R.

Proposition 7.2.1 *If G is a finite rank Butler group, then* (End G)$^+$ *is a finite rank Butler group.*

PROOF. By Corollary 3.2.4, there are completely decomposable groups C and D such that G is a pure subgroup of D and an epimorphic image of C, say $f : C \to G$. Then $\phi :$ End $G \to$ Hom (C, D), defined by $\phi(g) = gf$, is a monomorphism with pure image. To see that ϕ has pure image, suppose p is a prime, $h \in$ Hom (C, D), and $ph = gf$ for some $g \in$ End G. If $x \in C$, then $ph(x) = gf(x) \in pD \cap G = pG$. It follows that $(g/p) \in$ End G and $h = (g/p)f$. But Hom(C, D) is a completely decomposable group, since C and D are completely decomposable groups. This shows that (End G)$^+$ is a finite rank Butler group.

Proposition 7.2.2 *Assume that R^+ is a reduced torsion-free abelian group of finite rank. Then*

(a) R^+ *is a Butler group if and only if* $(NR)^+$ *is a Butler group and* $(R/NR)^+$ *is almost completely decomposable; and*

(b) *if* $(R/NR)^+$ *is a Butler group and* $\mathbb{Q}R/J\mathbb{Q}R$ *is a simple algebra, then* $(R/NR)^+$ *is homogeneous completely decomposable.*

PROOF. (a) The ring R is a subring of $\mathbb{Q}R$, a finite-dimensional $\mathbb{Q}$-algebra. Recall that $NR = R \cap J\mathbb{Q}R$, $J\mathbb{Q}R$ the Jacobson radical of $\mathbb{Q}R$.

First, assume that R^+ is a Butler group. Then $(R/NR)^+$, a homomorphic image of R^+, and $(NR)^+$, a pure subgroup of R, are Butler groups by Corollary 3.2.4. As in Lemma 2.2.7, there is a nonzero integer n with $nR^* \subseteq R/NR \subseteq R^* = R_1 \times \cdots \times R_m$, each R_i a finitely generated free S_i-module, S_i a Dedekind domain, and $\mathbb{Q}R_i$ a simple $\mathbb{Q}$-algebra. Since S_i is a domain, S_i^+ is homogeneous, so that R_i^+ is homogeneous. But $(R^*)^+$, hence each R_i^+, is a Butler group, since $(R^*)^+$ is quasi-isomorphic to the Butler group $(R/NR)^+$. Furthermore, each $(R_i)^+$ is homogeneous completely decomposable, because homogeneous Butler groups are completely decomposable by Corollary 3.2.7(a). This shows that $(R/NR)^+$ is almost completely decomposable.

Conversely, assume that $(NR)^+$ is a Butler group and $(R/NR)^+$ is almost completely decomposable. Then R^+ is quasi-isomorphic to $(NR)^+ \oplus (R/NR)^+$ by Theorem 2.4.7. Thus, R^+ is a Butler group by Corollary 3.2.4.

(b) In this case, as in (a), $(R/NR)^+$ is quasi-isomorphic to a homogeneous completely decomposable group, hence homogeneous completely decomposable.

Not every R with R^+ a Butler group is the endomorphism ring of a Butler group. For example, if R is the ring of Hamiltonian quaternions over $\mathbb{Z}_p$, then R^+ is a free $\mathbb{Z}_p$-module that is q-divisible for each prime $q \neq p$, and $\mathbb{Q}R$ is not a field, as required by (a) of the next corollary.

Corollary 7.2.3 [Arnold, Vinsonhaler 89] *Suppose $\mathbb{Q}R$ is a division algebra and R is the endomorphism ring of a finite rank Butler group.*

(a) *If R^+ is p-reduced for at most 4 primes p of $\mathbb{Z}$, then $\mathbb{Q}R$ is a field isomorphic to $\mathbb{Q}[x]/\langle g(x)^e\rangle$ for some irreducible $g(x) \in \mathbb{Q}[x]$.*
(b) *If R^+ is p-reduced for at most 3 primes p of $\mathbb{Z}$, then R is a subring of $\mathbb{Q}$.*

PROOF. (a) Since R is a subring of a division algebra, $NR = 0$ and R^+ is homogeneous completely decomposable. Hence, R^+ is not p-reduced if and only if R^+ is p-divisible. Thus, typeset $R^+ \subseteq T_4$, the finite Boolean algebra of types generated by {type $\mathbb{Z}_p : R$ is p-reduced}, since R^+ is p-reduced for at most 4 primes, say p_1, p_2, p_3, p_4.

Let G be a finite rank Butler group with End $G = R$. Then G is strongly indecomposable, since $\mathbb{Q}R$ is a division algebra. Moreover, typeset $G \subseteq T_4$. This is so because if q is a prime with $q \neq p_i$ for each i, then $qR = R$, and so $qG = G$. In particular, $G \in B(T_4)$. By Theorem 3.3.2, $B(T_4)_{\mathbb{Q}}$ is category equivalent to rep(S_4, $\mathbb{Q}$). Moreover, each indecomposable in rep(S_4, $\mathbb{Q}$) has endomorphism ring $\mathbb{Q}[x]/\langle g(x)^e\rangle$ for some irreducible $g(x) \in \mathbb{Q}[x]$ by Example 6.2.7. Thus, $\mathbb{Q}R = \mathbb{Q}$End G must also be of this form.

(b) In this case, just as in (a), if $R = $ End G for some finite rank Butler group G, then $G \in B(T_3)$ and G is strongly indecomposable. Thus, $\mathbb{Q}R = \mathbb{Q}$End $G = \mathbb{Q}$ by Example 3.3.3.

Theorem 7.2.4 [Arnold, Vinsonhaler 87] *Assume that R is a ring with additive group a finite rank Butler group. If R is p-reduced for at least five primes p, then R is the endomorphism ring of a finite rank Butler group.*

PROOF. Let $p_1, \ldots, p_5$ be five primes for which R is p-reduced. For each $1 \leq i \leq 5$, define X_i to be the subring of $\mathbb{Q}$ generated by $\{1/p_j | j \neq i\}$. By Example 1.1.7, there is

$$U = (\mathbb{Q}^m \oplus \mathbb{Q}^m, \mathbb{Q}^m \oplus 0, 0 \oplus \mathbb{Q}^m, (1+1)\mathbb{Q}^m, (1+A)\mathbb{Q}^m, (1+B)\mathbb{Q}^m) \in \text{rep}(S_5, \mathbb{Q})$$

for some $m \times m$ $\mathbb{Q}$-matrices A and B with $\mathbb{Q}R = C(A, B) = \text{End}(U)$.

Let $H = (X_1 R^+)^m \oplus \cdots \oplus (X_5 R^+)^m$, a finite rank Butler group, with $\mathbb{Q}H = (\mathbb{Q}R)^m \oplus \cdots \oplus (\mathbb{Q}R)^m$. Define $G = V \cap H$, V the row space of the $\mathbb{Q}$-matrix

$$M = \begin{pmatrix} I & 0 & I & A & B \\ 0 & I & I & I & I \end{pmatrix}.$$

Then G is a Butler group, being a pure subgroup of H. A routine matrix calculation shows that $\{(r, r) : r \in R\} \subseteq \text{End } G \subseteq \{(f, f) : f \in C(A, B)\} = (\{(r, r) : r \in \mathbb{Q}R\}$. Notice that each $X_i R^+$ is a fully invariant subgroup of H, since R is p_i-reduced for each $1 \leq i \leq 5$. It follows that if $(r, r) \in \text{End } G$, then $r \in \cap\{X_i R : 1 \leq i \leq 5\} = R$. Hence, End G is isomorphic to R, as desired.

Theorem 7.2.4 can be used to construct Butler groups with pathological direct sum decompositions just as for arbitrary torsion-free abelian groups of finite rank. For instance:

Example 7.2.5

(a) *There is a finite rank Butler group G and a prime p such that G is indecomposable at p but $(\text{End } G)_p$ is not a local ring.*

(b) *If $p \geq 5$ is a prime, then there is a finite rank Butler group H with a nonunique decomposition into indecomposables in the isomorphism at p category of finite rank Butler groups.*

(c) *There is a finite rank strongly indecomposable Butler group G with a subgroup H such that G is nearly isomorphic to H, H is a summand of $G \oplus G$, but G is not isomorphic to H.*

PROOF. (a) By Theorem 7.2.4, there is a finite rank Butler group G with End $G = \mathbb{Z}[i]$, the Gaussian integers, since $\mathbb{Z}[i]$ is a free abelian group of rank 2. The remainder of the proof is as in Example 2.1.5.

(b) A consequence of Theorem 7.2.4 and Example 2.1.11, in which case R is also a ring with finitely generated free additive group.

(c) By Theorem 7.2.4, there is a finite rank Butler group with End $G = \mathbb{Z}[(-5)^{1/2}] = R$, since R^+ is a free group of rank 2. Then R contains a non-principal ideal I with R/I finite, for example, $I = \langle 3, 1 + (-5)^{1/2} \rangle$. Since R is a Dedekind domain, I and R are in the same genus class, as in Lemma 2.2.8. Hence, $H = IG$ is a subgroup of G with G/H finite and H nearly isomorphic to G by Theorem 2.4.3.

7.3 Bracket Groups

This section consists of a brief introduction to a well-studied class of Butler groups. A *bracket group*, also called a $B^{(1)}$-*group* [Fuchs, Metelli 91] or a *Richman–Butler group* [Hill, Megibben 93], is a group $G = C/X$ with C a finite rank completely decomposable group and X a pure rank-1 subgroup of C. In this case, G is a finite rank Butler group with rank $G = \text{rank } C - 1$.

Example 3.2.1 is an example of a bracket group, as is C^+Y for a rank-1 group Y in $B(T)$; Exercise 7.1.3. In the latter case, $C^+Y = C/X$, where C is an injective group in $B(T)_Q$ and X is a rank-1 group, as a consequence of Example 6.3.6 and Proposition 7.1.3.

Bracket groups are intimately related to those almost completely decomposable groups containing a completely decomposable group of cyclic index. This relationship has yet to be investigated in any depth.

Proposition 7.3.1

(a) *If $G = C/X$ is a bracket group and n a nonzero integer, then $H = C + (1/n)X$ is an almost completely decomposable group with H/C a cyclic group isomorphic to X/nX.*

(b) *If G is an almost completely decomposable group and C is a completely decomposable subgroup of G with G/C cyclic, then G is a bracket group.*

(c) *If $G = C/X$ is a bracket group, then G is almost completely decomposable if and only if $T_{cr}(C)$ has a unique minimal element.*

PROOF. (a) The group H is almost completely decomposable. Moreover, H/C is isomorphic to $nH/nC = (nC + X)/X = X/(X \cap nC) = X/nX$, since X is a pure subgroup of C.

(b) Let $C = A_1 \oplus \cdots \oplus A_n$ be a completely decomposable subgroup of G with rank $A_i = 1$ for each i and $G = C + \mathbb{Z}x$. Write $x = (1/m)(a_1, \ldots, a_n)$ with $a_i \in A_i$ and m a nonzero integer. Then $G = (C \oplus \mathbb{Z}x)/Y$ is a bracket group, where Y is the pure rank-1 subgroup of G generated by $(mx, -a_1, \ldots, -a_n)$.

(c) Exercise 7.3.2.

Let $A = (A_1, \ldots, A_n)$ be an n-tuple of subgroups of $\mathbb{Q}$ with $1 \in A_i$ for each $i \in n^+ = \{1, 2, \ldots, n\}$. Define $G[A] = (A_1 \oplus \cdots \oplus A_n)/X$, where X is the pure rank-1 subgroup of $A_1 \oplus \cdots \oplus A_n$ generated by $(1, 1, 1, \ldots, 1)$. For example, $G[A_1, A_2] = A_1 + A_2$, since $0 \to A_1 \cap A_2 \to A_1 \oplus A_2 \to A_1 + A_2 \to 0$ is an exact sequence with $1 \in A_1 \cap A_2$.

Notice that type $X = \cap\{\text{type } A_i : 1 \leq i \leq n\}$, and if $G[A]$ is an almost completely decomposable group, then G must be completely decomposable. Call A *cotrimmed* if each $(A_i + X)/X$ is a pure subgroup of $G[A]$. For example, $G[A_1, A_2]$ is not cotrimmed unless $A_1 = A_2$.

Lemma 7.3.2 *Let* $G = G[A]$ *for some* $A = (A_1, \ldots, A_n)$.

(a) *Then A is cotrimmed if and only if* $\cap\{A_j : j \neq i\} \subseteq A_i$ *for each i.*

(b) *If* $B_i = A_i + \cap\{A_j : j \neq i\}$, *then* $B = (B_1, \ldots, B_n)$ *is cotrimmed with* $G[A] = G[B]$.

PROOF. Exercise 7.3.3.

Let $A = (A_1, \ldots, A_n)$, E a subset of n^+, and define $\tau_E = \text{type} \cap \{A_j : j \in E\}$. Given a partition $P = \{P(1), \ldots, P(m)\}$ of n^+, let $\tau(P) = \cap\{\tau_{P(i)} \cup \tau_{P(j)} : 1 \leq i < j \leq m\}$. For a type τ, two elements i and j of n^+ are τ-*equivalent* if either $i = j$ or there is a sequence $i(1), \ldots, i(n) \in n^+$ with $i(1) = i$, $i(n) = j$, and $\tau \not\leq \text{type } A_{i(k)} + A_{i(k+1)}$ for each $1 \leq k \leq m - 1$. It can be readily confirmed that τ-equivalence is an equivalence relation on n^+. For instance, if $\tau = \text{type } A_i$ for some i, then $\{i\}$ is a τ-equivalence class of n^+, since type $A_i \leq \text{type } A_i + A_j$ for each $j \in n^+$.

Lemma 7.3.3 *Let G be a bracket group. Then:*

(a) *G is quasi-isomorphic to $G[A] \oplus D$ for some completely decomposable group D and cotrimmed $A = (A_1, \ldots, A_n)$;*

(b) *typeset $G[A] = \{\tau(P) : P \text{ partition of } n^+\}$;*

(c) *$G[A]$ is quasi-isomorphic to $G[B(1)] \oplus \cdots \oplus G[B(m)]$, where each $B(i)$ is a subtuple of A_i and*

(d) *if $G = G[A]$ and $\tau \in \text{typeset } G$, then $G(\tau)$ is quasi-isomorphic to $G[B]$, where $B = (B_1, \ldots, B_r)$, B_i is a rank-1 group with type $B_i = \tau_{P_i}$ for a* *partition* $P_i = \{E(i), E(i)^c\}$, *and* $E(1), \ldots, E(r)$ *are the* τ-*equivalence classes of* n^+. *In particular,* rank $G(\tau) = r - 1$.

PROOF. (a) Write $G = C/Y$, where $C = X_1 \oplus \cdots \oplus X_n$ with rank $X_i = 1$ for each i and Y is a pure rank-1 subgroup of G. It is sufficient to assume that $0 \neq y = (x_1, \ldots, x_n) \in y$ with each $x_i \neq 0$. Let A_i be a subgroup of $\mathbb{Q}$ containing 1 and $f_i : X_i \to A_i$ an isomorphism.

Then $f = f_1 \oplus \cdots \oplus f_n : C \to C' = A_1 \oplus \cdots \oplus A_n$ is an isomorphism inducing an isomorphism $f : G \to C'/f(Y)$. If $0 \neq y = (q_1, \ldots q_n) \in f(Y)$, and X is the pure subgroup of C' generated by $(1, 1, \ldots, 1)$, then each $q_i \neq 0$ and $(q_1^{-1}, \ldots, q_n^{-1})f : G \to G[A]$ is a quasi-isomorphism. By Lemma 7.3.2, A may be assumed to be cotrimmed.

(b) Let $\pi : C_A \to G[A]$ be the canonical epimorphism with $C_A = A_1 \oplus \cdots \oplus A_n$ and ker $\pi = X = \langle(1, \ldots, 1)\rangle_*$. Recall from Section 3.2 that cosupport $G[A] = \{\text{cosupport} f : 0 \neq f \in \text{Hom}(G[A], \mathbb{Q})\}$, where cosupport $f = \{i : f\pi(A_i) = 0\}$. It follows that the maximal elements of cosupport $G[A]$ are the subsets I of n^+ with $|I| = n - 2$. For each such I write $n^+\setminus I = \{i, j\}$.

There is an epimorphism $f_I : G[A] \to G[A_i, A_j] = A_i + A_j \subseteq \mathbb{Q}$ induced by the projection of C_A onto $A_i \oplus A_j$. In particular, $f_I((a_1, \ldots, a_n) + X) = a_i - a_j$. There is a cobalanced monomorphism $\phi : G[A] \to \oplus\{f_I(G) : I \in S\} = \oplus\{A_i + A_j : 1 \leq i < j \leq n\}$ given by $\phi((a_1, \ldots, a_n) + X) = \oplus\{a_i - a_j : 1 \leq i < j \leq n\}$ (Lemma 3.2.3(b)). In particular, since the image of ϕ is pure, type $(a_1, \ldots, a_n) + X = \cap\{\text{type } a_i - a_j : 1 \leq i < j \leq n\}$.

Let $0 \neq y = (a_1, \ldots, a_n) + X$ with $a_i \in A_i$. There is an equivalence relation on n^+ defined by i y-*equivalent* to j if $a_i = a_j$. Let $P = \{P(1), \ldots, P(m)\}$ be the partition of n^+ consisting of y-equivalence classes. It follows from the definition of ϕ that type $y =$ type $\phi(y) = \tau(P)$, as desired.

(c) The proof is an induction on n. Let $\tau =$ type A_1 and $E(1) = \{1\}$, $E(2), \ldots, E(r)$ be the τ-equivalence classes of n^+. It suffices, by relabeling, to assume that $E(2) = \{2, 3, \ldots, m\}$. Define $B = (A_1, A_2, \ldots, A_m)$.

Now, $G[B]$ is a summand of $G[A]$. To see this, write $G[A] = C_A/X$ with $C_A = A_1 \oplus \cdots \oplus A_n$, $X = \langle(1, 1, \ldots, 1)\rangle_*$, and $G[B] = C_B/Y$ with $C_B = A_1 \oplus \cdots \oplus A_m$ and $Y = \langle(1, 1, \ldots, 1)\rangle_*$. A projection $f : C_A \to C_B$ induces an onto homomorphism $f : G[A] \to G[B]$. Next, define $g : G[B] \to G[A]$ by $g((a_1, \ldots, a_m) + Y) = (a_1, \ldots, a_m, a_1, \ldots, a_1) + X$. Since fg is the identity on $G[B]$, $G[B]$ is a quasi-summand of $G[A]$, provided that g is a quasi-homomorphism.

To this end, let $x \in G[B]$. By (b), type $x = \tau(P)$, where $P = \{P(1), \ldots, P(t)\}$ is the partition of m^+ given by the x-equivalence classes. The $g(x)$-equivalence classes in n^+ are $P' = \{P'(1), P(2), \ldots, P(n)\}$, where $P'(1)$ is the union of $P(1)$ and $\{i : m + 1 \leq i \leq n\}$. If $1 \leq i \leq m$ and $m + 1 \leq j \leq n$, then type $A_1 \leq$ type $A_i + A_j$, since $\tau =$ type A_1 and A_i and A_j are in different τ-equivalence classes. It follows that type $x = \tau(P) \leq$ type $g(x) = \tau(P')$. Consequently, g is a quasi-homomorphism from $G[B]$ to $G[A]$ by Theorem 3.3.2.

In particular, $G[A]$ is quasi-isomorphic to $G[B] \oplus$ ker f. However, ker $f = G[C]$, where $C = (A_{m+1}, \ldots, A_n)$. An induction on n completes the proof.

(d) Given B_i, define $f_i : B_i \to G[A]$ by $f_i(1) = (a_1, \ldots, a_n) + X$, where $a_j = 0$ if $j \notin E(i)$ and $a_j = 1$ if $j \in E(i)$. Then $f_i : B_i \to G[A](\tau)$ is a quasi-homomorphism as a consequence of (b), observing that type $f_i(1) = \text{type } B_i \geq \tau$. Now, $f = f_1 \oplus \cdots \oplus f_r : B_1 \oplus \cdots \oplus B_r \to G[A](\tau)$, and $f(1, 1, \ldots, 1) = 0$. Hence, f induces $f : G[B] \to G[A](\tau)$. If $y \in G[A](\tau)$, then type $y \geq \tau$. It follows from (b) that $y \in \text{image } f$, so that f is a quasi-epimorphism. Finally, image f has rank $r - 1$, since $E(1), \ldots, E(r)$ is a partition of n^+.

Theorem 7.3.4 *Let T be a finite lattice of types and $G = G[A]$ for some n-tuple $A = (A_1, \ldots, A_n)$ of subgroups of $\mathbb{Q}$. The following statements are equivalent:*

(a) *G is strongly indecomposable;*
(b) *rank $G(\tau) = 1$ for each $\tau = \text{type} A_i$;*
(c) *$\mathbb{Q}\text{End } G = \mathbb{Q}$.*

PROOF. (a) $\Rightarrow$ (b) If $n = 2$, then rank $G = 1$, and so rank $G(\tau) = 1$ for each $\tau = \text{type } A_i$. Now suppose $n \geq 3$, $\tau = \text{type } A_i$, and rank $G(\tau) > 1$. By Lemma 7.3.3(d), there are at least three τ-equivalence classes of n^+, $E(1) = \{\tau\}$, $E(2)$, and $E(3)$. In view of Lemma 7.3.3(c), $G[B]$ is a proper quasi-summand of G, where $B = (A_i : i \in E(1) \cup E(2))$. This is a contradiction to the assumption that G is strongly indecomposable.

(b) $\Rightarrow$ (c) If $f \in \text{End } G$, then $f(G(\tau_i)) \subseteq G(\tau_i)$ for each $\tau_i = \text{type } A_i$. Since rank $G(\tau_i) = 1$, $f : G(\tau_i) \to G(\tau_i)$ is multiplication by some $q_i \in \mathbb{Q}$. However, if $0 \neq x = (a_1, \ldots, a_n) \in X$, then $q_1 a_1 + \cdots + q_n a_n = f(x + X) = f(0) = 0 \in G$, and so $q_1 a_1 \oplus \cdots \oplus q_n a_n = q(a_1 \oplus \cdots \oplus a_n) \in C$ for some $q \in \mathbb{Q}$. Thus, $q = q_i$ for each i and $f = q \in \mathbb{Q}$, as desired.

(c) $\Rightarrow$ (a) is a consequence of Proposition 2.1.1(a), since $\mathbb{Q}\text{End } G = \mathbb{Q}$ has no nontrivial idempotents.

If $G[A]$ and $G[B]$ are strongly indecomposable, then they are isomorphic if and only if they are nearly isomorphic. This is a consequence of Theorem 2.4.3, since End $G[A]$ is a subring of $\mathbb{Q}$ by Theorem 7.3.4, hence a principal ideal domain. If $A = (A_1, \ldots, A_n)$ and $\tau \in \text{typeset } G[A]$ with rank $G(\tau) = 1$, then, by Lemma 7.3.3(d) and the distributivity of the lattice of all types, there is a subset E of n^+ with $\tau = \text{type } D_\tau$, where $D_\tau = \cap\{A_i : i \in E\} + \cap\{A_i : i \notin E\}$. Define $\delta(A) = (D_\tau : \text{rank } G(\tau) = 1)$.

Theorem 7.3.5 *Assume that T is a finite lattice of types, $A = (A_1, \ldots, A_n)$ and $B = (B_1, \ldots, B_m)$ are tuples of subgroups of $\mathbb{Q}$, and $G = G[A]$ and $H = G[B]$ are strongly indecomposable groups in $B(T)$. Then*

(a) *G and H are quasi-isomorphic if and only if rank $\Sigma\{G(\tau) : \tau \in M\} = \text{rank } \Sigma\{G(\tau) : \tau \in M\}$ for each nonempty subset M of T; and*
(b) *G and H are isomorphic if and only if they are quasi-isomorphic and there is $0 \neq q \in \mathbb{Q}$ with $q\delta(A) = \delta(B)$.*

PROOF. See [Arnold, Vinsonhaler 93] and references therein.

Theorem 7.3.6 [Goeters, Ullery, Vinsonhaler 94] *Assume that T is a finite lattice of types, $A = (A_1, \ldots, A_n)$ and $B = (B_1, \ldots, B_m)$ are tuples of subgroups of $\mathbb{Q}$, and $G = G[A]$ and $H = G[B]$ are strongly indecomposable groups in $B(T)$. Then G and H are quasi-isomorphic if and only if* rank $G(\tau) = $ rank $H(\tau)$ *for each $\tau \in T$.*

EXERCISES

1. [Goeters, Ullery 95] Let $A = (A_1, \ldots, A_n)$ for $n \geq 3$.
   (a) Prove that $G[A]$ is strongly indecomposable if and only if for each $i \in n^+$ and partition $\{\{i\}, E, F\}$ of n^+, type $A_i \nleq \tau_E \cap \tau_F$.
   (b) Prove that if for each 3-element subset $\{i, j, k\}$ of n^+, type $A_i + A_j$, type $A_j + A_k$, and type $A_j + A_k$ are pairwise incomparable, then $G[A]$ is strongly indecomposable.

2. [Fuchs, Metelli 91] Prove Proposition 7.3.1(c).

3. Prove Lemma 7.3.2.

4. Prove that if $A = (A_1, \ldots, A_n)$ is an n-tuple of subgroups of $\mathbb{Q}$, $G = G[A]$, T is a finite lattice of types, and $D : B(T)_{\mathbb{Q}} \to B(T^{op})_{\mathbb{Q}}$ is the duality given in Theorem 7.1.5, then $D(G[A]) = G(A)$, the kernel of the homomorphism $A_1 \oplus \cdots \oplus A_n \to \mathbb{Q}$ given by inclusion of the A_i's in $\mathbb{Q}$.

5. State and prove the duals of the results of Section 7.3 for groups of the form $G(A)$, as defined in the previous exercise.

NOTES ON CHAPTER 7

Projectives and injectives in $B(T)$ are characterized in [Butler 68] and [Lady 79]. Coxeter correspondences and almost split sequences for $B(T)_{\mathbb{Q}}$ have yet to find significant application. An alternative development of Coxeter correspondences is given in [Lady 83] for Butler modules over Dedekind domains.

There are scattered results on endomorphism rings of Butler groups of infinite rank; see [Arnold, Dugas 96], [Dugas, Göbel 97], and references therein. An interesting class of finite rank Butler groups, called Kravchenko groups and defined in terms of balanced projective resolutions, are investigated in [Nonxga, Vinsonhaler 96]. These classes of groups form a descending chain of subclasses of the class of finite rank Butler groups with intersection the class of finite rank completely decomposable groups. Representation analogues of Kravchenko groups are investigated in [Nonxga, Vinsonhaler 97].

There is an extensive body of literature on bracket groups, beginning with [Richman 83]. Although this is a restricted class of groups, it is rare in the theory of torsion-free abelian groups of finite rank to be able to classify groups by numerical invariants. A survey of this literature up to 1991, including a discussion of invariants and classification, can be found in [Arnold, Vinsonhaler 93]. More recent work includes [Höfling 93],

[Goeters, Ullery 91, 92, 94], [Goeters, Ullery, Vinsonhaler 94], [DeVivo, Metelli 96, 99], and [Yom 94, 98].

Cyclic regulating quotient groups, those almost completely decomposable groups G containing a regulating subgroup B with G/B cyclic, have been characterized, as noted in Notes on Chapter 5. In view of Proposition 7.3.1, cyclic regulating quotient groups are related to bracket groups. Implications of these characterizations of cyclic regulating quotient groups for the corresponding bracket groups, given in Proposition 7.3.1(a), remain unexplored.

8

Applications of Representations and Butler Groups

8.1   Torsion-Free Modules over Discrete Valuation Rings

The theory of Butler groups has, a priori, little relevance for p-local torsion-free groups. This is so because p-local Butler groups are completely decomposable. Nevertheless, isomorphism at p classes of finite rank Butler groups can be interpreted as torsion-free $\mathbb{Z}_p$-modules of finite rank (Corollary 8.1.9). Rank-1 groups correspond to purely indecomposable p-local groups, and bracket groups correspond to copurely indecomposable p-local groups.

Let R be a discrete valuation ring with prime p and quotient field Q, R^* the completion of R in the p-adic topology, and $k = R/pR$ a field. Then R^* is a discrete valuation ring with prime p and $R^*/pR^* = R/pR$ [Fuchs 70, 73]. For example, if $R = \mathbb{Z}_p$, then R^* is the ring of p-adic numbers with quotient field $\mathbb{Q}^*$, the field of p-adic rational numbers, and $k = \mathbb{Z}/p\mathbb{Z}$.

A torsion-free R-module M is *divisible* if $pM = M$ and *reduced* if M has no nonzero divisible submodules. There is a unique divisible submodule $d(M)$ of M with $M = d(M) \oplus N$ and N a reduced module. The *p-rank of M* is the R/pR-dimension of M/pM, and M^* is the completion of M in the p-adic topology. A pure submodule B of M is a *basic submodule* of M if B is a finitely generated free R-module with M/B divisible. Fundamental properties of completions [Fuchs 73] are listed in the following lemma.

Lemma 8.1.1 *If M is a reduced torsion-free R-module of finite rank, then:*

(a) *M is a pure R-submodule of M^*, M^* is a free R^*-module with R^*-rank $M^* = p$-rank M, and M^*/M is a divisible R-module;*

(b) *each R-endomorphism of M extends uniquely to an R^*-endomorphism of M^*;*
(c) *if B is a basic submodule of M, then $B^* = M^*$; and*
(d) *if N is a reduced torsion-free R-module of finite rank, then each $f \in \operatorname{Hom}_R(M, N)$ extends to a unique R^*-homomorphism $f^* : M^* \to N^*$.*

PROOF. Exercise 8.1.1.

Lemma 8.1.2 *Suppose M is a torsion-free R-module of finite rank.*

(a) *If M is not divisible, then M has a basic submodule B with* rank $B = p$-rank $M \leq$ rank M.
(b) *If p-rank $M = n$ and* rank $M = n + k$, *then M is free if and only if $k = 0$ and divisible if and only if $n = 0$.*
(c) *If $0 \to K \to M \to N \to 0$ is an exact sequence of torsion-free R-modules, then p-rank $M = p$-rank $K + p$-rank N.*

PROOF. (a) Notice that M/pM is a nonzero finite-dimensional k-vector space. Let $\{x_1 + pM, \ldots, x_n + pM\}$ be a k-basis of M/pM and define $B = Rx_1 \oplus \cdots \oplus Rx_n$, a free submodule of M. Then M/B is divisible as $B + pM = M$. To see that B is pure in M, let $x \in M$ and $px = r_1x_1 + \cdots + r_nx_n$. Since the $(x_i + pM)$'s are k-linearly independent, p divides r_i for each i. Thus, $x \in B$, as desired.

(b) Observe that M is free if and only if $B = M$, since if M is free, then M/B is both free and divisible, hence zero. Thus, by (a), M is free if and only if p-rank $M =$ rank M. Apply (a) to see that M is divisible if and only if M does not have a basic submodule, equivalently, p-rank $M = 0$.

Statement (c) follows from the fact that $0 \to K/pK \to M/pM \to N/pN \to 0$ must be an exact sequence, since the image of K is a pure submodule of M.

Let M be a finite rank torsion-free R-module. If M is reduced with p-rank $M = 1$, then M is a *purely indecomposable* module. In this case, M^* is isomorphic to R^*, and so M is isomorphic to a pure submodule of R^*. If p-rank $M = 1$, then $M = d(M) \oplus N$ with N a purely indecomposable module.

For an example of a purely indecomposable module, choose $t \in R^* \backslash R$ and let R_t be the pure submodule of R^* generated by $\{1, t\}$. Then R is a basic submodule of R_t, $R_t^* = R^*$, p-rank $R_t = 1$, and rank $R_t = 2$.

Purely indecomposable R-modules have properties reminiscent of torsion-free abelian groups of rank 1 (Lemma 3.1.1):

Proposition 8.1.3 *Suppose M and N are two purely indecomposable R-modules.*

(a) *The following statements are equivalent:*
   (i) *M and N are isomorphic.*
   (ii) *M is quasi-isomorphic to N.*
   (iii) $\operatorname{Hom}_R(M, N) \neq 0$ *and* $\operatorname{Hom}_R(N, M) \neq 0$.
   (iv) $\operatorname{Hom}_R(M, N) \neq 0$ *and* rank $M =$ rank N.

(b) *The endomorphism ring of M is a discrete valuation ring isomorphic to a pure subring of R^*.*

PROOF. (a) It suffices to assume that M and N are pure submodules of R^*. The implications (i) $\Rightarrow$ (ii) $\Rightarrow$ (iii) are routine.

(iii) $\Rightarrow$ (iv) Let $0 \neq f \in \operatorname{Hom}_R(M, N)$. Then f extends to an R^*-endomorphism f^* of R^*, and so f is multiplication by some $\alpha \in R^*$. Hence, f is a monomorphism, and rank $M = \operatorname{rank} f(M) \leq \operatorname{rank} N$. Similarly, rank $N \leq \operatorname{rank} M$, since $\operatorname{Hom}_R(N, M) \neq 0$.

(iv) $\Rightarrow$ (i) Let $0 \neq f \in \operatorname{Hom}_R(M, N)$. Since rank $M = \operatorname{rank} N$ and f is a monomorphism, $N/f(M)$ is bounded, say $rN \subseteq f(M)$ with $0 \neq r \in R$. Write $r = p^i u$ with p-height $u = 0$. Then u is a unit of R, and $p^i N \subseteq f(M) \subseteq N$. Since p-rank $N = 1$, $N/p^i N$ is a cyclic R-module. It now follows that $f(M) = p^j N$ for some j, whence M and N are isomorphic.

(b) Exercise 8.1.1.

The next proposition gives analogues of the pure fully invariant subgroups $G(\tau)$, for G a torsion-free abelian group and τ a type. Here, an element t of R^* plays the role of a type. Let M be a torsion-free module of finite rank with a basic submodule B and assume $B \subseteq M \subseteq M^*$, a free R^*-module with an R-basis of B as an R^*-basis of M^*. Let $t \in R^*$ and define $M(t) = \{m \in M : tm \in M\}$. Notice that if $t \in R$, then $M(t) = M$.

Proposition 8.1.4 *Assume that M and N are reduced torsion-free R-modules of finite rank and $t \in R^*$. Then:*

(a) *$M(t)$ is a pure submodule of M, and if $f : M \to N$ is an R-homomorphism, then $f(M(t)) \subseteq N(t)$;*
(b) *$R_t M(t) = \operatorname{Hom}(R_t, M)R_t$;*
(c) *$M(t) \cap M(s) \subseteq M(t + s)$; and*
(d) *if $s \in R^*$, then $M(t)(s) = M(st) \cap M(t)$.*

PROOF. (a) Let $m \in M$ and $pm \in M(t)$. Then $ptm \in pM^* \cap M = pM$, so that $tm \in M$. Thus, $m \in M(t)$ and $M(t)$ is pure in M. The R-homomorphism $f : M \to N$ extends to an R^*-homomorphism $f^* : M^* \to N^*$. Hence, if $m \in M$ with $tm \in M$, then $f(tm) = f^*(tm) = tf^*(m) = tf(m) \in N$. This shows that $f(M(t)) \subseteq N(t)$.

(b) If $t \in R$, then $M(t) = M$, $R_t = R$, and $RM = \operatorname{Hom}(R, M)R$. Now assume $t \in R^* \backslash R$ and let $m \in M(t)$. Then $Rm \oplus Rtm \subseteq M$, noting that $Rm \cap Rtm = 0$, since $t \notin R$. Hence, $R_t m \subseteq M$, since $R_t = \langle R \oplus Rt \rangle_*$ and M is pure in M^*. Define $f : R_t \to M$ by $f(y) = ym$, an R-homomorphism with $m = f(1) \in$ image f. Thus, $M(t) \subseteq \operatorname{Hom}(R_t, M)R_t$, and so $R_t M(t) \subseteq \operatorname{Hom}(R_t, M)R_t$.

Conversely, let $0 \neq f : R_t \to M$. Then f is a monomorphism, since R_t is purely indecomposable. Hence, image $f = R_t m \subseteq M$ for $m = f(1) \in M$. In particular, $m \in M(t)$, and so image $f \subseteq R_t M(t)$, as desired.

(c) If $m \in M$ and tm, sm are in M, then $(t + s)m$ is in M.

(d) Let $m \in M(t)(s)$. Then $m \in M(t)$ with $sm \in M(t)$, and so $tsm \in M$. This shows that $M(t)(s) \subseteq M(st) \cap M(t)$. Conversely, if $m \in M(st) \cap M(t)$, then $stm \in M$ and $tm \in M$. Hence, $m \in M(t)(s)$.

Let TF_R denote the category of reduced torsion-free R-modules of finite rank. As a consequence of the next two theorems, there is a variety of embeddings of $\mathrm{rep}(S_n, R, j)$ into TF_R.

Theorem 8.1.5 Let $A = \{t_1, \ldots, t_n\}$ be a subset of $R^* \backslash R$ such that $\{1, t_j, t_i t_j : 1 \le j \le n\}$ is a Q-independent set for each i.

(a) For each integer $j \ge 0$, there is a fully faithful functor $F_A : \mathrm{rep}(S_n, R, j) \rightarrow \mathrm{TF}_R$.

(b) The image of F_A consists of all those $M \in \mathrm{TF}_R$ such that
   (i) M has a basic submodule B with $M/(B + R_{t_1} M(t_1) + \cdots + R_{t_n} M(t_n))$ torsion;
   (ii) B is a fully invariant submodule of M; and
   (iii) $p^j B \subseteq M(t_1) + \cdots + M(t_n) \subseteq B$.

PROOF. Let $U = (U_0, U_i : i \in S_n) \in \mathrm{rep}(S_n, R, j)$ and define $M = F_A(U)$ to be the pure submodule of U_0^* generated by $U_0 + U_1 t_1 + \cdots + U_n t_n$. Given a representation morphism $f : U \rightarrow V$, define $F_A(f) = f^* : F_A(U) \rightarrow F_A(V)$, noticing that $f^*(u_i t_i) = f(u_i)t_i \in V_i t_i$ for each i. Then F_A is a faithful functor.

Observe that U_0 is a basic submodule of M, since U_0 is a free R-module, U_0 is a pure submodule of U_0^*, hence of M, and M/U_0 is divisible, since M is a pure submodule of U_0^* and U_0^*/U_0 is divisible.

Morever, $M(t_i) = U_i$ for each i. To see this, first observe that U_i is contained in $M(t_i)$. Conversely, let $m \in M(t_i)$. Then $m = r_0 u_0 + r_1 t_1 u_1 + \cdots + r_n t_n u_n$ for some $r_i \in Q$ and $u_i \in U_i$. Moreover, $t_i m = r_0 t_i u_0 + r_1 t_i t_1 u_1 + \cdots + r_n t_i t_n u_n = q_0 v_0 + q_1 v_1 t_1 + \cdots + q_n v_n t_n \in M$ for some $q_i \in Q$, $v_i \in U_i$. Fix a basis of U_0, express each u_i and v_i as a linear combination of the elements of this basis, equate coefficients, and use the hypothesis that $\{1, t_j, t_i t_j : 1 \le j \le n\}$ is Q-independent for each i to see that $m \in U_i$.

Now, $M/(U_0 + R_{t_1} M(t_1) + \cdots + R_{t_k} M(t_n))$ is torsion, since $U_i t_i \subseteq R_{t_i} M(t_i) = R_{t_i} U_i$. Moreover, $p^j U_0 \subseteq M(t_1) + \cdots + M(t_n) \subseteq U_0$, since $U_i = M(t_i)$ and $U \in \mathrm{rep}(S_n, R, j)$. Each $M(t_i)$ is fully invariant in M by Proposition 8.1.4(a). Thus, if $f \in \mathrm{End}\, M$, then $p^j f(U_0) \subseteq U_0 \cap p^j M = p^j U_0$, since U_0 is a pure submodule of M. This shows that U_0 is a fully invariant basic submodule of M.

To see that F is a full functor, let $g : F_A(U) \rightarrow F_A(V)$ and f the restriction of g to U_0, a basic submodule of $F_A(U)$. Now, $f : U_0 \rightarrow V_0$ with $f(U_i) = f(F_A(U)(t_i)) \subseteq F_A(V)(t_i) = V_i$ for each i. Thus, $f : U \rightarrow V$ is a representation morphism with $F_A(f) = g$, as desired.

Finally, let M be in TF_R satisfying the above conditions (i)–(iii). Then $U = (B, M(t_i) : i \in S_n) \in \mathrm{rep}(S_n, R, j)$, since $M(t_i)$ is a pure submodule, hence a summand, of the finitely generated free R-module B and $p^j B \subseteq M(t_1) + \cdots + M(t_n) \subseteq B$. Moreover,

$$F_A(U) = \langle B + M(t_1)t_1 + \cdots + M(t_n)t_n \rangle_* = \langle B + R_{t_1} M(t_1) + \cdots + R_{t_n} M(t_n) \rangle_* = M.$$

As a consequence of (b) of the next corollary, F_A sends rank-1 representations to purely indecomposable modules. In view of (a), not all modules in TF$_R$ are in the image of F_A. For example, if p-rank $M = n \geq 2$ and rank $M = n + 1$, then M is not in the image of F_A.

Corollary 8.1.6 *Let $F_A : \mathrm{rep}(S_n, R, j) \to \mathrm{TF}_R$ be the functor of Theorem 8.1.5.*

(a) *If $F_A(U) = M$, with $U = (U_0, U_i : i \in S_n)$, then p-rank $M = \mathrm{rank}\ U_0$ and rank $M = \mathrm{rank}\ U_0 + \mathrm{rank}\ U_1 + \cdots + \mathrm{rank}\ U_n \geq 2(p\text{-rank } M)$.*
(b) *If $U = (U_0, U_i : i \in S_n)$ with rank $U_0 = 1$, then $M = F_A(U)$ is purely indecomposable and $\mathrm{End}\ M = R$.*

PROOF. Assertion (a) follows from the definition of $F_A(U)$, since $U_0 + U_1 t_1 + \cdots + U_k t_k = U_0 \oplus U_1 t_1 \oplus \cdots \oplus U_k t_k$ by the independence condition on A and p-rank $M = \mathrm{rank}\ U_0 \subseteq \mathrm{rank}\ U_1 + \cdots + \mathrm{rank}\ U_k$.

Statement (b) is a consequence of (a) and the fact that $R = \mathrm{End}\ U$ is isomorphic to $\mathrm{End}\ F_A(U)$, since F_A is a fully faithful functor.

Recall from Section 4.2 that $\Delta(S_n, R)$ is the full subcategory of $\mathrm{rep}(S_{n+1}, R, 0)$ with objects $V = (V_0, V_1, \ldots, V_n, V_*)$ such that $V_0 = V_1 \oplus \cdots \oplus V_n$, V_* is a pure submodule of V_0, and $V_* \cap V_i = 0$ for each $1 \leq i \leq n$. Define $V_*(t_1, \ldots, t_n)$ to be the set of elements $(t_1 v_1, \ldots, t_n v_n) \in V_0^* = V_1^* \oplus \cdots \oplus V_n^*$, with $(v_1, \ldots, v_n) \in V_* \subseteq V_0 = V_1 \oplus \cdots \oplus V_k$.

Let $M \in \mathrm{TF}_R$ with basic submodule $B = B_1 \oplus \cdots \oplus B_n$. Then M is a pure submodule of $B^* = B_1^* \oplus \cdots \oplus B_n^*$. Given $(t) = \{t_1, \ldots, t_n\}$, a subset of R^*, define $M_{(t)} = \{(v_1, \ldots, v_k) \in M : (t_1 v_1, \ldots, t_k v_k) \in M\}$.

There is another uncountable family of embeddings of $\mathrm{rep}(S, R, 0)$ into TF_R.

Theorem 8.1.7 *Let $A = \{t_1, \ldots, t_n\}$ be a subset of $R^* \setminus R$ such that $\{1, t_j, t_i t_j : 1 \leq j \leq n\}$ is a Q-independent set for each i. There is a faithful embedding*

$$G_A : \mathrm{rep}(S_n, R, 0) \to \mathrm{TF}_R$$

that is full for all those $U \in \mathrm{rep}(S_n, R, 0)$ with no rank-1 summands. The image of G_A consists of all those M such that:

(i) *$M/(B + M_{(t)}(t_1, \ldots, t_n))$ is torsion, where $B = B_1 \oplus \cdots \oplus B_n$ is a fully invariant basic submodule of M with each B_i fully invariant in M; and*
(ii) *$M_{(t)}$ is contained in B.*

PROOF. In view of Lemma 4.2.3(b), it suffices to define $G_A : \Delta(S_n, R) \to \mathrm{TF}_R$. Let $V = (V_0, V_i, V_* : i \in S_n) \in \Delta(S_n, R)$, recalling that $V_0 = V_1 \oplus \cdots \oplus V_n$

and V_* is a summand of V_0. Define $M = G_A(V) = \langle V_0 + V_*(t_1, \ldots, t_k)\rangle_* \subseteq V_0^*$. Given a representation morphism $f : V \to V'$, extend $f : V_0 \to V_0'$ to an R^*-homomorphism $f^* : V_0^* \to V_0'^*$, and let $G_A(f) : G_A(V) \to G_A(V')$ be the restriction of f^* to $G_A(V)$. It follows that G_A is a faithful functor.

Now, V_0 is a basic submodule of M, since V_0 is free, V_0 is pure in V_0^*, hence in M, and M/V_0 is divisible, since M is pure in V_0^* and V_0^*/V_0 is divisible. Moreover, $M_{(t)} \subseteq B$, as a consequence of the independence conditions on A and the fact that $V_0^* = V_1^* \oplus \cdots \oplus V_n^*$.

To see that G_A is full, assume that V and V' have no proper rank-1 summands and let $g : M = G_A(V) \to M' = G_A(V')$ be an R-homomorphism. Extend g to an R^*-homomorphism $g^* : M^* = V_0^* \to (M')^* = (V_0')^*$. Then $g : M_{(t)} \to M_{(t)}'$: with $M_{(t)} = V_*$. Since V has no proper rank-1 summands, each $V_i/\pi_i(V_*)$ must be torsion; otherwise, there is a cyclic summand of V_i that gives rise to a representation summand of V. It now follows that $g : V_i \to V_i'$ for each i, since $g : V_* \to V_*'$. Now, $V_0 = V_1 \oplus \cdots \oplus V_k$, so letting $f = g$ gives a representation morphism $V \to V'$ with $G_A(f) = g$, as desired.

Finally, let M be as described in the statement of the theorem. Then $V = (B, B_i, M_{(t)} : i \in S_n) \in \Delta(S_n, R)$, noting that $M_{(t)}$ is pure in B and $M_t \cap B_i = 0$ for each i. Hence, $G_A(V) = \langle B + M_{(t)}(t_1, \ldots, t_n)\rangle_* = M$, as desired.

A torsion-free R-module of finite rank with no free summands is *copurely indecomposable* if rank $M = p$-rank $M + 1$. Observe that if M is a finite rank torsion-free module with rank $M = p$-rank $M + 1$, then $M = L \oplus F$, where F is a free R-module and L is copurely indecomposable.

Corollary 8.1.8 *Let* $G_A : \mathrm{rep}(S_n, R, 0) \to \mathrm{TF}_R$ *be the functor of Theorem 8.1.7.*

(a) *If* $M = G_A(U)$, *then* p-rank $M = $ rank $U_1 + \cdots + $ rank U_n *and* rank $M = 2\,(p\text{-rank } M) - $ rank $U_0 \le 2(p\text{-rank } M)$.

(b) *Let* $U = (U_0, U_i : i \in S_n) \in \mathrm{rep}(S_n, R, 0)$ *with* $n \ge 2$, $U_0 = R^{n-1}$, $U_1 = R \oplus 0 \oplus \cdots \oplus 0, \ldots, U_{n-1} = 0 \oplus 0 \oplus \cdots \oplus R$, *and* $U_n = R(1, 1, \ldots, 1)$. *Then* $M = G_A(U)$ *is copurely indecomposable with* End $M = R$.

PROOF. Assertion (a) is a consequence of the definition of G_A.

(b) Apply the functor of Lemma 4.2.3(b) to see that $U \to V$, where $V = (R^n = R \oplus \cdots \oplus R, R \oplus 0 \oplus \cdots \oplus 0, \ldots, 0 \oplus \cdots \oplus R, R(1, \ldots, 1)) \in \Delta(S_n, R)$. By (a), $M = G_A(V) = \langle R^n, R(t_1, \ldots, t_n)\rangle_*$ has p-rank n, rank $n+1$, and endomorphism ring R. In particular, M is copurely indecomposable.

Corollary 8.1.9 *Let* T *be a finite lattice of types with* $S_T = S_n$ *and* $A = \{t_1, \ldots, t_k\}$ *a subset of* $\mathbb{Z}_p^* \backslash \mathbb{Z}_p$ *such that* $\{1, t_j, t_i t_j, : 1 \le j \le k\}$ *is a* $\mathbb{Q}$-*independent set for each* i.

(a) *There is a fully faithful embedding* $F_A : B(T, j)_p \to \mathrm{TF}$ *given by*

$$F_A(G) = \langle G_p + G(\tau_1)_p t_1 + \cdots + G(\tau_k)_p t_k \rangle_* \subseteq G_p^*.$$

(b) *There is a faithful embedding $G_A : B(T, 0)_p \to$ TF that is full for groups H with no proper rank-1 summands given by*

$$G_A(H) = \langle H(\tau_1)_p \oplus \cdots \oplus H(\tau_k)_p + H_{*p}(t_1, \dots, t_k) \rangle_*$$

$$\subseteq H(\tau_1)_p^* \oplus \cdots \oplus H(\tau_k)_p^*,$$

where H_* is the kernel of $H(\tau_1) \oplus \cdots \oplus H(\tau_k) \to H$.

PROOF. A consequence of Corollary 4.3.2, Theorem 8.1.7, and Corollary 8.1.8.

Indecomposable modules in the image of F_A and G_A have wild modulo p representation type if $|A| \geq 3$, since rep$(S_3, R, 0)$ has wild modulo p representation type by Example 4.2.4.

Following is a complete list of indecomposable modules in the image of F_A and the image of G_A for the case that $|A| = 2$. This list arises from a complete list of indecomposables in rep$(S_2, \mathbb{Z}_p, j)$; Example 4.1.2.

Corollary 8.1.10 *Let T be a finite lattice of types with $S_T = \{\tau_1, \tau_2\} = S_2$ and $A = \{t_1, t_2\}$ a subset of $\mathbb{Z}_p^* \backslash \mathbb{Z}_p$ such that $\{1, t_1, t_2 t_1, t_2^2, t_1^2\}$ is a $\mathbb{Q}$-independent set for each i.*

(a) *If $G \in B(T, j)$ and $M = F_A(G)$ is indecomposable, then M is isomorphic to:*

  (i) *$\mathbb{Z}_p$ if M has p-rank 1 and rank 1;*

  (ii) *$\langle \mathbb{Z}_p, \mathbb{Z}_p t_1 \rangle_*$ or $\langle \mathbb{Z}_p, \mathbb{Z}_p t_2 \rangle_*$ if M has p-rank 1 and rank 2;*

  (iii) *$\langle \mathbb{Z}_p, \mathbb{Z}_p t_1, \mathbb{Z}_p t_2 \rangle_* \subseteq \mathbb{Z}_p^*$ if M has p-rank 1 and rank 3; or*

  (iv) *$\langle B, \mathbb{Z}_p t_1 \oplus 0, 0 \oplus \mathbb{Z}_p t_2 \rangle_* \subseteq B^*$ if M has p-rank 2 and rank 4, where $B = \mathbb{Z}_p \oplus \mathbb{Z}_p + \mathbb{Z}_p(1, 1)/p^j \subseteq \mathbb{Q} \oplus \mathbb{Q}$.*

(b) *If $H \in B(T, 0)$ and $M = G_A(H)$ is indecomposable, then M is isomorphic to:*

  (i) *$\mathbb{Z}_p$ if M has p-rank 1 and rank 1; or*

  (ii) *$\langle \mathbb{Z}_p \oplus \mathbb{Z}_p + \mathbb{Z}_p(t_1, t_2) \rangle_* \subseteq \mathbb{Z}_p^* \oplus \mathbb{Z}_p^*$ if M has p-rank 2 and rank 3.*

PROOF. Exercise 8.1.2.

The next corollary demonstrates a relationship between rank-1 groups in $B(T)$ and purely indecomposable p-local groups and between bracket groups and copurely indecomposable p-local groups.

Corollary 8.1.11 *Assume that T is a finite lattice of types with $S_T = S_n = \{\tau_1, \dots, \tau_n\}$.*

(a) *If $G \in B(T, j)$ has rank 1, then $F_A(G)$ is purely indecomposable.*

(b) *If $H = G[X_1, \dots, X_n]$ is a bracket group, where each X_i is a rank-1 group with type $X_i = \tau_i$, then $G_A(H)$ is copurely indecomposable.*

PROOF. Exercise 8.1.2.

Open Question: *Can the results of this section be extended to functorial embeddings of* rep(S, R, j) *for an arbitrary finite poset S? Can each reduced p-local torsion-free group of finite rank be so represented?*

EXERCISES

1. Prove Lemma 8.1.1 and Proposition 8.1.3(b).

2. Confirm Corollaries 8.1.10 and 8.1.11.

3. (a) Show that if M is a purely indecomposable R-module of rank 2, then M is isomorphic to M_t for some $t \in R^*$.
    (b) Compute the endomorphism ring of M_t.
    (c) [Richman 68] Find necessary and sufficient conditions in terms of s and t for M_s and M_t to be isomorphic.

4. [Arnold 72] Let M be a copurely indecomposable module with p-rank n and rank $n + 1$.
    (a) Prove that QEnd M is a subfield of the p-adic completion of the quotient field Q of R.
    (b) Give an example to demonstrate that quasi-isomorphic copurely indecomposable modules need not be isomorphic. Why must such an example have rank greater than or equal to 3?
    (c) Prove that if N is another copurely indecomposable module, then M is quasi-isomorphic to N if and only if the nth exterior power of M is isomorphic to the nth exterior power of N.

8.2   Finite Valuated Groups

There are correspondences between finite valuated p-groups and representations of finite posets over the field $\mathbb{Z}/p\mathbb{Z}$ that can be used to determine the representation type of some categories of finite valuated groups. These correspondences also give rise to correspondences between indecomposable finite rank Butler groups and indecomposable finite valuated p-groups with torsion-free groups of rank 1 corresponding to cyclic valuated p-groups.

Let p be a prime. An abelian p-group G is a *valuated p-group* if there is a mapping v from G to the ordinals together with ∞ such that $v(0) = \infty$, $v(px) > v(x)$, $v(rx) = v(x)$ if r is an integer prime to p, and $v(x + y) \geq \min\{v(x), v(y)\}$ for each $x, y \in G$. Here, it is agreed that $\infty > \infty$.

As an example of a valuated p-group, let G be a subgroup of a p-group H and define $v(x)$ to be the p-height of x in H for each $x \in G$. Then G is a valuated p-group, as a consequence of properties of the p-height function on H. In fact, this is the only example, since if G is a valuated p-group with valuation v, there is a p-group H such that G is a subgroup of H and $v(x)$ is the p-height of x in H for each $x \in G$ [Richman, Walker 79].

The class of finite valued p-groups forms a category. Morphisms, called *valuated homomorphisms*, are group homomorphisms $f : G \rightarrow H$ with $v(x) \leq v(f(x))$ for each $x \in G$. This category is an additive category; the direct sum of G and H is the group direct sum $G \oplus H$ with $v(x, y) = \min\{v(x), v(y)\}$ for $x \in G, y \in H$ [Richman, Walker 79]. As will be seen, indecomposable valuated groups need not be indecomposable as groups.

Let G be a finite valuated p-group and $x \in G$. The set $T(x, G) = \{y \in G : p^n y = x,$ some $n \geq 0\}$ is a finite poset, where $y \leq y'$ if $y = p^m y'$ for some $m \geq 0$. Then $T(x, G)$ is a tree with unique minimal element x, called a *root* of T. If $y, y' \in T(x, G)$ with $y < y'$, then $v(y') < v(y)$. Hence, $T(x, G)$ is an example of a *finite valuated tree*, a rooted tree T with each vertex y of T labeled by $v(y)$, an ordinal, or ∞, such that if $y, y' \in T$ with $y < y'$, then $v(y') < v(y)$.

Valuated trees play the same role for valuated p-groups as height sequences do for torsion-free groups. The point is that divisibility by integers is unique for a torsion-free group G, so that there is no need to record those y for which $ny = x$. However, divisibility by p is not unique for p-groups. Thus, vertices of the finite valuated tree $T(x, G)$ record the elements y of G for which $p^n y = x$; in addition, these vertices are labeled by their values.

For finite valuated trees S and T, define $S \leq T$ if there is a function $f : S \rightarrow T$ such that if $y < y'$ in S, then $f(y) < f(y')$ in T and $v(y) \leq v(f(y))$ for each $y \in S$. Two finite valuated trees S and T are *equivalent* if $S \leq T$ and $T \leq S$. If S is a finite valuated tree and $y \in S$, then length y is the length of a chain of elements from the root of S to s.

Lemma 8.2.1 *The set of equivalence classes $[T]$ of finite valuated trees forms a lattice. Specifically:*

(a) $[S] \leq [T]$ *if there is an S' equivalent to S and a T' equivalent to T with $S' \leq T'$.*
(b) $[S] \cap [T] = [U]$, *where $U = \{(s, t) \in S \times T : \text{length}(s) = \text{length}(t)\}$ is a finite valuated tree with* root $U = (\text{root } S, \text{root } T)$ *and $v(s, t) = \min\{v(s), v(t)\}$.*
(c) $[S] \cup [T] = [V]$, *where V, the disjoint union of S and T with the roots of S and T identified, is a finite valuated tree with $v(\text{root } V) = \max\{v(\text{root } S), v(\text{root } T)\}$ and the values of all other elements unchanged.*

PROOF. Exercise 8.2.1.

If x is an element of a finite valuated group G, then v-height x is the equivalence class of the valuated tree $T(x, G)$. For an equivalence class $[T]$ of valuated trees, define $G(T) = \{x \in G : v\text{-height } x \geq [T]\}$.

Lemma 8.2.2 *Suppose G and H are finite valuated p-groups.*

(a) *If $f : G \rightarrow H$ is a valuated homomorphism, then $f(G(T)) \subseteq H(T)$ for each finite valuated tree T.*
(b) *If $[S]$ and $[T]$ are equivalence classes of finite valuated trees with $[S] \leq [T]$, then $G(T) \subseteq G(S)$.*

PROOF. Statement (a) follows from the fact that $v(x) \leq v(f(x))$ for each $x \in G$, and (b) is clear.

The machinery is now in place for the finite valuated p-group analogue of the category of finite rank Butler groups with typeset contained in a finite lattice of types. Let L be a finite sublattice of the lattice of equivalence classes of valuated trees. An L-*group* is a finite valuated p-group G with v-height $x \in L$ for each $x \in G$.

Each finite valuated p-group G is an L-group for L the finite lattice generated by $\{v$-height $x : x \in G\}$. Define J_L to be the poset of join irreducible elements of L with reverse order. For a p-group G, and integer $j \geq 1$, $G[p^j] = \{x \in G : p^j x = 0\}$. Notice that if $f : G \to H$ is a group homomorphism, then $f(G[p^j])$ is contained in $H[p^j]$.

Theorem 8.2.3 *Let L be a finite sublattice of the lattice of equivalence classes of finite valuated trees and $j \geq 1$ an integer.*

(a) *There is a functor F_j from the category of L-groups to $\mathrm{rep}(J_L, \mathbb{Z}/p^j\mathbb{Z})$ given by $F_j(G) = (G[p^j], G(T)[p^j] : [T] \in J_L)$.*
(b) *If $U = (U_0, U_i : i \in J_L) \in \mathrm{rep}(J_L, \mathbb{Z}/p\mathbb{Z})$, then there is an L-group G with $F_1(G) = U$.*

PROOF. (a) For L-groups G and H and $f : G \to H$ a valuated homomorphism, define $F_j(f) : G[p^j] \to H[p^j]$. Then $F_j(f)$ is a representation morphism by Lemma 8.2.2(b). It now follows that F_j is a functor.

(b) Let $n = |S|$ and m the $\mathbb{Z}/p\mathbb{Z}$-dimension of U_0. Define G to be the direct sum of m copies of $\mathbb{Z}/p^n\mathbb{Z}$ with $U_0 = G[p]$. Write $S = \{1, 2, \ldots, n\}$, so that $U = (U_0, U_1, \ldots, U_n)$.

There is a valuation v of G given by $v(0) = \infty$, $v(x) = 2(n-k) + 1$ if order $x = p^k$ with $k \geq 1$ and $p^{k-1}x \in U_k$, and $v(x) = 2(n-k)$ if order $x = p^k$ with $k \geq 1$ and $p^{k-1}x \notin U_k$. To see that v is a valuation, notice that if $x, y \in G$ with order $y <$ order x, then $v(y) > v(x)$. Thus, $v(px) > v(x)$, since order $px <$ order x. Moreover, $v(x+y) \geq \min\{v(x), v(y)\}$, since order $x+y \leq \max\{$order x, order $y\}$.

If order $x + y =$ order $x =$ order $y = k$, then $v(x), v(y)$, and $v(x + y)$ are either $2(n - k)$ or $2(n - k) + 1$. If $v(x) = v(y) = 2(n - k)$, then $p^{k-1}x$ and $p^{k-1}y$ are elements of U_k, whence $p^{k-1}(x + y) \in U_k$ and $v(x + y) = 2(n - k)$. The other cases are routine. Finally, if $\gcd(r, p) = 1$, then $v(rx) = v(x)$, since $p^{k-1}x \in U_k$ if and only if $p^{k-1}rx \in U_k$.

For each $k \in S$, define a chain $T_k = \{t_1 > \cdots > t_n\}$ with $v(t_i) = 2i - 2$ if $U_k \cap U_{n-i+1} = 0$ and $v(t_i) = 2i - 1$ if $U_k \cap U_{n-i+1} \neq 0$. It remains to prove that $U_k = G(T_k)[p]$ for each k, in which case $F(G) = U$.

First let $0 \neq x \in U_k$. Then order $x = p$ so that $v(x) = 2n - 2$ if $x \notin U_1$ and $v(x) = 2n - 1$ if $x \in U_1$. Thus, $v(t_n) \leq v(x)$. Similarly, $T(x, G) \supseteq T_k$, so that v-height $x \geq T_k$ and $x \in G(T_k)[p]$. Therefore, $U_k \subseteq G(T_k)[p]$.

Conversely, let $x \in G(T_k)[p]$. Since $T(x, G) \supseteq T_k$, there is $y \in G$ with $p^{k-1}y = x$ and $v(y) \geq v(t_{n-k+1})$. Then $x \in U_k$; otherwise, $v(y) = 2(n - k) <$

$v(t_{n-k+1}) = 2(n - k + 1) - 1 = 2(n - k) + 1$, since $U_k \cap U_{n-(n-k+1)+1} = U_k \neq 0$. This shows that $G(T_k([p])$ is contained in U_k, as desired.

If U is an indecomposable representation and G is an L-group with $F_j(G) = U$, then G must be an indecomposable as a valued group. This observation allows an explicit construction of indecomposable valued L-groups from a given representation.

Example 8.2.4 *Assume $J_L = S_3$. The only indecomposable L-groups constructed in Theorem 8.2.3(b) are:*

(i) $G = (\mathbb{Z}/p^3\mathbb{Z})x$ with $(v(x), v(px), v(p^2x)) = (0, 2, 4), (1, 2, 4), (0, 3, 4), (0, 2, 5), (1, 3, 4), (1, 2, 5), (0, 3, 5), \text{ or } (1, 3, 5)$; and

(ii) $G = (\mathbb{Z}/p^3\mathbb{Z})a \oplus (\mathbb{Z}/p^3\mathbb{Z})b$ with $(v(a), v(pa), v(p^2a)) = (1, 2, 4)$, $(v(b), v(pb), v(p^2b)) = (0, 3, 4)$ and $(v(a + b), v(p(a + b)), v(p^2(a + b))) = (0, 2, 5)$.

PROOF. Via Example 1.1.5, the only indecomposable representations in rep(S_3, k), $k = \mathbb{Z}/p\mathbb{Z}$, are the 1-dimensional representations and a 2-dimensional representation. The groups in (i) correspond to the 1-dimensional representations $(k, 0, 0, 0)$, $(k, k, 0, 0), (k, 0, k, 0), (k, 0, 0, k), (k, k, 0), (k, k, 0, k), (k, 0, k, k)$, and (k, k, k, k), respectively, Similarly, the group in (ii) corresponds to the 2-dimensional representation $(k \oplus k, k \oplus 0, 0 \oplus k, (1 + 1)k)$.

If G is an L-group with $F_1(G) = U = (U_0, U_i : i \in S)$, as constructed in Example 8.2.4, then $p^n G = 0$, where n is the number of elements of S. Moreover, $v(x) \leq 2n + 1$ for each $x \in G$.

Example 8.2.5 *There are infinitely many isomorphism classes of indecomposable p^4-bounded finite valued groups G with $v(x) \leq 9$ for each $x \in G$.*

PROOF. Apply Theorem 8.2.3 and the fact that rep(S_4, $\mathbb{Z}/p\mathbb{Z}$) has infinite representation type; Example 1.1.6.

Example 8.2.6 *The category of p^5-bounded finite valued groups G with $v(x) \leq 11$ for each $x \in G$ has wild representation type.*

PROOF. A consequence of Theorem 8.2.3 and the fact that rep(S_5, $\mathbb{Z}/p\mathbb{Z}$) has wild representation type; Example 1.4.1.

The connection between finite rank Butler groups and finite valued p-groups is intriguing. A group $G \in B(T)_p$ is sent to $U_G = (G_p, G(\tau)_p : \tau \in S_T) \in$ rep(S_T, $\mathbb{Z}_p$). If L is a finite lattice of equivalence classes of finite valued trees with $J_L = S_T$ as posets, then $G' = (G/pG, (G(\tau) + pG)/pG : \tau \in S_T) \in$

rep(J_L, $\mathbb{Z}/p\mathbb{Z}$). Moreover, there is an L-group H with $F_1(H) = G'$ by Theorem 8.2.3. Consequently, $G \to H$ is a correspondence from $B(T)$ to L-groups such that if G is indecomposable at p, then H is indecomposable as a valuated group. Moreover, rank $G = 1$ if and only if H is a cyclic group.

Open Questions:
1. *Do the functors of Theorem 8.2.3 and Proposition 8.2.4 reflect isomorphisms? preserve indecomposables?*
2. *Is Theorem 8.2.3(b) true for all $j \geq 1$?*

EXERCISES

1. Prove Lemma 8.2.1.

2. Prove that there are infinitely many isomorphism classes of finite valuated p-groups G that are p^3-bounded with $v(x) \leq 13$ for each $x \in G$.

NOTES ON CHAPTER 8

There is a duality on the quasi-homomorphism category of finite rank torsion-free modules over a discrete valuation ring that sends purely indecomposable modules to copurely indecomposable modules [Arnold 72]. A more abstract version of this duality is given in [Lady 83] for Butler modules over a Dedekind domain R.

The representation type for subcategories of TF_R, defined in terms of the Q-dimension of a splitting field of a finite rank torsion-free R-module, is partially computed in [Lady 77B, 80A]. These computations employ representations of species [Ringel 76] instead of representations of partially ordered sets. Other papers on torsion-free modules of finite rank over discrete valuation rings include [Turgi 77] and [Franzen 83]. There is an extensive and evolving body of literature on torsion-free modules over valuation domains; see [Fuchs, Salce 85] and references therein.

Section 8.2 is a very elementary introduction to valuated groups. The seminal paper on the subject is [Richman, Walker 79]. Other embeddings from categories of representations of finite posets over $\mathbb{Z}/p\mathbb{Z}$ into the category of finite valuated p-groups are given in [Arnold, Richman, Vinsonhaler 91]. These embeddings are functorial and send rank-1 groups to simply presented valuated groups, but the functors are not additive functors.

It is evident in [Beers, Hunter, Walker 83] that classification of finite valuated p-groups is intimately related to the solution of matrix problems for representations of finite partially ordered sets over $\mathbb{Z}/p^j\mathbb{Z}$. A classification of valuated p-groups with values less than or equal to 4 is given in [Richman, Walker 99]. Each such group is the valuated direct sum of a short list of indecomposable finite valuated p-groups. This parallels the fact that if rep(S, k) has finite representation type, then a representation of arbitrary dimension is a direct sum of finite-dimensional indecomposable representations.

On the other hand, there are arbitrarily large valuated p-groups with values less than or equal to 5. The question of whether or not the category of finite valuated p-groups with values less than or equal to 5 has wild representation type remains unresolved. Just as for representations of finite posets over discrete valuation rings, a suitable definition of tame representation type remains elusive.

References

[Albrecht 83] Albrecht, U. Endomorphism rings and A-projective torsion-free abelian groups, *Abelian Group Theory*, Lect. Notes in Math. 1006, Springer-Verlag, New York, 1983, 209–252.

[Albrecht 89] Albrecht, U. Endomorphism rings and a generalization of torsion-freeness and purity, *Comm. in Alg.* 17 (1989), 1101–1135.

[Albrecht 91] Albrecht, U. On the quasi-splitting of exact sequences, *J. Alg.* 144 (1991), 344–358.

[Albrecht, Goeters 89] Albrecht, U. and Goeters, P. A dual to Baer's lemma, *Proc. Amer. Math. Soc.* 105 (1989), 817–826.

[Albrecht, Goeters 94] Albrecht, U. and Goeters, P. Pure subgroups of A-projective groups, *Acta. Math. Hungar.* 65 (1994), 217–227.

[Albrecht, Goeters, Faticoni, Wickless 97] Albrecht, U; Goeters, P.; Faticoni, T.; and Wickless, W., Subalgebras of rational matrix algebras, *Acta Math Hungar* 74 (1997).

[Albrecht, Goeters 98] Albrecht, U. and Goeters, P. Butler theory over Murley groups, *J. Alg.* 200 (1998), 118–133.

[Albrecht, Hill 87] Albrecht U. and Hill, P. Butler groups of infinite rank and axiom 3, *Czech. Math. J.* 37 (1987), 293–309.

[Anderson, Fuller 74] Anderson, F.W. and Fuller, K.R. *Rings and Categories of Modules*, Graduate Texts in Mathematics, Springer-Verlag, New York, 1974.

[Ara, Goodearl, O'Meara, Pardo 98] Ara, P., Goodearl, K., and O'Meara, K. Separative cancellation for projective modules over exchange rings, *Israel J. Math.* 105 (1998), 105–137.

[Arnold 72] Arnold, D.M. A duality for torsion free modules of finite rank over a discrete valuation ring, *Proc. London Math. Soc.* (3) 24 (1972), 204–216.

[Arnold 73] Arnold, D.M. A class of pure subgroups of completely decomposable abelian groups, *Proc. Amer. Math. Soc.*, 41 (1973), 37–44.

[Arnold 77] Arnold, D.M. Genera and decompositions of torsion free modules, *Abelian Groups*, Lect. Notes in Math. 616, Springer-Verlag, New York, 1977, 197–218.

[Arnold 81] Arnold, D.M. Pure subgroups of finite rank completely decomposable groups, *Abelian Group Theory*, Lect. Notes in Math. 874, Springer-Verlag, New York, 1981, 1–31.

[Arnold 82] Arnold, D.M. *Finite Rank Torsion-Free Abelian Groups and Rings*, Lect. Notes in Math. 931, Springer-Verlag, New York, 1982.

[Arnold 86] Arnold, D.M. Notes on Butler groups and balanced extensions, *Bull. U.M.I.* (6) 5-A (1986), 175–184.

[Arnold 89] Arnold, D.M. Representations of partially ordered sets and abelian groups, *Abelian Group Theory*, Contemporary Math. 87 (1989), 91–109.

[Arnold 95] Arnold, D.M. Near isomorphism of Butler groups and representations of finite posets, *Abelian Groups and Modules*, Kluwer, Boston, 1995, 29–40.

[Arnold, Dugas 93A] Arnold D.M. and Dugas, M. Block rigid almost completely decomposable groups and lattices over multiple pullback rings, *J. Pure and Appl. Alg.* 87 (1993), 105–121.

[Arnold, Dugas 93B] Arnold D.M. and Dugas, M. Butler groups with finite typesets and free groups with distinguished subgroups, *Comm. Alg.* 21 (1993), 1947–1982.

[Arnold, Dugas 95A] Arnold D.M. and Dugas, M. Locally free finite rank Butler groups and near isomorphism, *Abelian Groups and Modules*, Kluwer, Boston, 1995, 41–48.

[Arnold, Dugas 95B] Arnold, D.M. and Dugas M. Representations of finite posets and near-isomorphism of finite rank Butler groups, *Rocky Mountain J. Math.* 25 (1995), 591–609.

[Arnold, Dugas 96] Arnold, D.M. and Dugas, M. A survey of Butler groups and the role of representations, *Abelian Groups and Modules*, Lect. Notes in Pure and Appl. Math. 182, Marcel Dekker, New York, 1996, 1–13.

[Arnold, Dugas 97] Arnold, D.M. and Dugas, M. Representation type of finite rank Butler groups, *Colloquium Math.* 74 (1997), 299–320.

[Arnold, Dugas 98] Arnold D.M. and Dugas, M. Representation type of posets and finite rank almost completely decomposable groups, *Forum Math.* 10 (1998), 729–749.

[Arnold, Dugas 99] Arnold, D.M. and Dugas, M. Finite rank Butler groups with small typeset, *Abelian Groups and Modules*, Birkhäuser, Boston, 1999, 107–120.

[Arnold, Dugas, Rangaswamy 98] Arnold, D.M., Dugas, M., and Rangaswamy, K.M. Torsion-free abelian groups with semi-perfect quasi-endomorphism rings, preprint.

[Arnold, Lady 75] Arnold, D.M., and Lady, L. Endomorphism rings and direct sums of torsion-free abelian groups, *Trans. Amer. Math. Soc.*, 211 (1975), 225–237.

[Arnold, Hunter, Richman 80] Arnold, D.M., Hunter, R., and Richman, F. Global Azumaya theorems in additive categories, *J. Pure and Appl. Algebra*, 16 (1980), 223–242.

[Arnold, Murley 75] Arnold, D.M., and Murley, C.E. Abelian groups A, such that Hom(A, −) preserves direct sums of copies of A, *Pac. J. Math.*, 56 (1975), 7–20.

[Arnold, Richman 92] Arnold D. M. and Richman, F. Field independent representations of partially ordered sets, *Forum Math.* 4 (1992), 349–357.

[Arnold, Richman, Vinsonhaler 91] Arnold, D.M., Richman, F., and Vinsonhaler, C. General p-valuations of abelian groups, *Comm. in Alg.* 19 (11) (1991), 3075–3088.

[Arnold, Vinsonhaler 81] Arnold, D.M. and Vinsonhaler, C. Pure subgroups of finite rank completely decomposable groups II, *Abelian Group Theory*, Lect. Notes in Math. 1006, Springer Verlag, 1981, 97–143.

[Arnold, Vinsonhaler 87] Arnold D. and Vinsonhaler, C. Endomorphism rings of Butler groups, *J. Austral. Math. Soc.* 42 (1987), 322–329.

[Arnold, Vinsonhaler 89] Arnold, D. and Vinsonhaler, C. Quasi-endomorphism rings for a class of Butler groups, *Abelian Group Theory*, Contemporary Math. 87 (1989), 91–110.

[Arnold, Vinsonhaler 92] Arnold, D.M. and Vinsonhaler, C. Invariants for classes of indecomposable representations of finite posets, *J. Alg.* 147 (1992), 245–264.

[Arnold, Vinsonhaler 93] Arnold D.M. and Vinsonhaler, C. Finite rank Butler groups: A survey of recent results, *Abelian Groups*, Lect. Notes in Pure and Applied Math. 146, Marcel Dekker, New York, 1993, 17–41.

[Auslander 74] Auslander, M. Representation theory of Artin algebras II, *Comm. in Alg* 1 (1974), 269–310.

[Auslander, Reiten 75] Auslander, M. and Reiten, I. Representation theory of Artin algebras III: Almost split sequences, *Comm. in Alg.* 3 (1975), 239–294.

[Auslander, Reiten 77A] Auslander, M. and Reiten, I. Representation theory of Artin algebras IV: Invariants given by almost split sequences, *Comm. in Alg.* 5 (1977), 443–518.

[Auslander, Reiten 77B] Auslander, M. and Reiten, I. Representation theory of Artin algebras V: Methods for computing almost split sequences and irreducible morphisms, *Comm. in Alg.* 5 (1977), 519–554.

[Auslander, Smálo 81] Auslander, M. and Smálo, S.O. Almost split sequences in subcategories, *J. Alg.* 69 (1981), 426–454.

[Baer 35] Baer, R. Types of elements and characteristic subgroups of abelian groups, *Proc. London Math. Soc.* 39 (1935), 481–514.

[Baer 37] Baer, R. Abelian groups without elements of finite order, *Duke Math. J.* 3 (1937), 68–122.

[Bass, 68] Bass, H. *Algebraic K-Theory*, Benjamin, New York, 1968.

[Bautista, Martinez 79] Bautista, R. and Martinez, R. Representations of partially ordered sets and 1-Gorenstein Artin algebras, *Proc. of 1978 Antwerp Conf. on Ring Theory*, Marcel Dekker, New York, 1979, 385–433.

[Beaumont, Pierce 61A] Beaumont, R.A., and Pierce, R.S. Torsion-free rings, *Illinois. J. Math. 5* (1961), 61–98.

[Beaumont, Pierce 61B] Beaumont, R.A., and Pierce, R.S. Torsion free groups of rank two, *Mem. Amer. Math. Soc.* 38 (1961).

[Beaumont, Pierce 61C] Beaumont, R.A., and Pierce, R.S. Subrings of algebraic number fields, *Acta Sci. Math. (Szeged)* 22 (1961), 202–216.

[Beers, Hunter, Walker 83] Beers, D., Hunter, R., and Walker E.A. Finite valuated p-groups, *Abelian Group Theory*, Lect. Notes in Math. 1006, Springer-Verlag, 1983.

[Benabdallah, Mader 98] Benabdallah, K. and Mader, A. Almost completely decomposable groups and integral linear algebra, *J. Alg.* 204 (1998), 440–482.

[Bican 70] Bican, L. Purely finitely generated abelian groups, *Comm. Math. Univ. Carol.* 11 (1970), 1–8.

[Bican 78] Bican, L. Splitting in abelian groups, *Czech Math. J.* 28 (103) (1978), 356–364.

[Bican, Rangaswamy 95] Bican L. and Rangaswamy K.M. A result on B_1-groups, *Rend. Sem. Univ. Padova* 94 (1995), 95–98.

[Bican, Salce 83] Bican L. and Salce, L. Butler groups of infinite rank, *Abelian Group Theory*, Lect. Notes in Math. 1006, Springer-Verlag, New York, 1983.

[Blazhenov 96] Blazhenov, A.V. The genera and cancellation of torsion-free modules of finite rank, *St. Petersburg Math. J.* 7 (1996), 891–924.

[Brenner 67] Brenner, S. Endomorphism algebras of vector spaces with distinguished sets of subspaces, *J. Alg.* 6 (1967), 100–114.

[Brenner 74A] Brenner, S. Decomposition properties of some small diagrams of modules, *Symposia Mathematica* XIII, Academic Press, New York, 1974, 127–141.

[Brenner 74B] Brenner, S. On four subspaces of a vector space, *J. Alg.*, 29 (1974), 587–599.

[Bünermann 81] Bünermann, D. Auslander–Reiten quivers of exact one-parameter partially ordered sets, *Representations of Algebras*, Lect. Notes in Math. 903, Springer-Verlag, New York, 1981, 55–61.

[Bünermann 83] Bünermann, D. Hereditary torsion theories and Auslander-Reiten sequences, *Arch. der Math.* 41 (1983), 304–309.

[Burkhardt 84] Burkhardt, R. On a special class of almost completely decomposable groups I, *Abelian Groups and Modules*, CISM Courses and Lectures 287, Springer-Verlag, New York, 1984, 141–150.

[Burkhardt, Mutzbauer 96] Burkhardt, R. and Mutzbauer O. Almost completely decomposable groups with primary cyclic regulating quotient, *Rend. Sem. Mat. Univ. Padova* 95 (1996), 81–93.

[Butler 65] Butler, M.C.R. A class of torsion-free abelian groups of finite rank, *Proc. London Math. Soc.* (3) 15 (1965), 680–698.

[Butler 68] Butler, M.C.R. Torsion-free modules and diagrams of vector spaces, *Proc. London Math. Soc.* (3) 18 (1968), 635–652.

[Butler 87] Butler, M.C.R. Some almost split sequences in torsion-free abelian group theory, *Abelian Group Theory*, Gordon and Breach, New York, 1987, 291–302.

[Charles 68] Charles, B. Sous-groupes fonctoriels et topologies, *Studies on Abelian Groups*, Springer-Verlag Dunod, Paris, 1968, 75–92.

[Corner 61] Corner, A.L.S. A note on rank and direct decompositions of torsion-free abelian groups, *Proc. Cambridge Philos. Soc.* 57 (1961), 230–233.

[Corner 63] Corner, A.L.S. Every countable reduced torsion-free ring is an endomorphism ring, *Proc. London Math. Soc.* 13 (1963), 687–710.

[Corner 69] Corner, A.L.S. A note on rank and direct decompositions of torsion-free abelian groups, *Proc. Cambridge Philos. Soc.* 66 (1969), 239–240.

[Corner, Göbel 85] Corner, A.L.S. and Göbel, R. Prescribing endomorphism algebras: a unified treatment, *Proc. London Math. Soc.* (3) 50 (1985), 447–479.

[Crawley-Boevey 88] Crawley-Boevey, W. On tame algebras and bocses, *Proc. London Math. Soc.* (3) 56 (1988), 451–483.

[Crawley-Boevey 91] Crawley-Boevey, W. Tame algebras and generic modules, *Proc. London Math. Soc.* (3) 63 (1991), 241–265.

[Crawley-Boevey 92] Crawley-Boevey, W. Modules of finite length over their endomorphism rings, *London Math. Soc. Lecture Notes* 168, Cambridge University Press, 1992, 127–184.

[Cruddis 70] Cruddis, T.B. On a class of torsion free abelian groups, *Proc. London Math. Soc.* 21 (1970), 243–276.

[DeVivo, Metelli 96] DeVivo C. and Metelli, C. $B^{(1)}$-groups: Some counterexamples, *Abelian Groups and Modules*, Lect. Notes in Pure and Appl. Math. 182, Marcel Dekker, New York, 1996, 227–232.

[DeVivo, Metelli 99] DeVivo, C. and Metelli, C. Admissible matrices as base changes of $B^{(1)}$-groups, *Abelian Groups and Modules*, Birkhäuser, Boston, 1999, 135–148.

[Dittmann, Mader, Mutzbauer 97] Dittmann, U., Mader, A., and Mutzbauer, O. Almost completely decomposable groups with a cyclic regulating quotient, *Comm. in Alg.* 25 (1997), 769–784.

[Dlab, Ringel 76] Dlab, V. and Ringel, C. Indecomposable representations of graphs and algebras, *Mem. Amer. Math. Soc.* 173 (1976).

[Drozd 72] Drozd, Y. Representations of commutative algebras, *Funct. Anal. and Appl.* 6 (1972), 286–288.

[Drozd 74] Drozd, Y. Coxeter transformations and representations of partially ordered sets, *Funct. Anal. and App.* 8 (1974), 219–225.

[Dugas, Göbel 82] Dugas, M. and Göbel, R. Every cotorsion-free ring is an endomorphism ring, *Proc. London Math. Soc.* 45 (1982), 319–336.

[Dugas, Göbel 97] Dugas, M. and Göbel, R. Endomorphism rings of B_2-groups of infinite rank, *Israel J. Math.* 101 (1997), 141–156.

[Dugas, Oxford 93] Dugas, M. and Oxford, E. Near isomorphism invariants for a class of almost completely decomposable groups, *Abelian Groups*, Lect. Notes in Pure and Applied Math 146, Marcel Dekker, New York, 1993, 129–150.

[Dugas, Thomé 91] Dugas, M. and Thomé B. Countable Butler groups and vector spaces with four distinguished subspaces, *J. Alg.* 138 (1991), 249–272.

[Faticoni 87] Faticoni, T. Each countable reduced torsion-free commutative ring is a pure subring of an E-ring, *Comm. in Alg.* 15 (12), 1987, 2545–2564.

[Faticoni, Schultz 96] Faticoni, T. and Schultz, P. Direct decompositions of almost completely decomposable groups with primary regulating index, *Abelian Groups and Modules*, Marcel Dekker, New York, 1996, 233–242.

[Files, Göbel 98] Files, S. and Göbel, R. Gauβ' theorem for two submodules, *Math. Zeit.* 228 (1998), 511–536.

[Franzen 83] Franzen, B. Exterior powers and torsion-free modules over almost maximal valuation domains, *Abelian Group Theory*, Lect. Notes in Math. 1006, Springer-Verlag, New York, 1983, 599–606.

[Fuchs 58] Fuchs, L. *Abelian Groups.*, Publ. House of the Hungar. Acad. Sci., Budapest, 1958.

[Fuchs 70A] Fuchs, L. *Infinite Abelian Groups*, Vol. I, Academic Press, 1970.

[Fuchs, 70B] Fuchs, L. The cancellation property for modules, *Rings and Modules*, Lect. Notes in Math. 246, Springer-Verlag, New York, 1970, 192–213.

[Fuchs 71] Fuchs, L. Note on direct decompositions of torsion-free abelian groups, *Comment. Math. Helv.* 46 (1971), 87–91.

[Fuchs 73] Fuchs, L. *Infinite Abelian Groups*, Vol. II, Academic Press, New York, 1973.

[Fuchs 94] Fuchs, L. A survey of Butler groups of infinite rank, *Abelian Group Theory*, Contemporary Math. 171 (1994), 121–139.

[Fuchs, Grabe 75] Fuchs, L. and Grabe, P. Number of indecomposable summands in direct decompositions of torsion free abelian groups, *Rend. Sem. Mat. Univ. Padova* 53 (1975), 135–148.

[Fuchs, Loonstra 70] Fuchs, L., and Loonstra, F. On direct decompositions of torsion-free abelian groups of finite rank, *Rend. Sem. Mat. Univ. Padova* 44 (1970), 75–83.

[Fuchs, Loonstra 71] Fuchs, L. and Loonstra F., On the cancellation of modules in direct sums over Dedekind domains, *Indag. Math.* 33 (1971), 163–169.

[Fuchs, Metelli 91] Fuchs, L. and Metelli, C. On a class of Butler groups, *Manuscripta Math.* 71 (1991), 1–28.

[Fuchs, Metelli 92] Fuchs, L. and Metelli, C. Countable Butler groups, *Abelian Groups and Non-commutative Rings*, Contemporary Math. 130 (1992), 133–143.

[Fuchs, Salce 85] Fuchs, L. and Salce, L. *Modules over valuation domains*, Lect. Notes in Pure and Appl. Math. 97, Marcel Dekker, New York, 1985.

[Gabriel 72] Gabriel, P. Unzerlegbare Darstellung I, *Manuscripta Math.* 6 (1972), 71–103.

[Gabriel 73A] Gabriel, P. Représentations indécomposable des ensembles ordonnés, *Séminaire P. Dubriel*, Paris, 1972–1973, 1301–1304.

[Gabriel 73B] Gabriel, P. Indecomposable representations II, *Symposia Mathematica XI*, Academic Press, Orlando, 1973, 81–104.

[Gelfand, Ponomarev 70] Gelfand, I. and Ponomarev, I. Problems of linear algebra and classification of quadruples in a finite dimensional vector space, *Colloq. Math. Soc. Bolyai*, Tihany, 1970, 163–237.

[Göbel, Shelah 95] Göbel. R. and Shelah, S. On the existence of rigid aleph-1 free abelian groups of cardinality aleph-1, *Abelian Groups and Modules*, Kluwer, Boston, 1995, 227–237.

[Goeters 99] Goeters, H.P. Butler modules over 1-dimensional Noetherian domains, *Abelian Groups and Modules*, Birkhäuser, Boston, 1999, 149–166.

[Goeters, Ullery 90] Goeters, H.P. and Ullery W. Butler groups and lattices of types, *Comm. Math. Univ. Carol.* 31 (1990), 613–619.

[Goeters, Ullery 91] Goeters, H.P. and Ullery W. Homomorphic images of completely decomposable finite rank torsion-free groups, *J. Alg.* 140 (1991), 1–11.

[Goeters, Ullery 92] Goeters, H.P. and Ullery W. Almost completely decomposable torsion-free groups, *Rocky Mtn. J. Math.* 22 (1992), 593–600.

[Goeters, Ullery 94] Goeters, H.P. and Ullery W. The Arnold–Vinsonhaler invariants, *Houston J. Math.* 20 (1994), 17–25.

[Goeters, Ullery 95] Goeters, H.P. and Ullery W. On quasi-decompositions of $B^{(1)}$-groups, *Comm. in Alg.* 23 (1995), 131–137.

[Goeters, Ullery, Vinsonhaler 94] Goeters, H.P., Ullery W., and Vinsonhaler, C. Numerical invariants for a class of Butler groups, *Abelian Group Theory*, Contemporary Math. 171 (1994), 159–172.

[Goodearl 76] Goodearl, K.R. Power cancellation of groups and modules, *Pacific J. Math.*, 64 (1976), 387–411.

[Griffith 67] Griffith, P. Purely indecomposable torsion-free groups, *Proc. Amer. Math Soc.* 18 (1967), 738–742.

[Gustafson 82] Gustafson, W. The history of algebra and their representations, *Representations of Algebras*, Lect. Notes in Math. 944, Springer-Verlag, New York, 1982, 1–28.

[Hahn 97] Hahn, R. *Representations of Partially Ordered Sets*, Master's Thesis, Baylor University, 1997.

[Heller, Reiner 61] Heller, A. and Reiner, I. Indecomposable representations, *Ill. J. Math.* 5 (1961), 314–323.

[Hill, Megibben 93] Hill, P. and Megibben, C. The classification of certain Butler groups, *J. Alg.* 160 (1993), 524–551.

[Höfling 93] Höfling, B. On direct summands of $B^{(1)}$-groups, *Comm. in Alg.* 21 (1993), 1849–1856.

[Hungerford, 74] Hungerford, T. *Algebra*, Graduate Texts in Mathematics, Springer-Verlag, New York, 1974.

[Hunter, Richman, Walker 84] Hunter, R., Richman, F., and Walker, E.A. Subgroups of bounded abelian groups, *Abelian Groups and Modules*, CISM Courses and Lectures 287, Springer-Verlag, New York, 1984, 17–36.

[Jacobinski 68] Jacobinski, H. Genera and decompositions of lattices over orders, *Acta. Math.* 121 (1968), 1–29.

[Jónsson 57] Jónsson, B. On direct decompositions of torsion-free abelian groups, *Math. Scand.* 5 (1957), 230–235.

[Jónsson 59] Jónsson, B. On direct decomposition of torsion-free abelian groups, *Math. Scand.* 7, (1959) 361–371.

[Kaplansky 69] Kaplansky, I. *Infinite Abelian Groups*, University of Michigan Press, Ann Arbor, Michigan, 1954, 1969.

[Kerner 81] Kerner, O. Partially ordered sets of finite representation type, *Comm. in Alg.* 9 (1981), 783–809.

[Kleiner 75A] Kleiner, M. Partially ordered sets of finite type, *J. Soviet Math.* 23 (1975), 607–615.

[Kleiner 75B] Kleiner, M. On exact representations of finite type, *J. Soviet Math.* 23 (1975), 616–628.

[Koehler 64] Koehler, J. Some torsion-free rank two groups, *Acta. Sci. Math. Szeged* 25 (1964), 186–190.

[Koehler 65] Koehler, J. The type set of a torsion-free group of finite rank, *Illinois J. Math.* 9 (1965), 66–86.

[Krapf, Mutzbauer 84] Krapf, K.J. and Mutzbauer, O. Classification of almost completely decomposable groups, *Abelian Groups and Modules*, CISM Courses and Lectures 287, Springer-Verlag, New York, 1984, 151–162.

[Krugljak 64] Representations of the (p, p) group over a field of characteristic p, *Soviet Math. Dokl.* 4 (1964), 1809–1813.

[Lady 74A] Lady, E.L. Summands of finite rank torsion free abelian groups, *J. Alg.* 32 (1974), 51–52.

[Lady 74B] Lady, E.L. Almost completely decomposable torsion free abelian groups, *Proc. A.M.S.* 45 (1974), 41–47.

[Lady 75] Lady, E.L. Nearly isomorphic torsion free abelian groups, *J. Alg.* 35 (1975), 235–238.

[Lady 77A] Lady, L. On classifying torsion free modules over discrete valuation rings, *Abelian Group Theory*, Lecture Notes in Math. 616, Springer-Verlag, New York, 1977, 168–172.

[Lady 77B] Lady, L. Splitting fields for torsion-free modules over discrete valuation rings I, *J. Alg.* 49 (1977), 261–275.

[Lady 79] Lady, L. Extension of scalars for torsion-free modules over Dedekind domains, *Symposia Math.* XXIII, Academic Press, New York, 1979, 287–305.

[Lady 80A] Lady, L. Splitting fields for torsion-free modules over discrete valuation rings II, *J. Alg.* 66 (1980), 281–306.

[Lady 80B] Lady, L. Splitting fields for torsion-free modules over discrete valuation rings III, *J. Alg.* 66 (1980), 307–320.

[Lady 83] Lady, L. A seminar on splitting rings for torsion-free modules over Dedekind domains, *Abelian Groups*, Lect. Notes in Math. 1006, Springer-Verlag, New York, 1983, 1–48.

[Lambek 66] Lambek, J. *Lectures on Rings and Modules*, Blaisdell, Waitham, 1966.

[Lee 89] Lee, W.Y. Co-diagonal Butler groups, Chinese J. Math. 17 (1989), 259–271.

[Levi 17] Levi, F.W. Abelsche Gruppen mit abzähleren Elementen, Habilitationschrift (Leipzig, 1917).

[Lewis 93] Lewis, W.S. Almost completely decomposable groups with two critical types, *Comm. in Alg.* 21 (1993), 607–614.

[Loupias 75] Loupias, M. Indecomposable representations of finite ordered sets, *Representations of Algebras*, Lect. Notes in Math. 488, Springer-Verlag, New York, 1975, 201–209.

[Mader 93] Mader, A. Friedrich Wilhelm Levi, 1888–1966, *Abelian Groups*, Lect. Notes in Pure and Appl. Math. 146, Marcel Dekker, New York, 1993, 1–14.

[Mader 95] Mader, A. Almost completely decomposable groups, *Abelian Groups and Modules*, Kluwer, Boston, 1995, 343–366.

[Mader 00] Mader, A. *Almost Completely Decomposable Groups*, Gordon and Breach, 2000.

[Mader, Mutzbauer 93] Mader, A. and Mutzbauer O. Almost completely decomposable groups with cyclic regulating quotient, *Abelian Groups*, Lect. Notes in Pure and Appl. Math. 146, Marcel Dekker, 1993, 193–200.

[Mader, Mutzbauer, Rangaswamy 94] Mader, A., Mutzbauer, O., and Rangaswamy, K.M., A generalization of Butler groups, *Abelian Group Theory*, Contemporary Mathematics 171 (1994), 257–275.

[Mader, Nonxga 99] Mader, A. and Nonxga, L. Completely decomposable summands of almost completely decomposable groups, *Abelian Groups and Modules*, Birkhäuser, Boston, 1999, 167–190.

[Mader, Vinsonhaler 94] Mader, A. and Vinsonhaler, C. Classifying almost completely decomposable groups, *J. Alg.* 170 (1994), 754–780.

[Mader, Vinsonhaler 95] Mader, A. and Vinsonhaler, C. Almost completely decomposable groups with a cyclic regulating quotient, *J. Alg.* 177 (1995), 463–492.

[Mader, Vinsonhaler 97] Mader, A. and Vinsonhaler, C. The idempotent lifting property for almost completely decomposable groups, *Colloq. Math.* 72 (1997), 305–317.

[Mines, Vinsonhaler 92] Mines, R. and Vinsonhaler, C. Butler groups and Bext: a constructive approach, *Contemporary Math.* 130 (1992), 289–299.

[Mitchell 65] Mitchell, B. *Theory of Categories*, Academic Press, New York, 1965.

[Mouser 93] Mouser, M.A. On Z/p^2Z-modules with distinguished submodules, master's thesis, Baylor University, 1993.

[Mutzbauer 93] Mutzbauer, O. Regulating subgroups of Butler groups, *Abelian Groups*, Lect. Notes in Pure and Appl. Math. 146, Marcel Dekker, New York, 1993, 209–217.

[Mutzbauer 99] Mutzbauer, O. Normal forms of matrices with applications to almost completely decomposable groups, *Abelian Groups and Modules*, Birkhäuser, Boston, 1999, 121–134.

[Nazarova 75] Nazarova, L.A. Partially ordered sets of infinite type, *Math. USSR Izvestija* 9 (1975), 911–938.

[Nazarova 81] Nazarova, L.A. Poset representations, *Integral Representations and Applications*, Lect. Notes in Math. 882, Springer-Verlag, New York, 1981, 345–356.

[Nazarova, Roiter 75] Nazarova, L.A. and Roiter, A.V. Representations of partially ordered sets, *J. Soviet Math.* 23 (1975), 585–607.

[Nonxga, Vinsonhaler 95] Nonxga, L. and Vinsonhaler, C. Completely decomposable subgroups and factor groups of finite rank completely decomposable groups, *Abelian Groups and Modules*, Kluwer, Boston, 1995, 385–394.

[Nonxga, Vinsonhaler 96] Nonxga, L. and Vinsonhaler, C. Balanced Butler groups, *J. Alg.* 180 (1996), 546–570.

[Nonxga, Vinsonhaler 97] Nonxga, L. and Vinsonhaler, C. Balanced and cobalanced representations of posets, *Comm. in Alg.* 25 (12) (1997), 3735–3749.

[O'Meara, Vinsonhaler 99] O'Meara, K. and Vinsonhaler, C. Separative cancellation and multiple isomorphism in torsion-free abelian groups, *J. Alg.* 221 (1999), 536–550.

[Ould-Beddi, Strüngmann 99] Ould-Betti, M.A., and Strüngmann, L. Stacked bases for a pair of homogeneous completely decomposable groups with bounded quotient, *Abelian Groups and Modules*, Birkhäuser, Boston, 1999, 199–210.

[Pierce 60] Pierce, R.S. Subrings of simple algebras, *Michigan Math. J.*, 7 (1960), 241–243.

[Plahotnik 76] Plahotnik, V.V. Representations of partially ordered sets over commutative rings, *Math. USSR Isvestija* 10 (1976), 497–514.

[Pontryagin 34] The theory of topological commutative groups, *Ann. Math.* 35 (1934), 361–368.

[Rangaswamy, Vinsonhaler 94] Rangaswamy, K.M. and Vinsonhaler, C. Butler groups and representations of infinite rank, *J. Alg.* 167 (1994), 758–777.

[Reid 62] Reid, J. On quasi-decompositions of torsion-free abelian groups, *Proc. Amer. Math. Soc.* 13 (1962), 550–554.

[Reid 63] Reid, J. On the ring of quasi-endomorphisms of a torsion-free group, *Topics in Abelian Groups*, Scott-Foresman, Chicago, 1963, 51–68.

[Reid 81] Reid, J. Abelian groups finitely generated over their endomorphism rings, *Abelian Group Theory*, Lect. Notes in Math. 874, Springer-Verlag, New York, 1981, 41–52.

[Reid 83] Reid, J. Abelian groups cylic over their endomorphism rings, *Abelian Group Theory*, Lect. Notes in Math. 1006, Springer-Verlag, 1983, 190–203.

[Reid 99] Reid, J. Some matrix rings associated with acd-groups, *Abelian Groups and Modules*, Birkhäuser, Boston, 1999, 191–198.

[Reiner 75] Reiner, I. *Maximal Orders*, Academic Press, New York, 1975.

[Reiten 80] Reiten, I. The use of almost split sequences in the representation theory of Artin algebras, *Representations of Algebras*, Lect. Notes in Math. 944, Springer-Verlag, New York, 1980, 163–237.

[Richman 68] Richman, F. A class of rank-2 torsion-free groups, *Studies on Abelian Groups*, Springer-Verlag Dunod, Paris, 1968, 327–333.

[Richman 77] Richman, F. A guide to valuated groups, *Abelian Groups*, Lect. Notes in Math. 616, Springer, 1977, 73–86.

[Richman 83] Richman, F. An extension of the theory of completely decomposable torsion-free abelian groups, *Trans. A.M.S.* 279 (1983), 175–185.

[Richman 84] Richman, F. Butler groups, valuated vector spaces, and duality, *Rend. Sem. Mat. Univ. Padova* 72 (1984), 13–19.

[Richman 95] Richman, F. Isomorphism of Butler groups at a prime, *Abelian Group Theory*, Contemporary Math. 171 (1995), 333–337.

[Richman, Walker 79] Valuated groups, *J. Alg.* 56 (1979), 145–167.

[Richman, Walker 98] Richman, F. and Walker, E. Filtered modules over discrete valuation domains, *J. Alg.* 199 (1998), 618–645.

[Richman, Walker 99] Richman, F. and Walker, E. Subgroups of p^5-bounded groups, *Abelian Groups and Modules*, Birkhäuser, Boston, 1999, 55–74.

[Ringel 76] Ringel, C. Realizations of K-species and bimodules, *J. Alg.* 41 (1976), 269–302.

[Ringel 78] Ringel, C. The spectrum of a finite dimensional algebra, *Proceedings of 1978 Antwerp Conference on Ring Theory*, Marcel Dekker, New York, 1979, 535–597.

[Ringel 80] Ringel, C. Report on the Brauer-Thrall conjectures, *Representation Theory I*, Lect. Notes in Math. 831, Springer-Verlag, New York, 1980, 104–136.

[Ringel 84] Ringel, C. *Tame Algebras and Integral Quadratic Forms*, Lect. Notes in Math. 1099, Springer-Verlag, New York, 1984.

[Ringel 91] Ringel, C. Recent advances in the representation theory of finite dimensional algebras, *Progress in Math.* 95 (1991), 141–191.

[Ringel, Tachikawa 75] Ringel, C. and Tachikawa, H. QF-3 rings, *J. Reine Angew. Math.* 272 (1975), 49–72.

[Rojter 80] Rojter, A.V. Matrix problems and representations of bocses, *Representation Theory I*, Lect. Notes in Math. 831, Springer-Verlag, New York, 1980, 288–324.

[Schultz 87] Schultz, P. Finite extensions of torsion-free groups, *Abelian Group Theory*, Gordon and Breach, New York, 1987, 333–350.

[Simson 74] Simson, D. Functor categories in which every finite object is projective, *Bull. Acad. Polon. Sci.* 22 (1974), 375–380.

[Simson 92] Simson, D. *Linear Representations of Partially Ordered Sets and Vector Space Categories*, Gordon and Breach, Philadelphia, 1992.

[Simson 93] Simson, D. On representation types of module subcategories and orders, *Bull. Acad. Pol. Sci. Math.* 41 (1993), 77–93.

[Simson 96] Simson, D. Socle projective representations of partially ordered sets and Tits quadratic forms with applications to lattices over orders, *Abelian Groups and Modules*, Marcel Dekker, New York, 1996, 73–112.

[Stelzer 85] Stelzer, J. A cancellation criterion for finite rank torsion-free abelian groups of finite rank, *Proc. Amer. Math. Soc.* 94 (1985), 363–368.

[Stelzer 87] Stelzer, J. Ring theoretical criteria for cancellation, *Abelian Group Theory*, Gordon and Breach, New York, 1987, 175–196.

[Swan 62] Swan, R.G. Projective modules over group rings and maximal orders, *Ann. of Math.* (2) 76 (1962), 60–69.

[Turgi 77] Turgi, M. A sheaf-theoretic interpretation of the Kurosch theorem, *Abelian Group Theory*, Lect. Notes in Math. 616, Springer-Verlag, New York, 1977, 168–196.

[Turnbull, Aitken 61] Turnbull, H. and Aitken, A. *An Introduction to the Theory of Canonical Matrices*, Dover, New York, 1961.

[Vinsonhaler 95] Vinsonhaler, C. Invariants for almost completely decomposable groups, *Abelian Groups and Modules*, Kluwer, Boston, 1995, 485–490.

[Walker 64] Walker, E.A. Quotient categories and quasi-isomorphisms of abelian groups, *Proc. Colloq. Abelian Groups*, Budapest, 1964, 147–162.

[Warfield 68] Warfield, R.B. Jr. Homomorphisms and duality for torsion-free groups, *Math. Zeit.* 107 (1968), 189–200.

[Warfield 72] Warfield, R.B. Jr. Exchange rings and decompositions of modules. *Math. Ann.* 199 (1972), 31–36.

[Warfield 80] Warfield, R.B. Jr. Cancellation for modules and stable range of endomorphism rings, *Pac. J. Math.* 91 (1980), 457–485.

[Yakolev 91] Yakolev, A. Torsion-free abelian groups of finite rank and their direct decompositions, *J. Soviet Math.* 57 (1991), 3524–3533.

[Yom 94] Yom, P. A characterization of a class of Butler groups II, *Abelian Group Theory*, Contemporary Math. 171 (1974), 419–432.

[Yom 98] Yom. P. A characterization of a class of Butler groups, preprint.

List of Symbols

$A = (A_1, \ldots, A_n)$	- n-tuple of subgroups of $\mathbb{Q}$
$a^{\lhd}$	- subset of a poset S with $a \in S$
$A(m, \lambda)$	- Jordan block matrix
$\mathbf{A}_p$	- isomorphism at p category of torsion-free abelian groups
$\mathbf{A}_{\mathbb{Q}}$	- quasi-homomorphism category of torsion-free abelian groups
$b^{\rhd}$	- subset of poset S with $b \in S$
$b_\tau(G)$	- Burkhardt invariant of group G
$B(T)$	- category of finite rank Butler groups, T a lattice of types
$B(T, j)$	- subcategory of $B(T)$, $j \geq 0$
$B(T, j)_p$	- isomorphism at p category of $B(T, j)$
$B(T)_{\mathbb{Q}}$	- quasi-homomorphism category of $B(T)$
$C(A)$	- centralizer of matrix A
$C(A, B)$	- centralizer of two matrices A and B
cdn H	- coordinate vector of a torsion-free abelian group H
cdn U	- coordinate vector of representation U
C_G	- subring of the center of the center of End U_G, $U_G \in \text{rep}(S_T, \mathbb{Q})$
C^-	- Coxeter correspondence
C_n	- chain with n elements
$C(n)$	- a category of almost completely decomposable groups with typeset contained in an antichain
coker f	- the cokernel of an R-module homomorphism
Coker f	- the cokernel of a representation morphism
$\equiv (\text{mod } p)$	- congruence of integers modulo a prime p

$c(f)$	- content of polynomial $f(x) \in \mathbb{Z}[x]$
C^+	- Coxeter correspondence
CR	- center of ring R
$C\mathrm{rep}(S_T, \mathbb{Z}_p, j)$	- the subcategory of $\mathrm{rep}(S_T, \mathbb{Z}_p, j)$ corresponding to $C(T, j)_p$
$C\mathrm{rep}_{\mathrm{crit}}(S_T, \mathbb{Z}_p, j)$	- the subcategory of $\mathrm{rep}(S_T, \mathbb{Z}_p, j)$ corresponding to $C_{\mathrm{crit}}(T, j)$
$CT_{cr}(G)$	- critical cotypeset of G
$C(T, j)$	- almost completely decomposable groups in $B(T, j)$
$C(T, j)_p$	- isomorphism at p category of $C(T, j)$
$C_{\mathrm{crit}}(T, j)$	- almost completely decomposable groups in $B(T, j)$ with critical typeset $\subseteq S_T$
$\delta(A)$	- n-tuple of subgroups of $\mathbb{Q}$
$\delta_{(a,b)}S$	- derivative of poset S with respect to suitable pair (a, b)
$d(G)$	- maximal divisible subgroup of torsion-free abelian group G
$d(M)$	- maximal divisible submodule of torsion-free module over a discrete valuation ring
$\Delta\mathrm{rep}(S_T, \mathbb{Z}/p^j\mathbb{Z})$	- a subcategory of $\mathrm{rep}(S_T, \mathbb{Z}/p^j\mathbb{Z})$ of representations in matrix problem form
$\Delta(S_n, R)$	- a subcategory of $\mathrm{rep}(S_{n+1}, R)$ of representations in matrix problem form over a ring R
$\dim U$	- dimension of a representation U
$\oplus$	- direct sum
e_G	- exponent of $G/R(G)$ for a finite rank Butler group G
$\in$	- element of
$\notin$	- not element of
$\mathrm{End}\, U$	- ring of endomorphisms of representation U
$\mathrm{End}_R(M)$	- R-endomorphism ring of R-module M
$E(n, k)$	- a subcategory of $\mathrm{rep}(S_{n+2}, k)$
fg	- composition of morphisms f and g defined by $fg(x) = f(g(x))$
f^n	- composition of a morphism f repeated n times
$\mathrm{fr}\mathbf{A}$	- category of torsion-free abelian groups of finite rank
$\mathrm{fr}\mathbf{A}/p$	- mod p category of torsion-free abelian groups of finite rank
$\mathrm{fr}\mathbf{A}_p$	- isomorphism at p category of torsion-free abelian groups of finite rank
$\mathrm{fr}\mathbf{A}_{\mathbb{Q}}$	- quasi-homomorphism category of $\mathrm{fr}\mathbf{A}$
$G[A]$	- bracket group
$\gcd$	- greatest common divisor
G/H	- group of cosets of a subgroup H of an abelian group G
G^n	- direct sum of n copies of abelian group G
G_p	- localization of torsion-free abelian group at a prime p
$G[p^j]$	- p^j-socle of an abelian group

G_S	- localization of a torsion-free abelian group at a set of primes S
$G(T)$	- socle of a valuated abelian group for a tree T
$G[\tau]$	- τ-radical
$G^*[\tau]$	- subgroup containing τ-radical
$G(\tau)$	- τ-socle of torsion-free abelian group G
$G^*(\tau)$	- subgroup of G generated by $\{x : \text{type } X > \tau\}$
$G^\#(\tau)$	- pure subgroup generated by $G^*(\tau)$
G_τ	- τ-homogeneous completely decomposable subgroup of Butler group G with $G(\tau) = G_\tau \oplus G^\#(\tau)$
$h(x)$	- height sequence of an element x in an abelian group
$h_p(x)$	- p-height of an element x in an abelian group
$H + K$	- subgroup of group G generated by H and K
$\text{Hom}(A, B)$	- morphism group of objects A and B in additive category
$\text{Hom}(G, H)$	- group of homomorphisms of abelian groups G and H
$\text{Hom}(G, H)_p$	- localization of $\text{Hom}(G, H)$ at a prime p
$\text{Hom}_R(M, N)$	- group of R-homomorphisms of R-modules M and N
(i)	- chain of length i
$i(G)$	- regulating index of an almost completely decomposable group G
IM	- submodule of R-module M generated by $\{xm : x \in I, m \in M\}$ for ideal I of R
I^n	- nth power of ideal I
$\text{Ind}(\text{mod } R)$	- isomorphism classes of indecomposables in mod R
$\text{Ind}(n, k)$	- isomorphism classes of indecomposable k-representations of an antichain
$\text{Ind}(S, k)$	- isomorphism classes of indecomposable representations of poset S
$\text{Ind}(S, R, j)$	- isomorphism classes of indecomposables in $\text{rep}(S, R, j)$
$I(t)$	- 1-dimensional injective in $\text{rep}(S, k)$
$\text{IT}(G)$	- inner type of G
i_U	- injection for direct sum
$I(U)$	- injective envelope of $U \in \text{rep}(S, k)$
$\cap$	- intersection of sets, greatest lower bound of types
$JI(T)$	- join-irreducible elements of a poset T
J_L	- opposite of the poset of a lattice of v-heights
JR	- Jacobson radical of ring R
$\ker f$	- the kernel of an R-homomorphism f
$\text{Ker } f$	- the kernel of a representation morphism f
k^n	- n-dimensional k-vector space for field k
$k[x]$	- polynomial ring in indeterminate x with coefficients in k
$k(x)$	- quotient field of $k[x]$
$k\langle x, y \rangle$	- polynomial ring in two noncommuting variables x and y
$k(U)$	- a representation determined by $U \in \text{rep}(S, k)$

lcm	- least common multiple
$L(U)$	- representation determined by $U \in \mathrm{rep}(S, K)$
$\mathrm{Mat}_m(k)$	- ring of all $m \times m$ matrices
$M \approx N$	- equivalence of matrices M and N under row operations and block column operations
M^{-1}	- inverse of square matrix M
M/N	- module of cosets of submodule N of module M
$\mathrm{mod}\ R$	- category of finitely generated R-modules
M_P	- a localization of $M \in Q\mathrm{rep}(S, R)$ at an up-set P containing M
M^P	- a localization of $M \in Q\mathrm{rep}(S, R)$ at an up-set P contained in M
$M(t)$	- a module over R, $t \in R^*$
$M_{(t)}$	- a module over R, (t) a subset of R^*
μ_r	- right multiplication by an element r of a ring
$M_U = (A_1 \vert \cdots \vert A_n)$	- matrix of representation U with block matrices A_i,
$n(G)$	- an integer determined by the endomorphism ring of a torsion-free abelian group of finite rank
n^+	- $\{1, 2, \ldots, n\}$
NR	- nil radical of ring R with finite rank torsion-free additive group
$\mathrm{OT}(G)$	- outer type of G
Π	- set of primes of the integers
$P(A)$	- category of summands of direct sums of copies of an abelian group A
$P_f(A)$	- category of summands of finite direct sums of copies of an abelian group A
$P_f(R)$	- category of finitely generated projective right R-modules
$P(n, j)$	- a poset
$p^\omega G$	- p-divisible subgroup of abelian group G
$P(R)$	- category of projective right R-modules
$P(t)$	- 1-dimensional projective in $\mathrm{rep}(S, k)$, in $Q\mathrm{rep}(S, R)$
p_U	- projective cover $P(U) \to U \in \mathrm{rep}(S, k)$
$P(U)$	- projective representation determined by $U \in \mathrm{rep}(S, k)$
$\mathbb{Q}$	- field of rational numbers
Q	- quotient field of discrete valuation ring
qG	- $\{qx : x \in G\}$ for q a rational and G a torsion-free abelian group
$\mathbb{Q}G$	- $\{qx : q \in \mathbb{Q}, x \in G\}$ for torsion-free abelian group G
$Q\mathrm{rep}(S, R)$	- category of quiver representations of a finite poset S over a principal ideal domain R
q_U	- injective envelope $U \to I(U) \in \mathrm{rep}(S, k)$
R	- ring, associative, with identity
R^*	- completion of a ring R
$\mathrm{rep}(S, k)$	- category of finite-dimensional k-representations of S

$\text{rep}(S, R)$	- category of finite rank free representations of poset S over ring R, over discrete valuation ring
$\text{rep}_f(S, R)$	- category of finitely generated representations containing $\text{rep}(S, R)$
$\text{rep}(S, R, j)$	- subcategory of $\text{rep}(S, R)$ for R a discrete valuation ring
$\text{Rep}(S, k)$	- category of countable representations of poset S over field k
$\text{Rep}(S, R)$	- category of countable representations of poset S over ring R
$R_1 \times R_2$	- ring product of rings R_1 and R_2
$R(G)$	- regulator subgroup of a finite rank Butler group G
R^+	- additive group of a ring R
$\text{RT}(G)$	- Richman type of G
R_U	- endomorphism ring of matrix M_U
$|S|$	- cardinality of set S
S^*	- disjoint union of a poset S and a point $*$
σ	- duality from $\text{rep}(S, k)$ to $\text{rep}(S^{\text{op}}, k)$
$(S_1, \ldots, S_n)$	- disjoint union of posets S_i
S_n	- antichain with n elements
$S(n, j)$	- poset
S^{op}	- opposite of poset S
$\Sigma\{U_i : i \in S\}$	- subspace of U_0 generated by subspaces $\{U_i : i \in S\}$ of $U_0 \in \text{rep}(S, k)$
$\Sigma\{G_i : i \in S\}$	- subgroup of G generated by subgroups G_i of G
S_T	- opposite of poset of join irreducible elements of T
$\langle S \rangle$	- subgroup of an abelian group generated by a subset S
$\langle S \rangle_*$	- pure subgroup of an abelian group generated by a subset S
$T_{\text{cr}}(G)$	- critical typeset of G
$\otimes$	- tensor product
τ_E	- meet of a set of types with indices in E
TF_R	- category of torsion-free modules of finite rank over a discrete valuation ring R
$T(M)$	- torsion-subrepresentation of $M \in Q\text{rep}(S, R)$
$\tau(P)$	- type of a partition of n^+
$T(x, G)$	- finite valuated tree for $x \in G$, a valuated abelian p-group
$U_1 + U_2$	- subspace of U_0 generated by U_1 and U_2
$U = (U_0, U_i : i \in S)$	- representation of poset S
$U_{\text{crit}}(T, j)$	- subcategory of $C_{\text{crit}}(T, j)$ of uniform almost completely decomposable groups
$U_i^{\#}$	- subspace of U_i generated by $\{U_j : j < i\}$
$\hat{U}_i$	- $\cap\{U_j : j > i\}$ for $U = (U_0, U_i) \in \text{rep}(S, k)$
U^n	- direct sum of n copies of representation U
$\cup$	- union of sets, least upper bound of types

$U(T, j)$	- subcategory of $C(T, j)$ of uniform almost completely decomposable groups
v	- valuation of an abelian p-group
V_A	- $k[x]$-module defined by matrix A
$w(S)$	- width of poset S
$W(U)$	- injective envelope of C^+U for $U \in \mathrm{rep}(S, k)$
$W'(U)$	- projective cover of C^-U for $U \in \mathrm{rep}(S, k)$
$\mathbb{Z}$	- ring of integers
$\mathbb{Z}[1/n]$	- subring of rationals generated by 1 and $1/n$ for nonzero integer n
$\mathbb{Z}(i)$	- $\mathbb{Z}/i\mathbb{Z}$ if i integer and $\mathbb{Z}_p/\mathbb{Z}$ if $i = \infty$
$\mathbb{Z}_p$	- localization of integers at a prime p

Index of Terms

abelian group
  almost completely decomposable, 51
  balanced projective, 198
  $B^{(1)}$-, 205
  bounded, 48
  bracket, 205
  Butler, 91, 105
  central special, 115
  cobalanced injective, 198
  completely decomposable, 51, 105
  divisible, 56
  exponent of finite, 147
  finitely Butler, 105
  homogeneous, 79
  indecomposable at p, 51
  injective in $B(T)$, 197
  L-, 220
  locally completely decomposable, 99
  p-local, 56
  p-primary regulating quotient, 169
  pregeneric, 123
  preprojective, 201
  preinjective, 201
  p-reduced, 52
  projective in $B(T)$, 197
  quasi-generic, 113
  reduced, 56
  Richman–Butler, 205
  self-cancellation, 68
  self-small, 69
  semilocal, 58
  special, 115
  strongly indecomposable, 50
  T-, 116
  τ-homogeneous, 79
  torsion-free, 47
  uniform almost completely decomposable, 164
  valued p-, 218
algebraically independent elements, 73
antichain, 10

Boolean algebra, 84
Burkhardt invariants, 148

cancellation
  self, 68
Cartan matrix, 177
category
  additive, 10
  definition, 10
  equivalence, 69
categories of abelian groups
  almost completely decomposable, 154, 155, 160
  Butler, 100, 140
  generically wild, 121
  isomorphism at p, 51
  Krull–Schmidt, 50
  modulo p, 53
  quasi-homomorphism, 47
  torsion-free, 47
  uniform almost completely decomposable, 164